EVAPOTRANSPIRATION - AN OVERVIEW

Edited by **Stavros G. Alexandris**
and **Ruzica Stricevic**

Evapotranspiration - An Overview

http://dx.doi.org/10.5772/3383
Edited by Stavros G. Alexandris and Ruzica Stricevic

Contributors

Robert Criss, Elizabeth Hasenmueller, Amin Elshorbagy, Zohreh Izadifar, Nakayama, Baburao Kamble, Ayse Irmak, Ian Ratcliffe, Gary Hergert, Derrel Martin, Kenneth Hubbard, Sunil Narumalani, Robert Oglesby, Nebo Jovanovic, José Carlos Mendonça, Jozsef Szilagyi, Zoubeida Kebaili Bargaoui, Koutsoyiannis, Aristoteles Tegos, Andreas Efstratiadis, Renata Duffková, Hanbo Yang, Ahmed Al-Busaidi, Mir Matin, Charles Bourque

Published by InTech

Janeza Trdine 9, 51000 Rijeka, Croatia

Publishing Process Manager Dejan Grgur
Typesetting InTech DTP Team
Cover InTech Design Team

Additional hard copies can be obtained from orders@intechopen.com

Evapotranspiration - An Overview, Edited by Stavros G. Alexandris and Ruzica Stricevic
p. cm.
ISBN 978-953-51-1115-3

For EU product safety concerns:
IN TECH d.o.o., Prolaz Marije Krucifikse Kozulić 3, 51000 Rijeka, Croatia,
info@intechopen.com or visit our website at intechopen.com.

Meet the editors

Stavros Alexandris holds a PhD in Agricultural sciences from the Agricultural University of Athens, Greece. His research interests include evapotranspiration, crop water conservation, techniques and applications in Agrometeorology. He has more than six years of teaching experience and has published a number of scientific and technical papers. Since 1983, he has made an active contribution in many European research projects (Ecological impact of the air pollution in Attica region, Monitoring of the sea water intrusion in the groundwater aquifers of Argolis Plain, Rational application of irrigation and drainage under Greek conditions, Validity, limits and possible trends of coastal South Mediterranean traditional groundwater irrigated agriculture, MEDEFLU, LOLaqua, DMCSEE, COROADO). He is a member of the Geotechnical Chambers of Greece, Greek Society of Soil Science and Hellenic Meteorological Society. Today, Dr. Alexandris is an Assistant professor in AUA.

Ruzica Sticevic is a full time professor at University of Belgrade, Faculty of Agriculture. Her research interests include evapotranspiration, crop water requirements, water use efficiency, irrigation scheduling, crop growth modeling, drainage. She has more than fifteen years experience of teaching. She has published a number of scienticif papers in peer reviewed scientific journals, symposium, congresses, as well as practicum and textbook for students. Since 1991 she has actively participated in 5 EU projects (JOULE, WATERWEB, CROPWAT, LOLAqua, EURO-AGRIWAT), one bilateral and 7 national scientific projects, regarding the irrigation scheduling, water savings, water quality and soil and water management. She is a member of International Society on Horticultural Science, and Serbian Society of Soil Science.

Contents

Preface

Evapotranspiration has been the subject of thousands of research articles during the last century. It is a complicated subject, involving a number of interrelated procedures and standard, engaging a complex set of processes, which themselves are influenced by many factors dependent on local or global spatial scales of natural regimes. Maxwell in 1965 estimated that about ten trillion cubic meters of water per day returned to the atmosphere by evaporation and transpiration from the surface of the United States.

Despite the large number of scientific works worldwide on evapotranspiration, the phase transition of water into the biosphere is still an important area of investigation for the Geosciences. This book is designed to further enhance presenting recent investigations of many scientists on evapotranspiration modeling around the world. Thus, the objective of this book is to provide timely and comprehensive coverage of the principles, modeling and methods of measuring evapotranspiration from a single typical experimental site to large study areas, into natural and rural ecosystems.

The collective effort that solidified in this book, is only a minor part of the long line of hard research work that devoted scientists have contributed to the investigation of physical reality, to the never-ending struggle of man to unravel the unknown through logic and reason. Can this quest end? "...Evaporation is the most desperate branch of the desperate science of Meteorology...", Symons said as early as 1867.

The first chapters refer to the more general subjects and are followed by chapters with a more specific focus. Each chapter, however, was individually considered within this context.

The topics describe the latest research on ET estimation. The Bowen Ratio (BR) method, the complementary relation method (CR), the Penman Montieth (PM) method for estimating reference evapotranspiration, other simplified methods such as the Priestley Taylor method, or novel approaches on the simplification of the PM method with more emphasis on the empirical approach are all illustrated in the chapters of the book, in the framework of projects which include basins, rivers, or the laboratory setting of an experiment in a controlled environment.

These methods are a part of, or used in conjunction with models like NICE (National Integrated Catchment-based Eco-hydrology), WREVAP which is based on the CR method, HYDRUS2 which elaborates on the movement of water in the ground, an arithmetic solution of the Richards equation, as well as models that use satellite data or even models that utilize Artificial Neural Networks (ANN).

An equally wide range can be seen in the types of input data for the methods and the models used.

The research work estimates the quantity of AET, its relation to soil moisture and soil type, vegetation cover, land use, topography, weather conditions, climatic associations and geology. It compares the various methods in terms of their ability to predict, investigates topics like the spatiotemporal patterns of AET, the underlying relation of these patterns with plant development, the succession of the seasons during a year, and the subsequent variation in temperature, moisture, etc. It explores the limitations of the methods used, quantifies alternative management scenarios, and the effects of irrigation treatments with different water quality on AET.

The objective of this book isn't only to provide the knowledge, competencies and skills to serve as a recipe on water conservation, but seeks to contribute to the understanding the role of evapotranspiration and latent heat on the sensitivity of the climate system by bringing together the work of modelers and others involved in the analysis of observational data.

We are hopeful that the book "Evapotranspiration – An Overview" will provide a wider knowledge and may be a useful tool for many complicated aspects related to hydrological and specific environmental phenomena, as well as will put a premium on accurate approach of the mechanism of the water movement on the earth climatic system.

The editor is grateful to the authors who participated in this book project for their scientific contributions, which consists a useful volume on the topic of the state of the art applications of remote sensing and ground-based methods for ET estimates.

Lastly, I express here my sincerest thanks to the co-editor Prof. Ruzica Stricevic and our partner, Dr. Costas Chatzithomas for the manifold suggestions which I have received from them, and for their constant input of this work.

Editor, Stavros Alexandris
Department of Natural Resources
Management & Agricultural Engineering, Sector of Water resources Management
Agricultural University of Athens, Greece

Co-editor, Prof. Ruzica Stricevic
Faculty of Agriculture, University of Belgrade, Serbia

Chapter 1

Water Balance Estimates of Evapotranspiration Rates in Areas with Varying Land Use

Elizabeth A. Hasenmueller and Robert E. Criss

Additional information is available at the end of the chapter

http://dx.doi.org/10.5772/52811

1. Introduction

In the continental United States, approximately 2/3 of all rainfall delivered is lost to evapotranspiration (ET; US Water Resource Council, 1978). It follows that the ET rate, representing the combined processes of physical evaporation and biological transpiration, is essential for predicting water yields, designing irrigation and supply projects, managing water quality, quantity, and associated environmental concerns, and negotiating disputes, contracts, or treaties involving water. Water fluxes in catchments are controlled by these physical and biological processes as well as by hydrogeologic properties that are complex, heterogeneous, and poorly characterized by field and laboratory measurements. As a result, practical theories of ET rates and their impact on runoff generation and catchment hydrology remain elusive.

Many methods have been used to determine ET rates in watersheds. Since atmospheric vapor flux is difficult to measure directly, most methods monitor the change of water in the system. Potential ET (PET), the amount of ET that would occur if unlimited water were available, can be measured using an evaporation pan or ET gauge. Pan data can also be used to estimate the actual ET, representing the ET that occurs when water is limited, for the vegetation of interest using relationships presented by Jensen et al. (1990). Lysimeters, soil water depletion, and the energy balance method have also been used to estimate ET (e.g., van Bavel, 1961), though measurements are difficult. Another approach, the water balance method, provides simple, effective estimates of ET rates if accurate stream gauging and precipitation data are available. This method is generally used for large watersheds, and compares water inputs (e.g., precipitation) and outflows (e.g., stream flow) for a given basin over long periods of time.

This paper uses the water balance approach to calculate ET rates by comparing precipitation and runoff data for watersheds. We expand on this traditional method to show how ET can be deconvolved into physical and biological components whose magnitudes vary both seasonally and with variations in basin properties, such as land use or hydrogeology. We also demonstrate how results for different basins can be directly compared to quantify ET disparities and out-of-basin gains or losses of water.

1.1. Previous work on water balance

The water balance approach entails determining the ET from the following equation:

$$ET = P - R \pm \Delta W \pm \Delta S \tag{1}$$

where P is precipitation, R is runoff, ΔW represents withdraws, diversions, or interbasin transfers, and ΔS is the change in groundwater and soil water storage (Ward & Trimble, 2004). The main assumptions of the model are: (1) the net groundwater flow across the watershed boundaries is zero and, hence, other than ΔW, the only inflow of water is P and the only outflows are ET and R; (2) the saturated and unsaturated storages are lumped in a single term (S); and (3) the entire water storage is accessible by plant roots in the watershed (Palmroth et al., 2010). For short intervals, ΔS is important and should be measured, but over an annual period it is normally small and to first order can be neglected, so:

$$ET = P - R \pm \Delta W \tag{2}$$

Previous workers assert that different land use practices have different and sizeable effects on ET and runoff rates (Dunn & Mackay, 1995; Gerten et al., 2004). In urban settings, impervious surfaces such as buildings, roads, parking lots, and other structures prevent rainfall from infiltrating; consequently, abnormally large fractions of runoff are directed into stream channels (Schilling & Libra, 2003). In contrast, in forest settings more water is evaporated directly off of tree leaves during throughfall, and once rainwater reaches the soil it is absorbed and transpired. Thus, in forested areas, ET is enhanced and runoff is reduced (Murakami et al., 2000). Crops and grasses can also intercept and transpire rainwater, but at significantly lower rates than in forests.

The following discussion tests and expands upon these concepts by comparing water balance calculations for several watersheds. Basins selected for analysis have long-term meteorological and discharge records as well as other special characteristics that simplify calculations or exemplify special processes and effects. Discharge and meteorological data are correlated with GIS data to determine the effects of land use on ET-runoff relationships.

1.2. Vegetation and transpiration

Vegetation and the water cycle, including the relationship between the ET rate and runoff generation, are intrinsically linked (Hutjes et al., 1998; Arora, 2002; Gerten et al., 2004). Basin water balance is a fundamental constraint on the productivity (Clark et al., 2001) and distribution (Stephenson, 1990) of terrestrial vegetation. Similarly, the plant community structure and geographic location are of primary importance for ET and runoff generation dynamics

(Dunn & Mackay, 1995). Transpiration accounts for the movement of water within a plant and the subsequent loss of water as vapor through stomata in its leaves, and the type and abundance of vegetation significantly affects the overall ET rate. Plant communities influence runoff processes in numerous ways, including phenology (Peel et al., 2001), plant maturity (Neilson, 1995), leaf area (Kergoat, 1998), stomatal behavior (Skiles & Hanson, 1994), and rooting strategy (Milly, 1997). In turn, processes such as albedo, interception (Eckhardt et al., 2003), percentage of soil cover, solar radiation, humidity, temperature, and wind (Swank & Douglass, 1974) affect plants transpiration rates. Herbaceous plants generally transpire less than woody plants because they typically have less extensive foliage. Conifer forests tend to have higher ET rates than deciduous forests, which is primarily due to the enhanced amount of precipitation intercepted, evaporated, and transpired by conifer foliage during the winter and early spring seasons.

It is well established that reduced forest cover decreases the ET rate and subsequently increases basin runoff, whereas reforestation typically lowers runoff (Bosch & Hewlett, 1982). For a large part of the southeastern Unites States, forested areas, over a period of decades, have transpired about 33 cm/y (area-depth) more than other land covers (Trimble et al., 1987). Vegetation also effects ET on the global scale; for instance, the variability of annual runoff between continents is controlled not only by differences in precipitation but also by the geographical distribution of various types of vegetation (e.g., evergreen vs. deciduous; Peel et al., 2001). In areas that are not irrigated, actual ET is usually no greater than precipitation, with some buffer in time depending on the soil's ability to hold water. Actual ET will usually be less than precipitation because some water will be lost due to percolation or surface runoff. An exception is areas with high water tables, where capillary action can cause water from the groundwater to rise through the soil matrix to the surface. If PET is significantly greater than precipitation, then soils will dry out if not irrigated.

The role of vegetation in the hydrologic cycle has been extensively studied (Horton, 1919; Wicht, 1941; Penman, 1963; Bosch & Hewlett, 1982; Turner, 1991), and these investigations have generally been split into two categories. The first involves "paired-catchment" experiments, and comparisons among > 90 catchments revealed large variations in runoff response attributable to differences in vegetative cover, with the catchments that have lower forest cover showing increased water yield due to lower ET (Hibbert, 1967; Bosch & Hewlett 1982). The second category, i.e., "single-catchment" water balance studies, is also used to determine the impact of vegetation on runoff. These studies were not designed to study the specific effects of land use on ET rates and water balance, but encompass a diverse group of catchments with different climates, vegetation, and soil types, and thus provide useful information about the role of vegetation in catchment water balance.

1.3. Hydrogeology and runoff

In addition to type and density of vegetation in a watershed, runoff volumes in the stream channel are controlled by the geography and hydrogeology of the basin. The size (Criss & Winston, 2008), shape (Hodge & Tasker, 1995), and orientation relative to the storm path (Ward & Trimble, 2004) of the watershed affect the rate of runoff delivery. Lithology and soil type significantly influence runoff volumes because permeability varies enormously for different geologic materials, with karstic rocks, gravels, and sand being most permeable, and igneous and metamorphic rocks, shales, and clay being least permeable (Bureau of Reclamation, 1977). If recent rainfall cannot penetrate into the subsurface, it will not be stored and subsequently transpired. In addition, basin topography, especially slope, exerts great influence on the transmittal of water to stream channels. Further, closed depressions can direct water to the groundwater system in karst areas or can cause ponding that enhances ET. Rainfall intensity and rainfall duration also influence runoff percentages and rates, and if they are high, are major causes of flash floods.

Anthropogenic activities also modify the hydrology of basins. Urban watersheds are particularly vulnerable to flash flooding due to the high percentage of impervious surface (low permeability), such as roads, roofs, sidewalks, and parking lots (Konrad, 2003). However, rainfall-runoff relationships in developed areas are highly complicated because of storm sewers and detention basins. Interbasin transfers are also possible, especially where storm and sanitary sewer systems are combined.

1.4. Data sources

Meaningful water balance calculations require accurate, long-term discharge and meteorological records. The US Geological Survey (USGS) currently maintains nearly 8,000 real-time gauging stations that monitor stage and/or discharge of streams and rivers in the United States (Wahl et al., 1995). The monitored watersheds vary greatly in size, climate, lithology, land use, engineering modifications, and other anthropogenic impacts, and thus a huge and diverse database is available for analysis. Long-term records of annual and monthly discharge are available online for many of these sites (USGS, 2012a). To avoid confusion and to simplify correlations with rainfall records, in the following discussion and diagrams we always use calendar years, not the USGS "water year."

Annual and monthly precipitation data were obtained from several National Oceanic and Atmospheric Administration weather stations (NOAA, 2012). The closest weather station to a given basin that had essentially complete records was used to calculate the average precipitation for the catchment. To evaluate the influence of land use and vegetative cover on ET, land use/land cover GIS data from the 2006 National Land Cover Database (USGS, 2012b) were used. The catchment area above each discharge gauging station was calculated and percentages of each type of land use/land cover were generated using ArcGIS 10 software.

1.5. Hydrologic setting of the meramec river basin

The Meramec River, which drains a 10,300 km^2 area of east-central Missouri, USA (Fig. 1), has many special characteristics that render it optimal for the study of ET rates and runoff generation processes, so special reference is made to it below. First, the Meramec River is one of the few remaining large, unimpounded rivers in the United States, as it has been spared from the engineering works and flood management practices found on practically all other waterways in the United States (Jackson, 1984; Ruddy, 1992). Second, water balance relationships are further simplified because the basin has a low population density and negligible withdrawals. Third, very long (46 to 90 years) discharge records are available for eight gauging stations in the basin, and additional sites were monitored for a shorter interval or intermittently. Finally, the high accuracy of the available discharge data can be quantitatively established, justifying their use in making reliable assessments of ET rates in the different subbasins. In particular, the sum of annual flows for the three major subbasins matches that for the downstream station, when a minor adjustment for the evident difference in areas is made; the error is less than 2% (Fig. 2). For all the sites, the average annual runoff is around 32%, indicating that 68% of the precipitation has been removed from the system. This is similar to the average annual ET rate for the continental United States, which is close to 66% of the annual rainfall (US Water Resource Council, 1978).

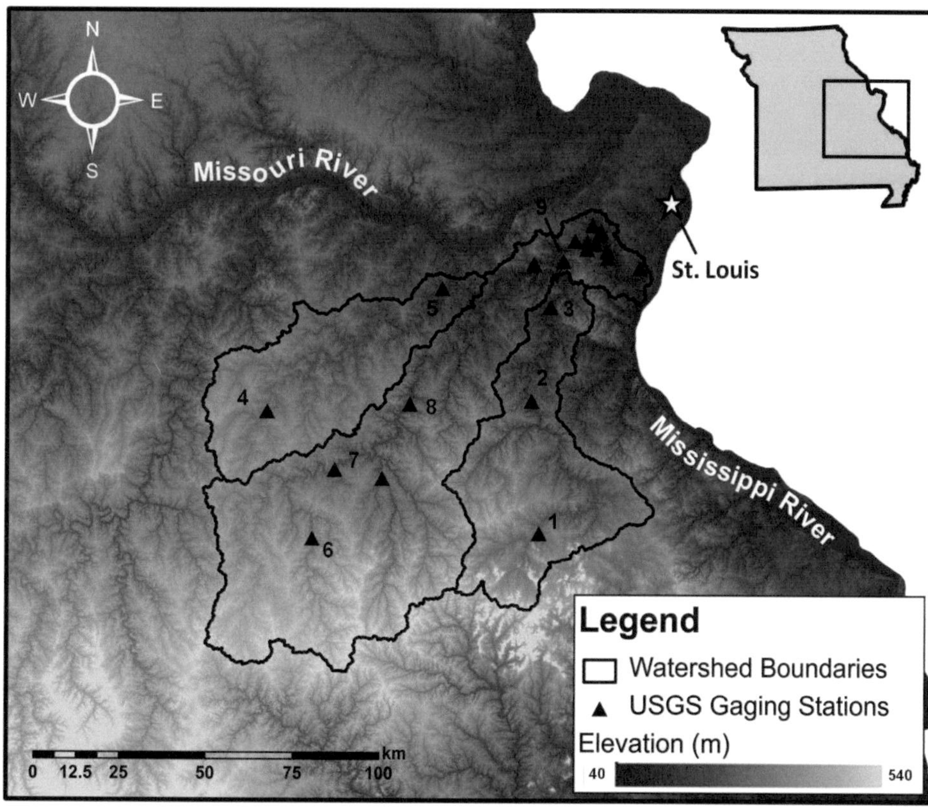

Figure 1. Shaded relief map of east-central Missouri, USA showing the whole Meramec River basin (10,300 km^2) and the Bourbeuse River (2,183 km^2) and Big River (2,513 km^2) subbasins; the City of St. Louis (open star) is shown for reference and the closed triangles are USGS gauging stations in the basin. Selected gauging stations are labeled (1 = Big River at Irondale; 2 = Big River near Richwoods; 3 = Big River at Byrnesville; 4 = Bourbeuse River near High Gate; 5 = Bourbeuse River at Union; 6 = Meramec River at Cook Station; 7 = Meramec River near Steelville; 8 = Meramec River near Sullivan; and 9 = Meramec River at Eureka). The elevation in the map area ranges from 390 m in the southwest to 110 m along the Mississippi River in the southeast; digital elevation model (DEM) base map data generated by the USGS are provided by MSDIS (2012).

The Meramec River and its two main tributaries, the Bourbeuse and Big Rivers, flow generally north and northeast until joining the Mississippi River south of St. Louis (Fig. 1). The flow pattern is asymmetrical as the basin lies on the northeastern flank of the Salem Plateau in east-central Missouri (Fenneman, 1938), and includes the foothills of the Ozark Mountains. Relief is greatest in the south and gradually decreases northward into the rolling hills of the Bourbeuse River subbasin. The unconfined Ozark aquifer crops out throughout the area and predominantly consists of lower Paleozoic dolostone and limestone units that underlie thin soils (Imes & Emmett, 1994). The basin features diverse karst topography including many springs, losing and gaining streams, 'swallow holes,' and sinkholes, which allow rapid connection between surface water and groundwater reservoirs (Vandike, 1995). Recharge occurs exclusively through infiltration of rainwater, with annual precipitation averaging ~ 100 cm and monthly totals for April and May usually exceeding 10 cm.

In addition to its proximity to the unimpounded Meramec River, the St. Louis region is optimal for the study of ET phenomena because discharge data are also available for a diverse suite of small streams. In particular, the USGS currently maintains 39 gauging stations in the City of St. Louis and St. Louis County (USGS, 2012a) that quantify discharge in watersheds that vary in area from 0.65 to 215 km^2, and include urban, industrial, commercial, residential, agricultural, and rural forested land use. Discharge records for most of these sites span 5 to 15 years.

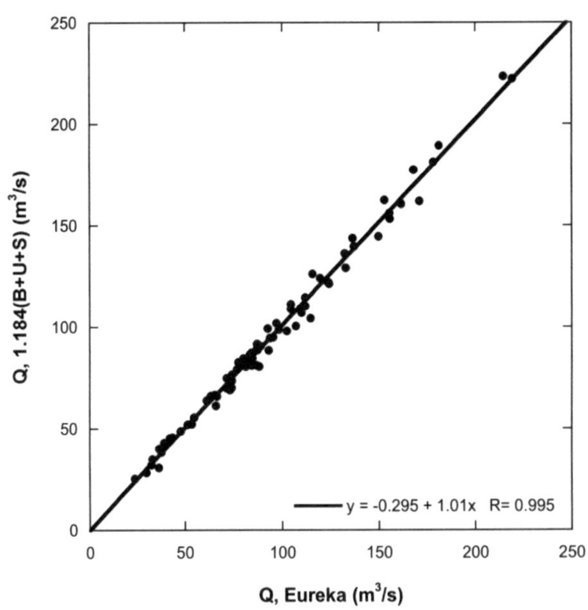

Figure 2. When multiplied by 1.184, the combined annual discharge for the Big River at Byrnesville (B; 2,375 km^2), Bourbeuse River at Union (U; 2,093 km^2), and upper Meramec River near Sullivan (S; 3,820 km^2) compares very closely with the discharge of the lower Meramec River measured at Eureka (9,811 km^2, see Fig. 1). The factor of 1.184 represents the quotient of the relevant respective basin areas, 9,811 km^2/(2,375 km^2 + 2,093 km^2 + 3,820 km^2). The unit slope, small y-intercept, and high correlation coefficient of the regression line attest to the high quality of these discharge data, which are continuous at all four sites since 1923 except for Sullivan from 1933 to 1943.

2. Results

The observed difference between rainfall delivered to the watershed and the resultant stream discharge can be used to determine the ET rate. For a river with negligible withdrawals or out-of-basin gains or losses, and over a sufficiently long interval, eq. 2 applies and can be rewritten as:

$$R = P - ET \tag{3}$$

This equation provides an important means of determining the average ET by simply subtracting the long-term mean value of runoff from that of precipitation. The equation also suggests a straightforward graphical procedure for determining ET, namely plotting observed discharge vs. observed precipitation for different months or years and determining ET from the y-intercept. Multiple complications and sources of confusion interfere with the latter approach.

2.1. Units of measure

Use of eq. 3 requires attention to the relevant units of measure. Because precipitation in the basin is measured as meters delivered over a specified interval of time, then both runoff and ET must be expressed in the same units for the same area and over the same time interval. Thus, the relevant runoff quantity, R, can be determined from the total discharge volume (in m^3) flowing out of the watershed over the specified time interval, divided by the contributing basin area (in m^2). The resultant unit, for example m/s, appears to be a "rate," but is actually a volumetric flux of water, $m^3/m^2/s$. All quantities in eq. 3 will therefore reduce to units of "rates," yet they actually all represent volumetric fluxes. Thus, the parameters in eq. 3 must all be expressed as cm/s, m/d, or m/y, etc., as is internally consistent and appropriate for any given case. This elementary matter has caused much confusion over the years, particularly because precipitation is normally reported and casually understood in length units, such as "cm" or "m" of rain. Less commonly precipitation is reported as a rate, (e.g., "cm/d," "m/y," etc.), which at least conveys acceptable units, yet precipitation is almost never considered as a volumetric flux, which is, in fact, what it actually represents.

2.2. Runoff vs. precipitation plots

Eq. 3 suggests that a graph of runoff vs. precipitation data will conform to a simple linear relationship with a negative y-intercept. A unit slope is expected for that relationship, because taken at face value, eq. 3 suggests that $\partial R/\partial P = 1$. In other words, in elementary algebraic parlance, where linear equations are understood to have the form $y = mx + b$, it appears that m in eq. 3 is equal to unity, giving $y = x + b$. Indeed, Fredrickson (1998) demonstrated that plots of discharge vs. precipitation for different parts of the Meramec basin define linear relationships with negative y-intercepts. However, when discharge is reported in appropriate flux units, the dimensionless slopes are only about 0.5 ± 0.2, depending on the site. Annual data for the Meramec River at Eureka and for a tributary near High Gate exemplify this result (Fig. 3).

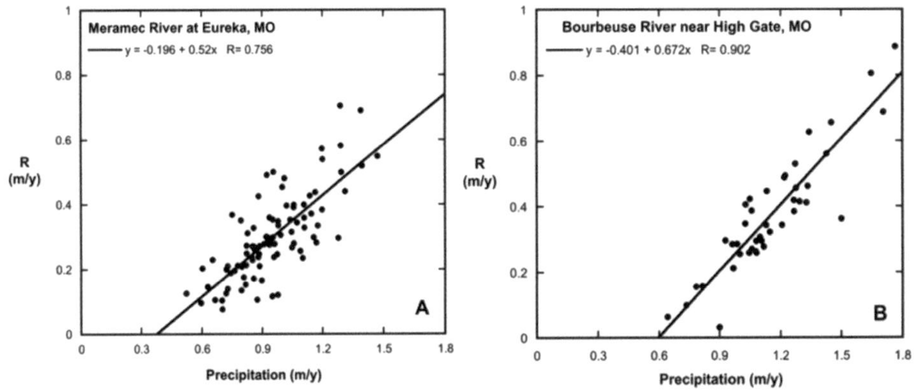

Figure 3. Graph of (A) runoff for the Meramec River at Eureka, Missouri vs. annual rainfall at St. Louis, Missouri for the years 1922 to 2011 and (B) runoff for the Bourbeuse River near High Gate, Missouri vs. annual rainfall at Rolla, Missouri for the years 1966 to 2010. Note that the slope and x-intercept are higher at the High Gate station than at the Eureka station.

In fact, for all basins we have examined, a linear correlation with a negative y-intercept is obtained when basin runoff (*R*) is plotted against precipitation (*P*), but unit slopes are never observed. Eq. 3 must be modified to:

$$R = mP + b \tag{4}$$

where *m* is the observed slope. Note that the dimensionless quantity *m* represents the fraction of the rainfall in excess of the x-intercept that becomes runoff. Moreover, because $m \neq 1$, *b* cannot represent negative ET.

Graphs of runoff vs. precipitation for numerous temperate-zone basins display a range of slopes (mostly 0.6 ± 0.25), and have positive x-intercepts that equal $-b/m$. The positive x-intercepts indicate that a certain amount of precipitation is "lost" every year before runoff is generated, equaling approximately 35 ± 20 cm in Missouri. Possible reasons for this "lost" rainfall are evaluated below and include ET processes and groundwater recharge. Once this fixed amount of water is removed from the basin, the river channel then receives a fixed fraction of the remaining rainfall budget, which is represented by the slope of the line. However, even this "excess" rainfall is still subject to additional ET losses, as there is not a 1:1 slope between *R* and *P*. This result shows that there are at least two components to ET.

2.2.1. *Physical meaning of b*

The fundamental reason that $m < 1$ is that ET is a function of *P*. For this reason, eqs. 1 – 3, although clearly correct and almost universally invoked in hydrologic literature, are highly misleading. Despite all appearances, eq. 3 does not indicate that $y = (1 \times x) + b$, rather it states that $y = x + f(x)$, where $f(x)$, representing the ET, is an unknown function of precipitation. Fortunately, this function is readily determined from available data.

Comparison of eqs. 3 and 4 shows that:

$$ET = (1-m)P - b \tag{5}$$

In effect, ET is seen to consist of two components, one that depends on the amount of precipitation delivered, equal to $(1-m)P$, and the second ($-b$) that is independent of precipitation. In other words, the second component represents the base ET that exists even when the precipitation delivered is zero. It is logical, though not perfectly accurate, to conceptualize the first ET component as physical evaporation effects on standing water, and the second as plant transpiration. Practically no physical evaporation can occur when wet surfaces are absent, yet even in times of little or no rainfall, plants would wilt and die if they were unable to extract large quantities of moisture from soils. Thus, a somewhat simplistic but conceptually useful interpretation of eq. 5 is:

$$ET = (1-m)P + ET_0 \tag{6}$$

where $(1-m)P$ mostly represents physical evaporation but includes "excess" transpiration, and ET_0 is the minimum allowable transpiration, numerically equal to $-b$.

In the general case where eq. 1 applies, the right hand side of eq. 6 would need to include the terms $\pm \Delta W$ and $\pm \Delta S$, and the latter quantities could both be functions of P, which would greatly complicate interpretations. In what follows, we find it convenient to refer to two equations that are only slightly more complicated than eq. 6, namely:

$$\begin{aligned} ET &= (1-m)P + (ET_0 \pm \Delta W) \quad \text{(a)} \\ &\text{and} \\ ET &= (1-m)P + (ET_0 \pm \Delta S) \quad \text{(b)} \end{aligned} \tag{7}$$

where $\pm \Delta W$ and $\pm \Delta S$ are not neglected but are viewed as constants. We will show that the form of $-b$ in eq. 7a is helpful in analyzing annual runoff and precipitation data, while the form of $-b$ in eq. 7b is helpful in interpreting monthly data.

Eqs. 5 – 7 suggest another procedure for estimating ET, which would be to graph the quantity $P - R$, representing the estimated ET, directly against P. Ideally, the slope m^* on this graph would equal the quantity $1-m$, and the y-intercept b^* would equal $-b$. Although this construct has certain advantages for visualizing ET, it masks an induced correlation between the x- and y-axes. This defect is considerable because the plotted variables on such a graph are not linearly independent. Among other problems, random errors in the measurements would lead to a systematic overestimation of the slope m^*. In contrast, when R is plotted directly against P as in Fig. 3, the y- and x-axes represent entirely different, independent variables.

2.2.2. *Mean annual ET*

Many direct measurements quantify evaporation rates from pans (Farnsworth & Thompson, 1982), and far fewer measurements quantify ET rates using lysimeters (e.g., van Bavel, 1961). Some germane examples are shown in Fig. 4; the data suggest that the annual ET rate in the eastern USA is ~ 0.8 m/y, while the pan rate is ~ 1.3 m/y for the indicated sites. Of course, these rates depend on location and they vary year to year, but as a rule of thumb, ET appears to be about 63% of the annual pan rate.

It is useful to compare the total, mean ET from lysimeters (Fig. 4) with the long-term average for the Meramec basin, using eq. 5 and the regressions given in Fig. 3. For Eureka, the mean ET is 0.65 m/y given the average rainfall of 0.95 m/y for the relevant interval, and at High Gate, ET is 0.78 m/y given the mean precipitation at Rolla of 1.15 m/y for the relevant interval. Thus, over many decades in the Meramec basin, ET is about 68% of total precipitation while the complementary runoff is about 32%.

It is reassuring that the mean ET values secured for the Meramec basin are in reasonable agreement with the mean annual ET data provided by van Bavel (1961; Fig. 4) for several sites in the eastern USA. More importantly, eq. 5 provides a means of showing how ET depends on the annual precipitation. In particular, in years with the lowest observed precipitation, ET is observed to be > 90% of *P*, while in years having the most rainfall, ET can be < 60% of *P*.

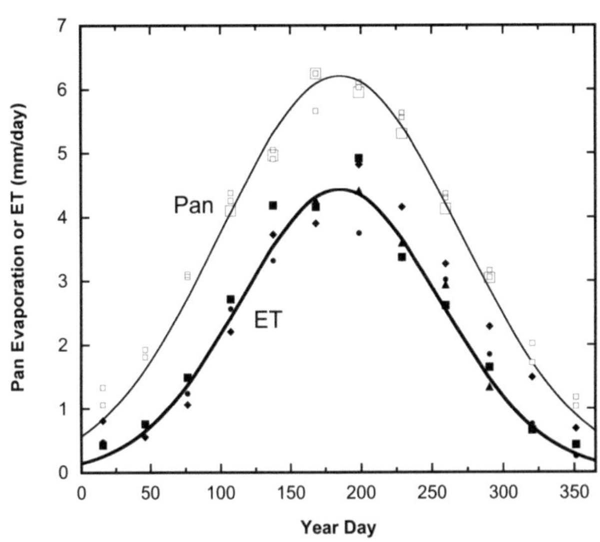

Figure 4. Pan evaporation (open symbols) from Weldon Spring, Missouri (large square) and from Russellville and Stuttgart, Arkansas (small squares), compiled in Farnsworth & Thompson (1982). Weldon Springs pan data were not collected during winter and early spring, likely due to freezing, but data for the available months is similar to that measured at the selected Arkansas sites, which offer complete annual coverage. ET data (solid symbols) are from van Bavel (1961) for Coshocton, Ohio (square); Seabrook, New Jersey (diamonds); Raleigh, North Carolina (triangles); and Waynesville, North Carolina (circles). The data approximate simple bell curves (solid lines), notably $E_{pan} \sim 6.22\ Exp\{-0.00007(Year\text{-}Day - 185)^2\}$, and $ET \sim 4.44\ Exp\{-0.0001(Year\text{-}Day - 185)^2\}$; modeled after Criss & Winston (2008).

2.2.3. *Seasonal behavior of ET*

Theoretically, the seasonal behavior of ET can be defined from graphs of runoff vs. precipitation constructed for each month; however, several matters interfere with this approach. The ET determinations require that the change in groundwater storage over the interval of interest is small, a condition much more likely to be realized over an annual cycle than over a short interval. Consequently, large basins are not well suited for monthly analysis. For example, the lower Meramec basin would be a poor choice, as it features considerable seasonal variations in groundwater storage, indicated by its hydrologic residence time of ~ 3 months determined from oxygen isotope data (Frederickson & Criss, 1999). Small basins are much more likely to have short storage constants, but few have long records and they tend to not be gauged as accurately as large basins; a result of their flashy nature. Due to such problems and short-term weather vagaries, the monthly regressions at individual sites were found to have rather low correlation coefficients, causing uncertainties in the slopes and y-intercepts.

Nevertheless, in an attempt to define monthly relationships between runoff and precipitation, we searched for small basins in Missouri and Illinois that have long-term records and proximal meteorological stations. Those selected for examination in Missouri include the Bourbeuse River near St. James (60 km^2; USGS #07015000) and at High Gate (350 km^2; USGS #07015720), Little Beaver Creek (16.6 km^2; USGS #06931500) near Rolla, and in Illinois include Indian Creek (95 km^2; USGS #05588000) at Wanda, Asa Creek (20.8 km^2; USGS #05591500) at Sullivan, and Farm Creek (71 km^2; USGS #05560500) at Farmdale.

Fig. 5 shows the monthly variations in slope and y-intercept for these particular sites. In each case, the variations are "noisy" over an annual cycle, but taken as a group, the variations show systematic behavior. The slopes (Fig. 5A) are steepest in the cold months, consistent with low physical evaporation and enhanced runoff due to frozen ground, and are much smaller during the hot, sunny, dry months when physical evaporation is high. When the data are inverted or graphed as $1 - m$, they feature a "humped" pattern, but one that is skewed compared to the pan and ET data plotted in Fig. 4. This shape probably reveals a key characteristic of physical evaporation disclosed by this analysis. In particular, due to leaf growth, the effective surface area of a basin late in the year is much greater than that early in the year. This effect could produce the asymmetry in the curve because physical evaporation rates depend on surface area, which changes over an annual cycle in natural settings, but not in an evaporation pan. Similar seasonal asymmetry has been calculated by Hoskins (2012).

Fig. 5B shows the monthly variations in y-intercept for these particular sites. The y-intercepts are near zero during winter, when plants are inactive, confirming the expectation that this term is related to ET_0 as implied by eqs. 6 and 7b. This expectation is also consistent with the strongly negative y-intercepts during spring, when plant growth is rapid. Surprisingly, the y-intercepts resume low values during the hottest months, followed by a second minimum in fall, then recover to low values with winter's approach, defining a "w-shaped" pattern over an annual cycle. In the latter half of the year, it is useful to interpret $-b$ using eq. 7b. During the hot summer months when large fractions of precipitation are lost to physical evaporation, plants strive to conserve water, and much of the transpired water is derived from soils, which subsequently dry out. In fact, it appears that ET_0 and ΔS approximately offset each other during summer, so the intercept representing their sum is small (eq. 7b). In fall when plant activity is greatly reduced and ET_0 is small, rainfall replenishes dry soils before runoff is generated, so this second minimum represents the "repayment" of soil moisture (ΔS) losses incurred during summer.

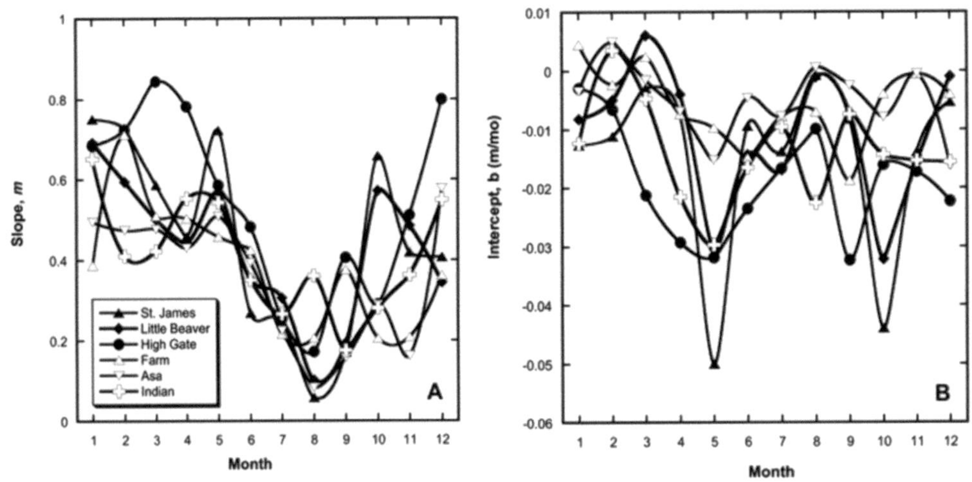

Figure 5. Seasonal variations of (A) slope *m* and (B) intercept *b* determined from monthly regressions using eq. 4, for several small watersheds in Missouri (solid symbols) and Illinois (open symbols) with extensive records..

3. Interbasin comparisons

On an annual basis, the total rainfall delivered to geographically proximal basins is similar, and changes in storage are rather small. This feature can be exploited to directly compare the runoff of proximal river basins to define differences attributable to land use disparities or due to out-of-basin transfers (eq. 7a). Consider that two proximal basins, denoted by subscripts 1 and 2, have annual runoff vs. precipitation regressions that can be expressed in the form of eq. 4:

$$R_1 = m_1 P_1 + b_1 \text{ and } R_2 = m_2 P_2 + b_2 \qquad (8)$$

If P_1 and P_2 are similar for these basins, a direct consequence is that:

$$R_2 = (m_2 / m_1)\, R_1 + \{(m_2 / m_1)\, b_1 - b_2\} \quad \text{(a)}$$

So: (9)

$$\partial R_2 / \partial R_1 = m_2 / m_1 \quad \text{(b)}$$

Eqs. 9a and 9b provide a means to directly compare basin runoff and to gather insights about their relative amounts of ET (Table 1). Importantly, note that the slopes and y-intercepts on such diagrams are *not* the same as *b* and *m* derived from runoff vs. precipitation plots (e.g., eqs. 9a and 9b).

Site	USGS Site Number	Area (km²)	Altitude‡ (m)	Intercept† (m/y)	Slope†	$r^*_{(m,b)}$	Years of Data
Irondale	07017200	453	229.6	-0.004	1.170	0.919	46
Richwoods	07018100	1,904	159.4	+0.002	1.079	0.967	62
Byrnesville	07018500	2,375	132.2	+0.002	1.079	0.970	89
High Gate	07015720	350	244.5	-0.016	1.248	0.938	46
Union	07016500	2,093	148.9	-0.009	1.057	0.961	90
Cook Station	07010350	515	263.5	-0.019	0.881	0.963	18
Steelville	07013000	2,023	207.8	-0.001	0.896	0.958	89
Sullivan	07014500	3,820	177.3	+0.003	0.951	0.978	79
Eureka	07019000	9,811	123.2	0.000	1.000	1.000	92

‡ Altitude of gauging station.

† Note that these slopes and y-intercepts are not the same as m and b (e.g., see eq. 9b).

r^* is the correlation coefficient.

Table 1. Regression lines for runoff in numerous subbasins of the Meramec River compared to runoff from the lower basin measured at Eureka.

3.1. Interpretation of slope on Runoff vs. Runoff plots

In theory, a runoff vs. runoff plot would provide a complete characterization of ET effects in basin 2 if those effects in basin 1 were fully characterized, i.e., if m_1 and b_1 are known. This supposition requires that P is highly similar in the basins being compared. For example, any systematic, proportional differences in P would directly factor into the slope on this diagram. Interpretations are best for small, proximal watersheds, and even this restriction may be insufficient in mountainous areas where P varies strongly with altitude, rain shadow effects, etc.

Given this caveat, on runoff vs. runoff plots, approximately unit slopes indicate that the physical ET losses in basin 2 are similar to those of basin 1, whereas low slopes (< 1) indicate that the losses for basin 2 exceed those of basin 1, and high (> 1) slopes indicate the opposite. An example is shown for two subbasins in the Meramec basin, one with considerable pastureland and the other dominated by forested land, which are compared to the main stem of the upper Meramec River that in all key aspects (see Table 2) has intermediate character (Fig. 6).

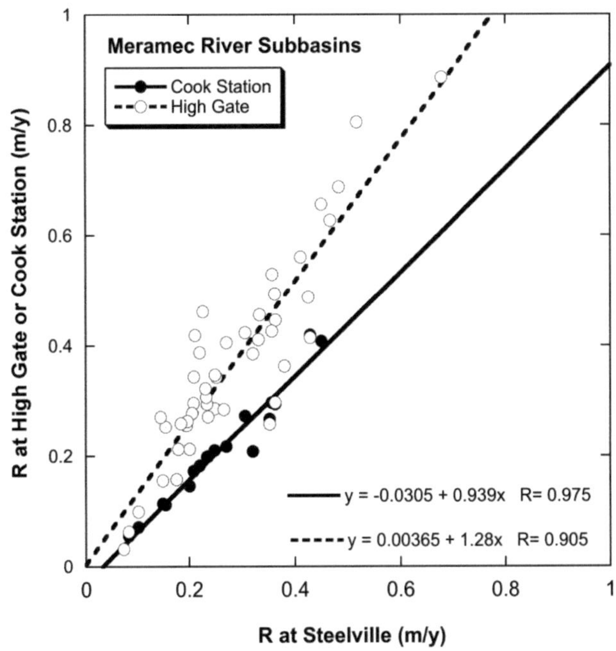

Figure 6. Comparison of runoff from three subbasins in the upper Meramec River basin. Compared to the reference basin near Steelville, the upper Bourbeuse River above High Gate (open circles) has much more pastureland, while the upper Meramec River above Cook Station (closed circles) is dominantly forested. Note that the slope for High Gate is > 1 but that for Cook Station is < 1, indicating that forests have high physical ET. The y-intercepts are small, but the intercept for Cook Station is clearly negative.

3.2. Interpretation of y-intercept in Runoff vs. Runoff plots

It is both expected and observed that y-intercepts are normally small on runoff-runoff plots (e.g., Fig. 6). In fact, the errors in the regression equations may normally overwhelm any small actual differences in the ET_0 values of the watersheds that would dominate the magnitude of this quantity (eqs. 6 and 9a). However, in cases where significant storage effects occur or where there are large transfers of water into or out of a watershed, the y-intercepts can be large and significant. In such a case, ΔW in eqs. 1, 2, and 7a cannot be neglected.

Transfers of water between proximal subbasins are affected by elevation. That is, high areas tend to lose water to the groundwater system that flows to regions of lower head and normally, but not necessarily, is discharged at lower elevation along the same stream. Data for numerous gauged sites in Missouri show the tendency for high altitude subbasins to have below average runoff, illustrating this effect. For example, compared to lower basin runoff near Eureka, the upper subbasins of the Meramec River (Irondale, High Gate, Cook Station, and Steelville) all have negative y-intercepts relative to the downstream site near Eureka (Table 1).

An extreme example of an interbasin transfer is provided by the Chicago Sanitary and Ship Canal in Illinois. Most surface waters and all wastewaters in the Chicago region flow away from, or are diverted away from, Lake Michigan into the Illinois River system. Ostensibly, the area of the contributing watershed at the canal gauging station is 1,914 km^2; however, this canal also receives wastewater discharge from the Stickney Treatment Plant, whose average output of ~ 6 million m^3/d ranks it as the world's largest. That output, representing an average of ~ 70 m^3/s, primarily represents water originally drawn from Lake Michigan that is subsequently treated to provide the municipal water supply of Chicago. Following use and then cleanup at Stickney, all this water is diverted from the Great Lakes watershed into the Mississippi River watershed, via the canal.

Fig. 7 compares runoff for the Chicago Sanitary and Ship Canal (USGS #05536995) to that of the Vermilion River in east-central Illinois. The regression is poor because approximately 75% of the flow in the canal is treated wastewater, derived from outside the basin. Nevertheless, that is the relevant point. The y-intercept in this case is huge, greatly exceeding the total rainfall normally delivered to this "watershed," and its value independently quantifies the total, average, man-made contribution to the canal's flow as ~ 80 m^3/s.

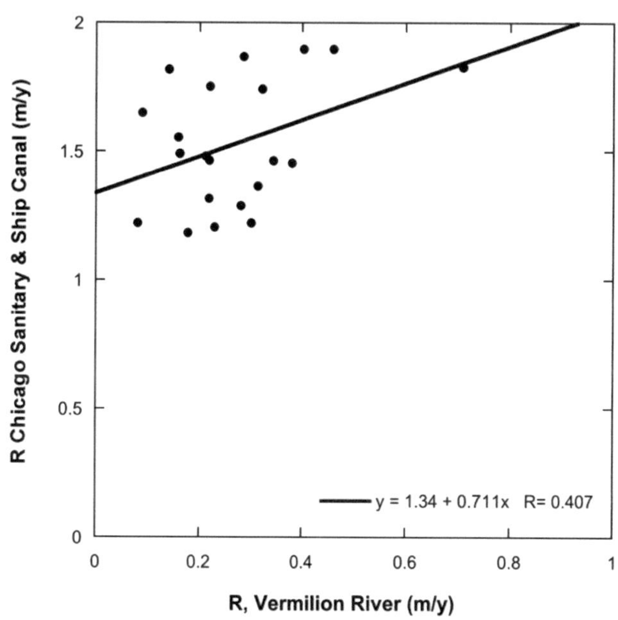

Figure 7. Comparison of runoff in the Chicago Sanitary and Ship Canal (USGS #05536890) to that in the Vermilion River at Pontiac, Illinois (USGS #05554500). The huge y-intercept represents the artificial transfer of ~ 80 m^3/s of water into the canal.

3.3. Land use and ET

Different land use practices have different effects on the ET and runoff rates. For example, it is well known that conversion of forested areas to urban or agricultural areas causes increased runoff (Bosch & Hewlett 1982). This effect is due to the reduction in ET in urban environments because of the decreased vegetative coverage. As noted above, in forested areas physical evaporation is also enhanced by leaf area, due to "interception" of precipitation. Crops and grasses can also intercept and transpire rainwater, but at lower rates than in forested ecosystems.

Land use data were compiled for all the subwatersheds in the Meramec basin (Table 2). The data confirm that the more heavily forested the subbasin, the lower the runoff (Fig. 8). Subbasins with higher percentages of pasture/hay and cultivated crops had increased runoff relative to forested areas; a result of the smaller surface area from which water is transpired by these grasses and smaller plants. However, urban land use had the largest impact on ET rates and strongly increased runoff when compared to forest.

Station	Open Water	Developed	Barren Land‡	Forest	Wetland	Grass-land	Scrub/ Shrub	Cultivated Crops	Pasture/ Hay
Irondale	0.50	3.10	0.02	70.99	0.21	1.81	0.43	0.02	22.92
Richwoods	0.56	6.03	0.34	72.19	0.40	2.71	0.30	0.14	17.33
Byrnesville	0.63	6.39	0.29	72.38	0.54	2.57	0.24	1.00	15.96
High Gate	0.43	7.72	0.08	50.74	0.30	2.44	0.00	0.58	37.71
Union	0.49	6.30	0.07	56.51	0.61	2.04	0.09	2.08	31.81
Cook Station	0.21	3.47	0.06	77.39	0.34	1.81	0.01	0.09	16.62
Steelville	0.18	4.95	0.04	65.98	0.34	2.25	0.00	0.12	26.14
Sullivan	0.22	4.45	0.11	73.16	0.33	2.16	0.88	0.08	18.61
Eureka	0.40	4.94	0.16	73.56	0.46	2.06	0.70	0.66	17.06
All Basins	0.56	7.35	0.21	67.01	0.56	2.18	0.35	1.33	20.45

‡Rock, sand, and clay.

Table 2. Percentage of various types of land use in the different subbasins of the Meramec River watershed.

Another important factor that dramatically affects ET is bedrock geology. Distinct trends in the slope (Table 1) and the land use (Table 2) were observed in the western, shale-rich subbasins and the eastern, carbonate-hosted subbasins (Fig. 8).

3.4. Runoff dynamics of small and urban watersheds

Many have argued that impervious surfaces such as buildings, roads, parking lots, and other structures enhance runoff because these structures prevent water from infiltrating. Consequently, surface runoff is directly conveyed into stream and river channels. St. Louis is ideal for such study given the large number of small watersheds that are gauged. Examination of runoff relationships in all 39 small, gauged basins in City of St. Louis and St. Louis County revealed many large and sometimes inexplicable differences. Surprisingly, the area-weighted average runoff from all these basins was only slightly higher (~ 35%) than that for the Meramec basin (~ 32%). This result reveals a major complication. In urban areas, ΔW in the water balance equations can be highly important, as storm sewers can cause large intrabasin and interbasin

http://dx.doi.org/10.5772/52811

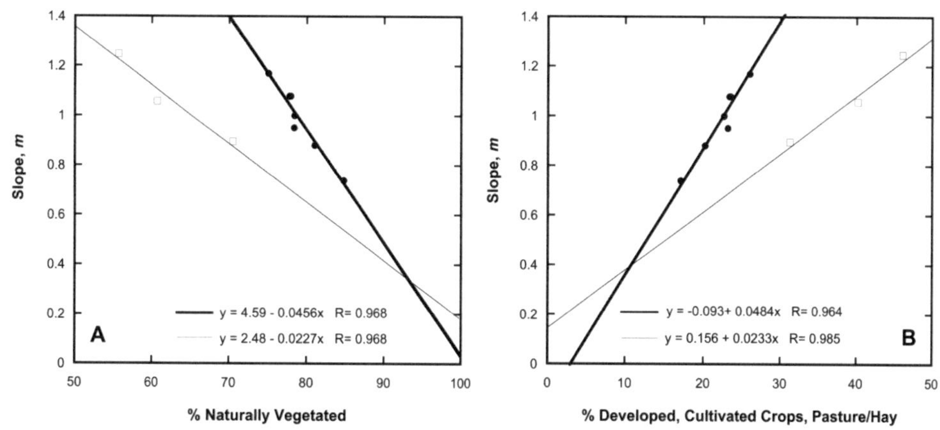

Figure 8. Plots of slopes (Table 1) vs. land use (Table 2) for the Meramec River basin (Fig. 1). (A) Demonstrates that the more heavily forested subwatersheds have the lowest runoff, while (B) shows that the urban development, cultivated crops, and pasture/hay increase the percentage of runoff. A distinct difference was observed between the behavior in the eastern (closed circles) and the western (open squares) portion of the watershed, attributed to lithologic differences.

transfers. This situation is particularly magnified by combined sewer systems, where stormwaters and sanitary discharges enter the same pipes, ultimately to be treated and released into major rivers. Moreover, in many residential areas, lawn irrigation can supply the equivalent of several inches of rainfall during summer months (Spronken-Smith & Oke, 1998).

We found the runoff-runoff plot to be particularly useful in interpreting discharge data in these small, moderately to intensely developed watersheds. The predominantly residential Creve Coeur watershed was selected as the reference basin as it was relatively large and behaved similarly to several other small watersheds in the area. Each graph in Fig. 9 contrasts runoff from Creve Coeur Creek (hereafter, CCC) to runoff from selected watershed pairs that display contrasting characteristics.

Fig. 9A compares runoff from the Kiefer Creek watershed and the Fishpot Creek watershed to CCC. The y-intercepts are large but one is positive and the other is negative. This feature exemplifies the large and opposite values for ΔW; in this case caused not by storm sewers, but by karst groundwater flow. That is, these contrasting sites respectively represent gaining and losing stream reaches, whereas no large springs occur in the CCC basin. This comparison is compelling because land cover is predominantly residential in all three watersheds. Fig. 9B compares runoff from the Black Creek watershed, which has extensive areas of impervious surface, and the mostly forested Williams Creek watershed to CCC. The profound difference in land cover is clearly reflected in the slopes of the regression lines. The y-intercepts are probably also significant; Williams Creek is gauged below several significant springs, while Black Creek contains several combined sewer lines and combined sewer overflows, and is probably losing stormwater runoff to the Mississippi River, where treated stormwaters are discharged. Fig. 9C compares runoff from the upper River des Peres watershed near University City and the Sugar Creek watershed to CCC. The upper River des Peres basin is residential and commercial, while the Sugar Creek basin has low-density residential development, so the former would be expected to have the highest slope, yet the opposite is seen. Both basins are underlain by shale-rich Pennsylvanian strata, so compared to the CCC watershed that has much more limestone, the slopes would be expected to be > 1. The likely cause of these disparate slopes is that the topographic relief in the upper River des Peres watershed is very low compared to CCC and Sugar Creek. Thus, the slopes of the trend lines suggest that the effect of topographic relief on runoff generation is more important than that of land cover in this case.

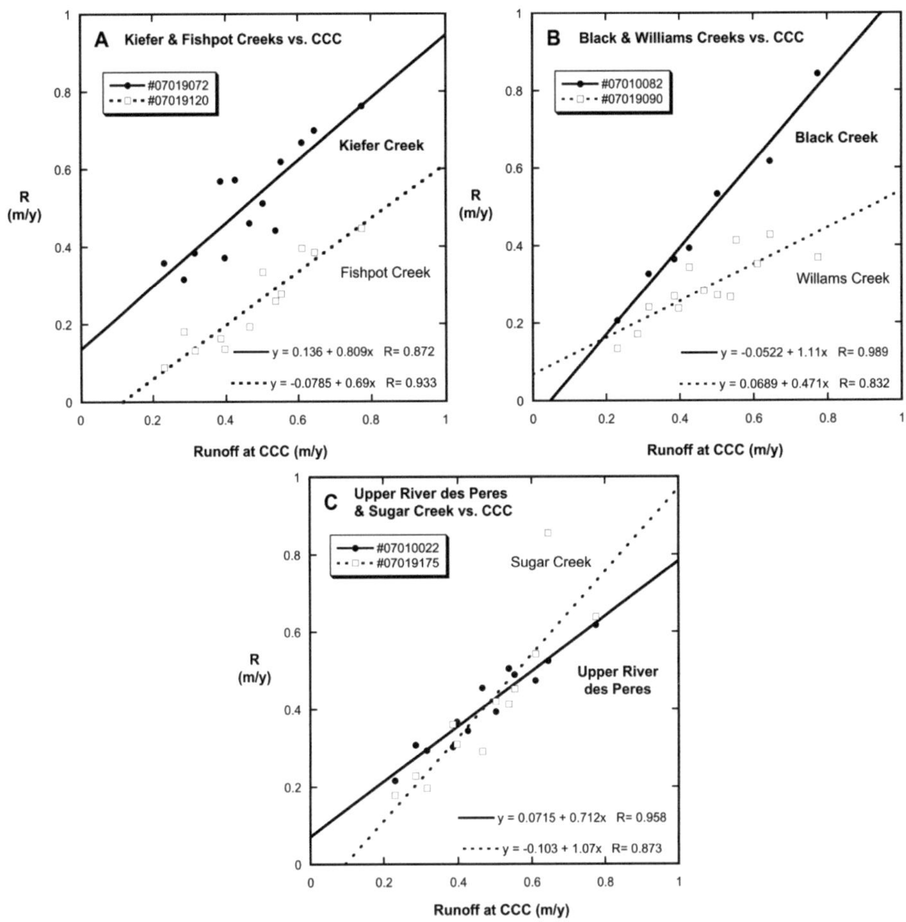

Figure 9. Comparison of runoff in several small watersheds to that of the 57 km^2 CCC watershed (USGS #06935890). (A) Runoff in the 10 km^2 Kiefer Creek watershed (closed circles) gauged just below a significant spring and that of the proximal 25 km^2 Fishpot Creek watershed (open squares) gauged just above a spring, both vs. CCC. Note the positive and negative effects of groundwater transfer on the y-intercept for these proximal watersheds. (B) Runoff in the 15 km^2 Black Creek watershed (closed circles) that has large areas of impervious surface and the mostly forested, 20 km^2 Williams Creek watershed (open squares) gauged below a spring, both vs. CCC. Note the effects of land cover on the slopes for these watersheds. (C) Runoff in the 23 km^2 watershed of the upper River des Peres watershed (closed circles) and that of the 13 km^2 Sugar Creek watershed (open squares), both vs. CCC, contrasting the effects of flat and steep topographic slopes.

4. Conclusions

The water balance equation provides an effective means to calculate the ET rate if long-term data precipitation and stream discharge are available for a given watershed. Simple estimates of ET can be made by subtracting the long-term mean values of runoff (*R*) from precipitation (*P*). Much richer information is provided by the fundamental graph, *R* vs. *P*, which depicts the relationship between these measured, independent variables over time intervals of interest. Annual data for *R* and *P* display good linear trends on this graph, but they show several surprising characteristics. First, a large x-intercept is seen on this graph, indicating that significant precipitation is "lost" before any runoff is generated; second, even after this demand is satisfied, the "excess" rainfall is subject to additional losses, so a 1:1 slope between *R* and *P* is not realized. We show how the y-intercept and slope of this plot can be used to deconvolve ET into different components that approximate physical evaporation and biological transpiration. We also show how the magnitudes of these quantities vary with basin character and throughout the year. We confirm standard expectations for runoff generation from different landscapes, such as high runoff fractions shed from impervious surfaces and low runoff generated by forested watersheds. However, over an annual cycle, we find that physical evaporation effects from natural basins are strongly skewed compared to the symmetrical, bell-shaped curves defined by pan data, an effect we attribute to changes in leaf area. Moreover, we show that short-term changes in soil moisture confuse monthly ET analyses.

Another graph, a direct graphical comparison of annual runoff from different, proximal basins, is very useful for estimating relative ET differences. More importantly, the y-intercepts on such plots both identify and quantify out-of-basin gains or losses of water. Such interbasin transfers can be very significant in karst areas due to groundwater flows, as well as in developed areas due to storm sewers, especially combined sewer systems.

Author details

Elizabeth A. Hasenmueller* and Robert E. Criss

Department of Earth and Planetary Sciences, Washington University, Saint Louis, USA

References

[1] Arora, V. (2002). Modeling vegetation as a dynamic component in soil-vegetation-atmosphere transfer schemes and hydrological models. *Rev. Geophys.* , 40, 3.

[2] Bosch, J. M, & Hewlett, J. D. (1982). A review of catchment experiments to determine the effect of vegetation changes on water yield and evapotranspiration. *J. Hydrol.* , 55, 3-23.

[3] Bureau of Reclamation(1977). *Ground Water Manual.* US Department of the Interior, Washington, DC, , 480.

[4] Clark, D. A, Brown, S, Kicklighter, D. W, Chambers, J. Q, Thomlinson, J. R, Ni, J, & Holland, E. A. (2001). Net primary production in tropical forests: an evaluation and synthesis of existing field data. *Ecol. Appl.* , 11, 371-384.

[5] Criss, R. E, & Winston, W. E. (2008). Discharge predictions of a rainfall-driven theoretical hydrograph compared to common models and observed data. *Water Resour. Res.* W10407., 44

[6] Dunn, S. M, & Mackay, R. (1995). Spatial variation in evapotranspiration and the influence of land use on catchment hydrology. *J. Hydrol.* , 171, 49-73.

[7] Eckhardt, K, Breuer, L, & Frede, H. G. (2003). Parameter uncertainty and the significance of simulated land use change effects. *J. Hydrol.* , 273, 164-176.

[8] Farnsworth, R. K, & Thompson, E. S. (1982). *Mean monthly, seasonal and annual pan evaporation for the United States.* NOAA Technical Report NWS 34, , 82.

[9] Fenneman, N. M. (1938). *Physiography of Eastern United States.* McGraw-Hill, New York, , 714.

[10] Frederickson, G. C. (1998). Relationship between the stable isotopes of precipitation and springs and rivers in east-central Missouri and southwestern Illinois. Master thesis, Washington University, St. Louis, Missouri, USA, , 235.

[11] Frederickson, G. C, & Criss, R. E. (1999). Isotope hydrology and residence times of the unimpounded Meramec River Basin, Missouri. *Chem. Geol.* , 157, 303-317.

[12] Gerten, D, Schaphoff, S, Haberlandt, U, Lucht, W, & Sitch, S. (2004). Terrestrial vegetation and water balance-hydrological evaluation of a dynamic global vegetation model. *J. Hydrol.* , 286, 249-270.

[13] Hibbert, A. R. (1967). Forest treatment effects on water yield. In: *Forest Hydrology*, eds. W.E. Sopper & H.W. Lull, Pergamon, Tarrytown, NY, , 527-543.

[14] Hodge, S. A, & Tasker, G. D. (1995). *Magnitude and frequency of floods in Arkansas.* US Geological Survey Water Resources Investigations Report, 95-4224, Little Rock, AR, , 275.

[15] Horton, R. E. (1919). Rainfall interception. *Mon. Weather Rev.* , 47, 603-623.

[16] Hoskins, J. (2012). Locally derived water balance method to evaluate realistic outcomes for runoff reduction in St. Louis, Missouri. *Watershed Sci. Bull.* , 3, 63-67.

[17] Hutjes, R. W. A, Kabat, P, Running, S. W, Shuttleworth, W. J, Field, C, Hoff, H, Jarvis, P. G, Kayane, I, Krenke, A. N, Liu, C, Meybeck, M, Nobre, C. A, Oyebande, L, Pitman, A, Pielke, R. A, Raupach, M, Saugier, B, Schulze, E. D, Sellers, P. J, Tenhunen, J.

D, Valentini, R, Victoria, R. L, & Vörösmarty, C. J. (1998). Biospheric aspects of the hydrological cycle. *J. Hydrol.* , 212, 1-21.

[18] Imes, J. L, & Emmett, L. F. (1994). *Geohydrology of the Ozark Plateaus Aquifer System in parts of Missouri, Arkansas, Oklahoma, and Kansas.* US Geological Survey Professional Paper 1414-D, , 127.

[19] Jackson, J. P. (1984). *Passages of a stream: a chronicle of the Meramec.* University of Missouri Press, Columbia, MO, , 138.

[20] Jensen, M. E, Burman, R. D, & Allen, R. G. (1990). *Evapotranspiration and Irrigation Water Requirements.* American Society of Civil Engineers, New York, , 332.

[21] Kergoat, L. (1998). A model for hydrological equilibrium of leaf area index on a global scale. *J. Hydrol.* , 212, 268-286.

[22] Konrad, C. P. (2003). *Effects of urban development on floods.* US Geological Survey Fact Sheet FS-076-03, , 4.

[23] Milly, P. C. D. (1997). Sensitivity of greenhouse summer dryness to changes in plant rooting characteristics. *Geophys. Res. Lett.* , 24, 269-271.

[24] Missouri Spatial Data Information Service (MSDIS)(2012). Data Resources. MSDIS: University of Missouri, http://www.msdis.missouri.edu.

[25] Murakami, S, Tsuboyama, Y, Shimizu, T, Fujieda, M, & Noguchi, S. (2000). Variations of evapotranspiration with stand age and climate in a small Japanese forested catchment. *J. Hydrol.* , 227, 114-127.

[26] National Oceanic and Atmospheric Administration (NOAA)(2012). National Weather Service (NWS) Weather: NWS, http://www.weather.gov/.

[27] Neilson, R. P. (1995). A model for predicting continental-scale vegetation distribution and water balance. *Ecol. Appl.* , 5, 362-386.

[28] Palmroth, S, Katul, G. G, Hui, D, Mccarthy, H. R, Jackson, R. B, & Oren, R. (2010). Estimation of long-term basin scale evapotranspiration from streamflow times series. *Water Resour. Res.* W10512., 46

[29] Peel, M. C, Mcmahon, T. A, Finlayson, B. L, & Watson, F. G. R. (2001). Identification and explanation of continental differences in the variability of runoff. *J. Hydrol.* , 250, 224-240.

[30] Penman, H. L. (1963). *Vegetation and hydrology.* Technical Communications 53, Commonwealth Bureau of Soils, Harpenden, England, UK, , 124.

[31] Ruddy, T. M. (1992). *Damming the dam: the St. Louis District Corps of Engineers and the controversy over the Meramec Basin project from its inception to its deauthorization.* US Army Engineer District, St. Louis, , 136.

[32] Schilling, K. E, & Libra, R. D. (2003). Increased baseflow in Iowa over the second half of the 20th century. *J. Am. Water Resour. As.* , 39, 851-860.

[33] Skiles, J. W, & Hanson, J. D. (1994). Responses of arid and semiarid watersheds to increasing carbon dioxide and climate change as shown by simulation studies. *Clim. Change.* , 26, 377-397.

[34] Spronken-smith, R. A, & Oke, T. R. (1998). The thermal regime of urban parks in two cities with different summer climates. *Int. J. Rem. Sens.* , 19, 3039-3053.

[35] Stephenson, N. L. (1990). Climatic control of vegetation distribution: the role of the water balance. *Am. Nat.* , 135, 649-670.

[36] Swank, W. T, & Douglass, J. E. (1974). Streamflow greatly reduced by converting deciduous hardwood stands to pine. *Science.* , 185, 857-859.

[37] Trimble, S. W, Weirich, F, & Hoag, B. L. (1987). Reforestation and the reduction of water yield on the Southern Piedmont since c. 1940. *Water Resour. Res.* , 23, 425-437.

[38] Turner, K. M. (1991). Annual evapotranspiration of native vegetation in a Mediterranean-type climate. *Water Resour. Bull.* , 27, 1-6.

[39] US Geological Survey (USGS). (2012a). USGS Real-time data for Missouri: USGS Real-time data for Missouri, http://waterdata.usgs.gov/mo/nwis/rt.

[40] US Geological Survey (USGS). (2012b). USGS Land Cover Institute (LCI): US Land Cover, http://landcover.usgs.gov/uslandcover.php.

[41] US Water Resources Council. (1978). *The Nation's Water Resources, 1975-2000: second national water assessment.* US Water Resources Council, Washington, DC.

[42] Van Bavel, C. H. M. (1961). Lysimetric measurements of evapotranspiration rates in the eastern United States. *Soil Sci. Soc. Am. Proc.* , 25, 138-141.

[43] Vandike, J. E. (1995). Missouri State Water Plan Series, Surface Water Resources of Missouri. *Water Resources Report*, Missouri Department of Natural Resources, , 1(45), 122.

[44] Wahl, K. L, Thomas, W. O, & Hirsch, R. M. (1995). *The stream-gaging program of the US Geological Survey*. US Geological Survey Circular 1123, , 22.

[45] Ward, A. D, & Trimble, S. W. (2004). *Environmental hydrology*, 2nd ed. CRC Press, Boca Raton, Florida, , 475.

[46] Wicht, C. L. (1941). Diurnal fluctuation in Jonkershoek streams due to evaporation and transpiration. *J. S. Aft. For. Assoc.* , 7, 34-49.

Chapter 2

Recent Updates of the Calibration-Free Evapotranspiration Mapping (CREMAP) Method

Jozsef Szilagyi

Additional information is available at the end of the chapter

http://dx.doi.org/10.5772/52392

1. Introduction

This study is the revision of an earlier evapotranspiration (ET) estimation technique [1], called Calibration-Free Evapotranspiration Mapping (CREMAP), applied for Nebraska (Figure 1). Major differences between the current and previous versions are

a. the ET rates of the winter months (December, January, February) are included in the current annual maps;

b. the method became largely automated and thus its application made simpler.

In principle, the recent modeling period could have been extended to include 2010 and 2011, however, the multi-institutional research project [2] that provided the monthly incoming global radiation values was terminated in 2010, thus no radiation data are available after June 2010 from that source. Rather than looking for alternative data sources, the original 10-year long modeling period, i.e., 2000-2009, was kept, thus ensuring that the same data types were employed throughout the study.

In the earlier version of the ET maps it was assumed that the ET rates in the winter time are negligible when one is concerned with the mean annual value. In the light of the present version of ET mapping, this assumption was found true only partially: there are regions within Nebraska for which winter ET indeed seems to be negligible (mostly the north-central and north-western parts) in comparison with water-balance data, while in other regions it is not so. These latter regions include parts of Nebraska (mostly the south and south-west portion) with the highest winter-time daily maximum temperatures and/or with most abundant precipitation (eastern, south-eastern portion of the state). As a result the precipitation (P) recycling ratio (i.e., ET / P) rose from a previously estimated mean annual value of 93% to 95%, leaving an estimated 5% of the precipitation to emerge as runoff (Ro) in the streams.

Naturally, as any estimation method, the current approach is not perfect. In the Pine Ridge Escarpment and in the Niobrara River Breaks regions (Figure 1) the ET estimates (similarly to the previous version of the ET map) had to be corrected via comparison with precipitation data because otherwise they would overestimate ET rates by about 10-20%. The reason is in the gross violation of the underlying hypotheses of the current ET estimation method in areas of very rough terrain. After the corrections in these distinct geomorphic regions, it is believed that the resulting ET rates are quite reasonable across the whole state. Overall, the method yields a state-representative ET rate value of 549 mm/yr, within 2% of the simplified water balance (P – Ro) derived rate of 538 mm/yr, employing the USGS [3] values of computed runoff for catchments with level-8 hydrologic unit codes (HUC), and explains 87% of the observed spatial variance of the water balance ET values among the HUC-8 catchments (there are 70 such watersheds within the state) for the same period.

2. Description of the current ET estimation method

An ET estimation method had been proposed by Bouchet [4], employing the complementary relationship (CR) of evaporation which was subsequently formulated for practical regional-scale ET applications by Brutsaert & Stricker [5] and Morton et al. [6]. In this study the WREVAP program of [6] was applied for the estimation of the regional-scale ET rates at monthly periods. Disaggregation of the regional ET value in space is based on the Moderate Resolution Imaging Spectroradiometer (MODIS) data [7] that have a nominal spatial resolution of about 1 km. The disaggregation is achieved by a linear transformation of the 8-day composited MODIS daytime surface temperature (T_s) values into actual ET rates on a monthly basis [1, 8] by first aggregating the composited T_s data into monthly mean values. Compositing is used for eliminating cloud effects in the resulting composite data by removing suspicious, low pixel-values in the averaging over each eight-day period. See [7] for more detail of data collection and characteristics.

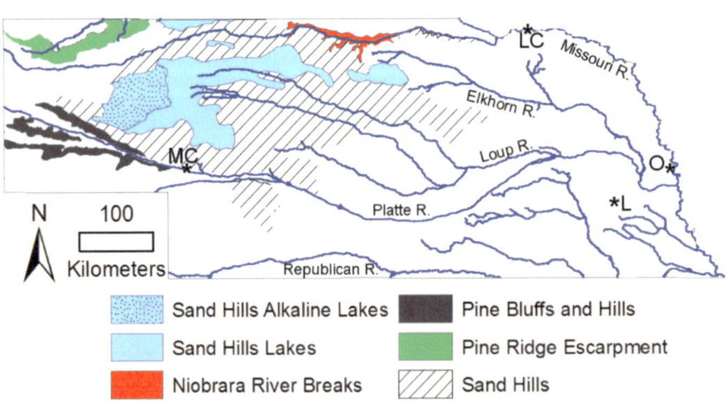

Figure 1. Stream network and selected geomorphic regions of Nebraska. MC: McConaughy Reservoir; LC: Lewis and Clark Reservoir; L: Lincoln; O: Omaha.

The transformation requires the specification of two anchor points in the T_s versus ET plane (Figure 2). The first anchor point is defined by the spatially averaged daytime surface temperature, $\langle T_s \rangle$, and the corresponding regionally representative ET rate, E, from WREVAP. (The original FORTRAN source code can be downloaded from the personal website of the author: snr.unl.edu/ szilagyi/szilagyi.htm). The second anchor point comes from the surface temperature, T_{sw}, of a completely wet cell and the corresponding wet-environment evaporation, E_w, (defined by the Priestley-Taylor [9] equation with a coefficient value of 1.2). The two points identify the linear transformations of the T_s pixel values into ET rates for each month. The resulting line is extended to the right, since in about half the number of the pixels ET is less than the regional mean, E. A monthly time-step is ideal because most of the watershed- or large-scale hydrologic models work at this time-resolution, plus a monthly averaging further reduces any lingering cloud effect in the 8-day composited T_s values. Wet cells within Nebraska were identified over Lake McConaughy and the Lewis and Clark Lake on the Missouri River (Figure 1). An inverse-distance weighting method was subsequently used to calculate the T_{sw} value to be assigned to a given MODIS cell for the linear transformation.

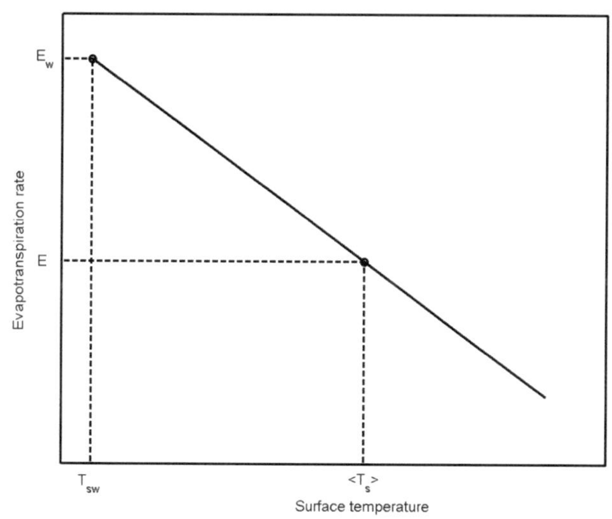

Figure 2. Schematics of the linear transformation of the MODIS daytime surface temperature values into ET rates (after [1]), applied in CREMAP.

The core assumption of CREMAP is that the surface temperature of any MODIS cell is predominantly defined by the rate of evapotranspiration due to the large value of the latent heat of vaporization for water and that the energy (Q_n) available at the surface for sensible (i.e., heat convection) and latent heat (i.e., evapotranspiration) transfers are roughly even among the cells of a flat-to-rolling terrain. Heat conduction into the soil is typically negligible over a 24-hour period, and here considered negligible over the daytime hours as well.

This last assumption is most likely true for fully vegetated surfaces where soil heat conduction is small throughout the day, and is less valid for bare soil and open water surfaces. While a spatially constant Q_n term at first seems to be an overly stringent requirement in practical applications due to spatial changes in vegetation cover as well as slope and aspect of the land surface, Q_n will change only negligibly in space provided the surface albedo (i.e., the ratio of in- and outgoing short-wave radiation) value also changes negligibly among the pixels over a flat or rolling terrain. For the study region, the MODIS pixel size of about 1 km may indeed ensure a largely constant Q_n value among the pixels since the observed standard deviation in the mean monthly (warm season) surface albedo value of 17% is only 1.2% among the MODIS cells.

A further assumption of the method is that the vertical gradient of the air temperature near the surface is linearly related to the surface temperature [10, 11], thus sensible heat (H) transfer across the land-atmosphere interface, provided changes in the aerodynamic resistance (r_a) among the MODIS pixels are moderate, can also be taken a linear function of T_s. This can be so because under neutral atmospheric conditions (attained for time-steps a day or longer) r_a depends linearly on the logarithm of the momentum roughness length, z_{0m} [11], thus any change in z_{0m} between pixels becomes significantly dampened in the r_a value due to the logarithm. Consequently, the latent heat (LE) transfer itself becomes a linear function of T_s under a spatially constant net energy (Q_n) term required by the CR, therefore $Q_n = H + LE$, from which $LE = mT_s + c$ follows, m and c being constants for the computational time step, i.e., a month here, within a region.

8-day composited MODIS daytime surface temperature data were collected over the 2000 – 2009 period. The 8-day composited pixel values were averaged for each month to obtain one surface temperature per pixel per month, except for December, January, and February. The winter months were left out of the linear transformations because then the ground may have patchy snow cover which violates the constant Q_n assumption since the albedo of snow is markedly different from that of the land. Therefore in the wintertime the WREVAP-derived regional ET rates were employed without any further disaggregation by surface temperatures but, rather, with a subsequent correction, discussed later.

Mean annual precipitation, mean monthly maximum/minimum air and dew-point temperature values came from the PRISM database [12] at 2.5-min spatial resolution. Mean monthly incident global radiation data at half-degree resolution were downloaded from the GCIP/SRB site [2]. While previously [1] the regions were defined by subdividing the state into eight distinct areas (a largely arbitrary process) for the calculation of the regionally representative values of the mean monthly air temperature, humidity and radiation data, required by WREVAP, now such a subdivision is not necessary. Instead, a "radius of influence" is defined over which the regional values are calculated separately for each designated MODIS cell, very similar to a temporal moving-average process, but now in two dimensions of space. In principle, such a spatial averaging could be performed for each MODIS cell, in practice however, it becomes computationally overwhelming on the PC, and it is also unnecessary, since the spatial averages form a 2-D signal of small gradient, making possible that "sampling" (i.e., the actual calculation of the spatial mean values including the WREVAP-calculated ET rate) is performed only in a selected set of points, which was chosen as each tenth MODIS cell in space (both row-, and column-wise). The remaining cells were then filled up with spatial mean values, linearly interpolated first by row among the selected MODIS-cell values, and then by column, involving the already interpolated values in the rows as well. Near the eastern and southern boundaries of the state any necessary spatial extrapolation was done by the gradient method (i.e., keeping the first two terms of the Taylor-expansion). This "sampling" sped up calculations by at least two orders of magnitude.

Care had to be exercised with the choice of the radius of influence. Rather than applying a constant radius, a spatially changing one was required because near the boundary of the state the "window" becomes asymmetrical around the MODIS cells, therefore the radius changed linearly with distance to these boundaries from a starting value of 25 cells up to a maximum of 125 cells (at a rate of 4/5 cell by each line or column) in the central portion of the domain. It was simpler to define a rectangular region around each designated MODIS cell, rather than a circular one, therefore the radius of influence is half the side-length of the resulting square.

Once the spatial mean values were available for each MODIS cell, the actual linear transformation of the T_s to ET values was performed for each month (except the winter months). The linear transformation of the T_s values into ET rates assumes a negligible change in the r_a value among the cells. As was mentioned above, r_a is directly proportional –up to a constant and with a negative slope— to the logarithm of z_{0m} under neutral stability conditions of the atmosphere, provided the wind speed at the blending height (about 200 m above the ground) is near constant within the region [11]. The momentum roughness height, z_{0m}, of each MODIS cell has been estimated over the state (Figure 3) with the help of a 1-km digital elevation model, as the natural logarithm of the standard deviation in the elevation values among the 25 neighboring cells surrounding a given cell, including the chosen cell itself. The minimum value of z_{0m} has been set to 0.4 m, so when the estimate became smaller than this lower limit (possible for flat regions), the value was replaced by 0.4 m. Note that the z_{0m} = 0.4 m value is the upper interval value for a "prairie or short crops with scattered bushes and tree clumps" in Table 2.6 of [13]. The rugged hills regions of Nebraska (e.g., the Pine Ridge and the Pine Bluffs, just to name a few) are characterized (Figure 3) by the largest z_{0m} values (larger than 3 m), covering the 3-4 m range for "Fore-Alpine terrain (200-300 m) with scattered tree stands" of [13]. Since the r_a estimates are proportional (up to a constant) to the logarithm of the z_{0m} values, their change among the MODIS cells is much subdued: about 67% of the time they are within 5% of their spatial mean and more than 94% of the time they remain within 15% (Figure 4), supporting the original assumption of the present ET mapping method.

Cells that had r_a values smaller than 95% of the mean r_a value (involved about 20% of all cells) were identified, and the corresponding ET values corrected by the relative change in r_a, considering that the sum of the latent (LE) and sensible heat (H) values are assumed to be constant among the cells (equaling Q_n) and that H is proportional to dT_z / r_a [11], where dT_z is the vertical gradient of the air temperature above the surface, itself taken proportional to T_s. The reason that only the "overestimates" of ET are corrected is that the linear transformation of the T_s values into ET rates seems to be more sensitive to more rugged-than-average terrain than to smoother one. That is why the most rugged part of Nebraska, i.e., the Pine Ridge, required an additional (to the above) 10% ET adjustment if no Ponderosa Pine was detected in the 3x3-cell region of the land use-land cover map around a given cell, and a 20% cut if it was. The assumption is that in this extremely rugged region cells with other than Ponderosa pine designation, may still contain scattered trees, if in the vicinity there are pine-forested areas plus the air turbulence, enhanced by the rugged terrain, may have a wake with a characteristic length of about a km. Within the Niobrara River Breaks region (less rugged than the Pine Ridge) only a 10% additional ET adjustment was applied without regard if the cell-surroundings are pine-covered or not. The underlying reasons of these deviations may be (after accepting that the PRISM precipitation values are correct) the way z_{0m} is estimated, perhaps a DEM with a finer resolution would yield better results.

Or maybe the type of vegetation, even at a 1-km resolution, has relevance (similar to plot-scale applications), in addition to the primary elevation variance. Or even, due to the enlarged surface area of the rugged terrain, the global radiation value should be reduced, which would lower ET. This topic certainly requires further research.

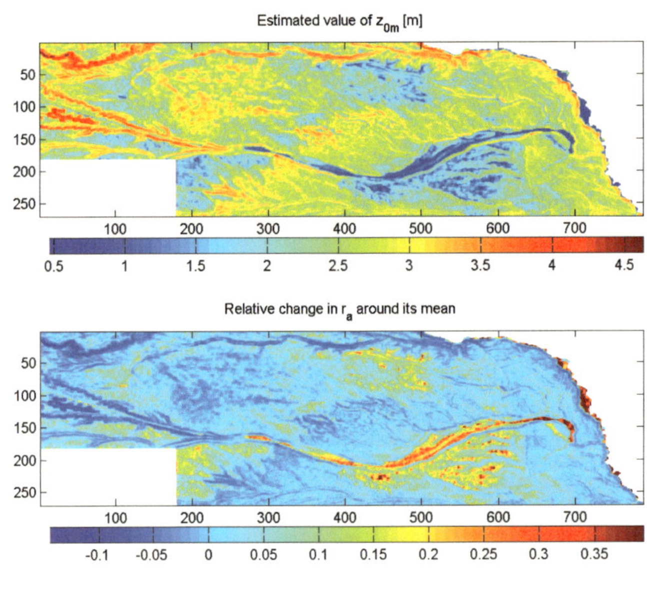

Figure 3. Estimated values of the a) momentum roughness height (z_{0m}), and; b) relative change in the aerodynamic resistance (r_a) around the state-wide mean. The numbers along the left and bottom edge of the panels are the MODIS cell numbers.

A final correction was applied for cells of "extreme" elevation. Namely, when the elevation of a cell differed from the regional mean value by more than 100 m, its surface temperature was corrected by 0.01 Kelvin per meter, reflecting the dry-adiabatic cooling rate of the air.

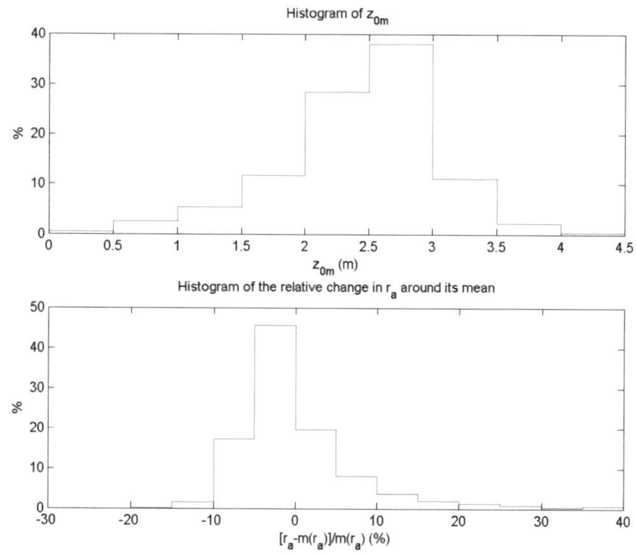

Figure 4. Relative histograms of the momentum roughness height (z_{0m}) and the relative change in the aerodynamic resistance (r_a) around its spatial mean value [$m(r_a)$] across Nebraska, estimated from a 1-km resolution digital elevation model.

The WREVAP model is based on the complementary relationship [4] which performs the worst in the cold winter months [8, 14, 15], thus the resulting WREVAP-obtained winter ET rates become the most uncertain. A yet unpublished study by the present author, involving the Republican River basin, indicated that inclusion of the winter ET rates of WREVAP improved the mean annual ET estimates in comparison with water balance derived [3, 12] data. Other studies [1, 16] also indicated that WREVAP somewhat overestimates ET rates in the Nebraska Sand Hills region even without inclusion of the winter months. Finally, a water balance based [3, 12] verification of the current ET estimates indicated that the WREVAP-provided winter ET rates are necessary in the most humid eastern, south-eastern part of the state. Based on these comparisons, the WREVAP winter months were fully included in the mean annual ET rates [besides the wettest part of the state, defined by (P_{sm} – ET_{WREVAP}) > 50 mm] for the Republican River basin, and for areas where the mean monthly daytime maximum temperature values exceeded 5 ºC. The latter area almost fully covers the Republican River basin, plus the south and south-western part of the panhandle region. P_{sm} designates the spatially smoothed precipitation values of PRISM, applying a 30-by-30-cell window, to filter out the unrealistic grainy structure of the PRISM precipitation field (Figure 5) due probably to its spatial interpolation method. No winter ET rates were included in the mean annual ET values wherever (P_{sm} – ET_{WREVAP}) < 10 mm; and a 50% reduction of the WREVAP winter ET rates were used for areas where [10 mm < (P_{sm} – ET_{WREVAP}) < 50 mm] held true.

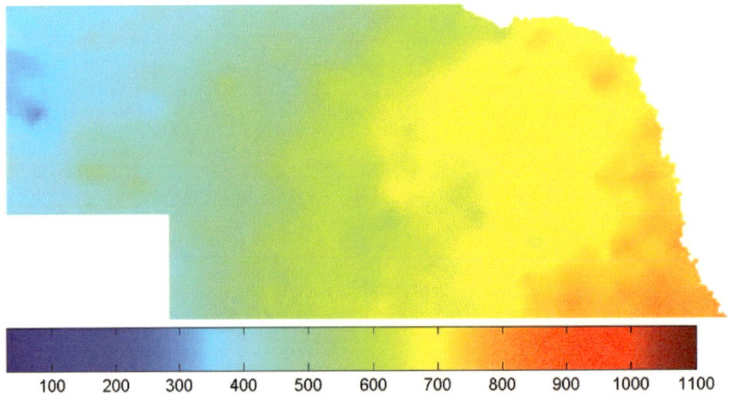

Figure 5. Distribution of the mean annual precipitation (mm) values across Nebraska from the PRISM data, 2000-2009. The state-wide mean annual precipitation rate is 577 mm.

3. Results and conclusion

The mean annual ET rates across Nebraska are displayed in Figure 6. By and large, ET follows the distribution of precipitation, as expected. Most of the ET values are between 250-500 mm in the panhandle, around

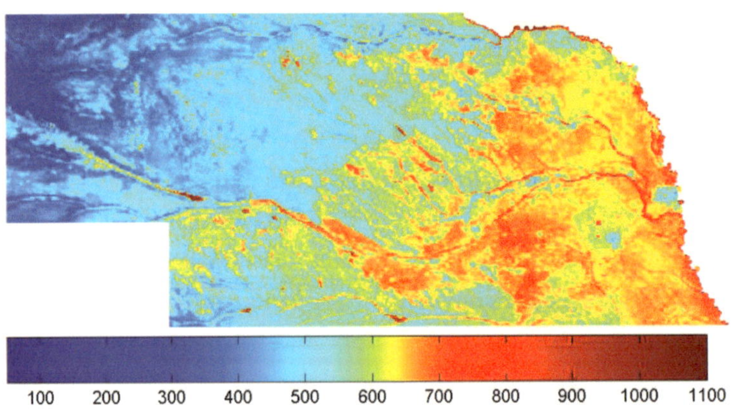

Figure 6. Estimated mean annual ET rates (mm) in Nebraska (2000-2009). The state-wide mean ET value is 549 mm/yr.

500-650 mm in the middle of the state and near 650 mm in the eastern portion of it. Locally, however, there are large differences due to land use and land cover variance. The sizeable reservoirs (McConaughy, Lewis and Clarke, Harlan County, Swanson, Calamus, etc.) large enough to fully accommodate a MODIS cell, display the largest ET rates, around 1000 mm annually. The reservoirs/lakes are followed by the wider rivers (i.e., Platte, Missouri, Loups,

Elkhorn, and partly the Republican) and their valleys in ET rates. The reason, beside the presence of the open water surface, is in the relatively small distance to the groundwater table in these river valleys, enabling the root system of the vegetation to tap into it, plus in the accompanying large-scale irrigation within the valleys. The river valleys on the ET scale are followed by areas of intensive irrigation, reaching 750 mm a year. The driest regions, with the smallest rate of ET in eastern Nebraska are the urban areas of Omaha and Lincoln (Figure 1), where the built in surfaces enhance surface runoff. The eastern outline of the Sand Hills is clearly visible, as well as the sandy areas (the elongated green-colored features) between the Loup and the Platte Rivers. The sandy soil, due to its large porosity favors deep percolation of the water often out of reach of the vegetation.

Figure 7 depicts the monthly ET rates from January through December. In July and August the irrigated plots in south-central Nebraska can evaporate as intensively as the open water surfaces. Even in September, when most of the produce has been harvested, the soil through its enhanced moisture due to summer irrigation, evaporates more than the surrounding, non-irrigated land. In November the distribution of ET rates becomes zonal and follows the precipitation distribution.

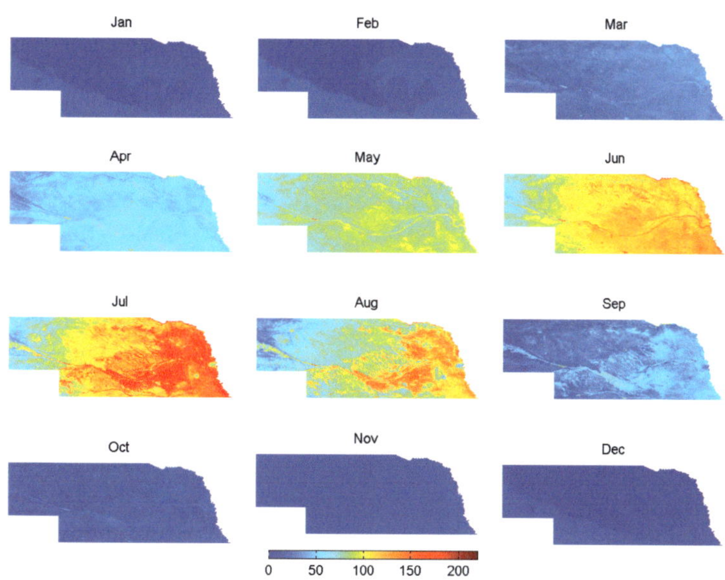

Figure 7. Estimated mean monthly ET rates (mm) in Nebraska., 2000-2009.

While in absolute numbers the south-central portion of the state produces the highest ET rates, the picture changes significantly, when one looks at the ET to precipitation (so-called precipitation recycling) ratios of Figure 8.

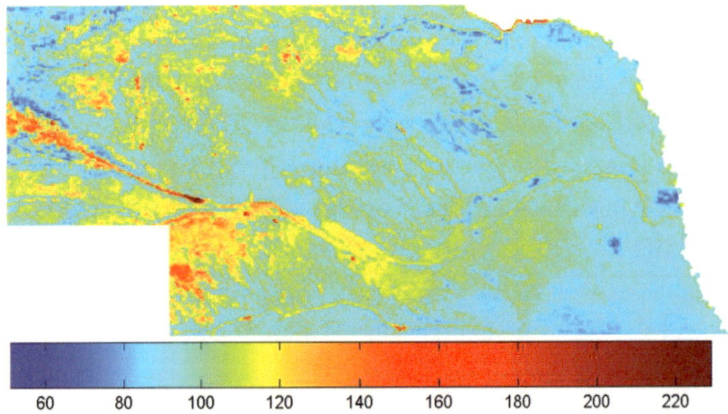

Figure 8. Estimated mean annual ET to precipitation ratios (%), 2000-2009. The state-wide mean ratio is 95%.

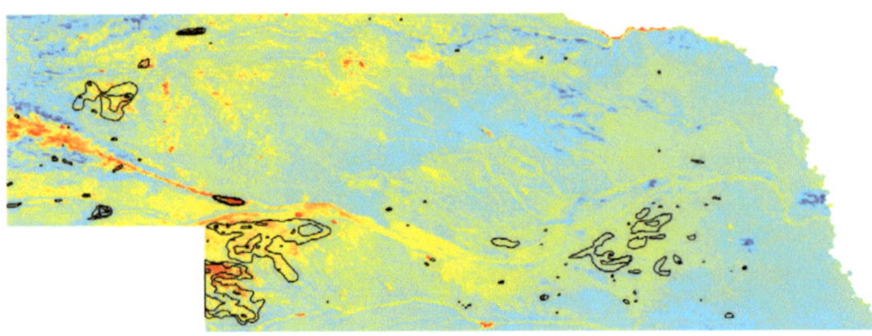

Figure 9. Distribution of areas with the largest observed groundwater decline (at least 3 m, to up to 8 m) over the 2000-2009 period, overlain the ET / P map. For the correct color to ratio correspondence, please, use the colors in Figure 8 instead of the current ones.

Lake McConaughy is the clear winner (followed by smaller lakes in the vicinity), evaporating about twice as much as it receives from precipitation. It does not mean, of course, that the other small lakes in the area would not evaporate as much as Lake McConaughy per unit area, they probably evaporate even more (the smaller the lake the larger typically its evaporation rate, provided other environmental factors are equal), but their size inhibits MODIS to detect their surface temperatures without "contamination" from the surrounding land. Note again the eastern outline of the Sand Hills and the elongated sandy areas between the Loup and Platte Rivers as areas of relatively low ET rates. The two urban areas of Omaha and Lincoln are clearly visible again.

Two large irrigated areas stand out clearly as the most intensive water users (relative to precipitation), one in the Republican River basin and the other in the North-Platte River valley of the panhandle. In these areas ET rates significantly exceed (up to 50%) precipitation rates. Another significant irrigated area in Box Butte County (at the western edge of the Sand Hills) plus the one in the Republican River basin coincide largely with regions of extensive groundwater depletions, displayed in Figure 9.

The lines in Figure 9 designate areas (after [17]) where groundwater decline was at least 3, 5, 8 m over the 2000-2009 period. Naturally, in heavily irrigated areas close to major streams (e.g., North- and South Platte, Platte River), such groundwater depletions are absent (but not around Lake McConaughy, where reservoir water levels have been below normal most of 2000-2009) since the chief source of the irrigation water is the stream itself. Figure 10 displays the distribution of irrigated land, overlain the ET / P values.

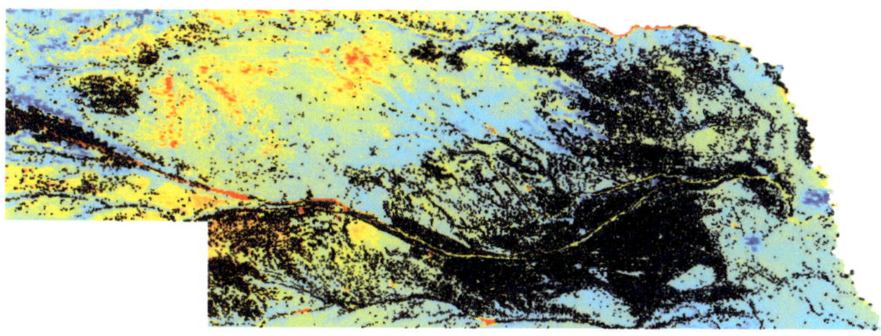

Figure 10. Irrigated land (marked in black) distribution [after 18, 19] in Nebraska, 2005.

As seen in Figure 10, not all land areas with larger than unity ratios are connected to irrigation, good examples are the Sand Hills wetlands. Similarly, not all areas that come up with values larger than 100% do actually evaporate more than they receive from precipitation. Such an area is the table-land just south of the western edge of Lake McConaughy, between the North- and South-Platte Rivers (please, refer to Figure 8 for corresponding precipitation recycling ratios, the colors of Figure 10 are slightly off because it was produced by a different software that enabled marking the irrigated areas on top of the ratios). In this area irrigation is largely absent (or at least it was in 2005, the date of the irrigation data), yet the ratio is between 100-120%. The error may be caused by several factors, namely a) the well-known, often 10% underestimation of precipitation; b) inaccuracy of the ET estimates, and; c) problems with the spatial interpolation of the measured precipitation values. The latter is well demonstrated in Figure 5, which shows that in the southern panhandle region there can be a difference of 125 mm (about 25-30% of the annual value) in the precipitation values within a distance of 30 km or less. Added to this uncertainty is the wide-spread underestimation of precipitation, especially in windy areas where a measurable portion of the raindrops (and especially snowflakes) is swept away from the rain gage. Finally, the present ET estimates have an error term (discussed further later) of about plus/minus (±) 5-10%. Employing a ±5% error in the latter, another -5% underestimation in the measured precipitation values together with yet another independent ±5% error in the interpolated values, the resulting ET / P ratio may contain an error of -5% to 16%, coinciding well with the error extent found in the table-land area. Therefore, the ratios in Figure 8 must be treated with this uncertainty in mind.

A comparison with the previous version [1] of the mean annual ET map (Figure 11) reveals that the largest differences are found in the Republican River basin and the southern panhandle region, where now the full values of the WREVAP-estimated winter-time ET rates were added to the warm-season values (March-November). Note that the procedures used for preparing the two maps are different (application of a radius of influence around each MODIS cell versus distinct geographic regions) as was explained above. The perceptible diagonal and level straight lines suggest some problems with the interpolation method employed in the previous ET map.

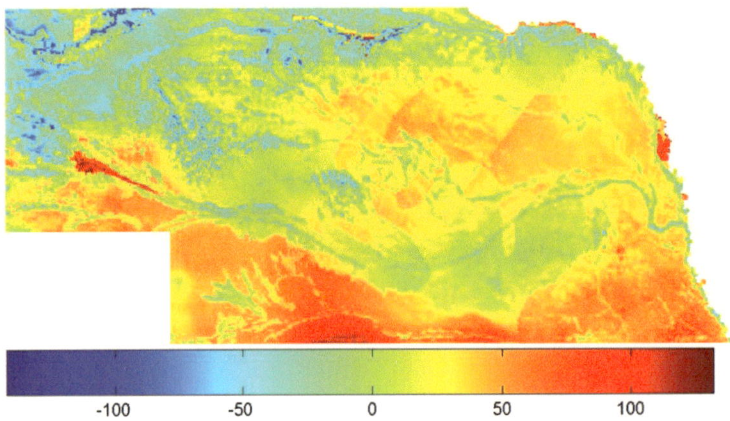

Figure 11. Differences in the present and previous [1] mean annual ET maps (mm). Mean is 18.5 mm.

Verification of the estimated mean annual ET rates can be best performed on a watershed-by-watershed basis by subtracting the stream discharge values (expressed in mm) from the mean precipitation values of the catchments, assuming that groundwater level changes are negligible over the study period, i.e., 2000-2009. As seen above, the latter is not true in many regions within Nebraska, but a transformation of these groundwater-level changes into water depth values would require the state-wide distribution of the specific yield value (also called drainable porosity, defined by dividing the drained water volume value with that of the control volume, fully saturated with water at the start of drainage) of the water bearing aquifer, a hydro-geological parameter not available for the whole state. Figure 12 displays the water-balance derived ET rates employing the PRISM precipitation values [12] and USGS-derived watershed-representative runoff values [3] for the HUC-8 watersheds within Nebraska, while Figure 13 displays the spatial distribution of estimation error (predicted minus water-balance derived) of the CREMAP ET values among the same catchments. As seen, in the majority of the watersheds the estimation error is within 30 mm of the "observed" value. The largest overestimation takes place for watersheds within the Missouri River basin, within the Sand Hills, and within and just north of the Republican River basin. The latter area corresponds to one of the most severe groundwater depletion regions within the state, where the "missing" water certainly contributes to elevated ET rates, not detectable by the simplified water balance approach.

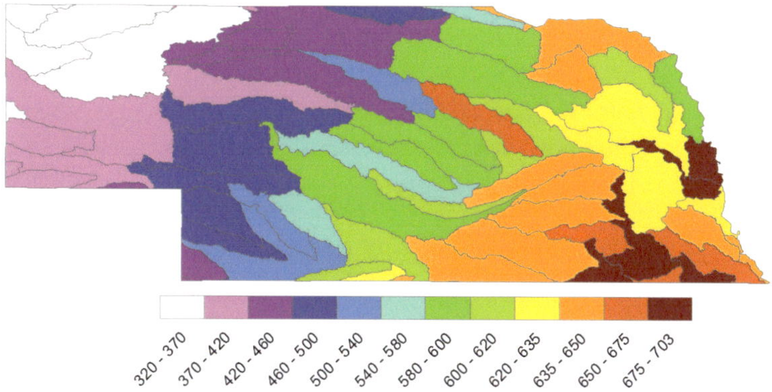

Figure 12. Water balance derived (P – Ro) mean annual ET rates (mm) of the USGS HUC-8 watersheds in Nebraska, 2000-2009.

Note also that a systematic underestimation of the precipitation rates automatically leads to a virtual overestimation of the ET rates by the present method during verification. Another problem with the verification is that the watershed area employed for the transformation of the discharge values into water depth, may also be somewhat uncertain, since the groundwater catchment does not always overlap perfectly with the surface-water catchment delineated by the help of surface elevation values. Probably that is why the largest over- and under-estimation of ET is found within the Sand Hills, in catchments very close to each other. Also, the USGS watershed runoff values employ simplifications that may cause serious errors in the estimated watershed runoff rate, such as found for the Lower Republican basin in Kansas (not shown here), where USGS reports a mean runoff rate of 5 mm/yr for 2000-2009, while a Kansas Geological Survey study [20] found a mean annual runoff rate of 106 mm/yr, an almost twenty-fold difference.

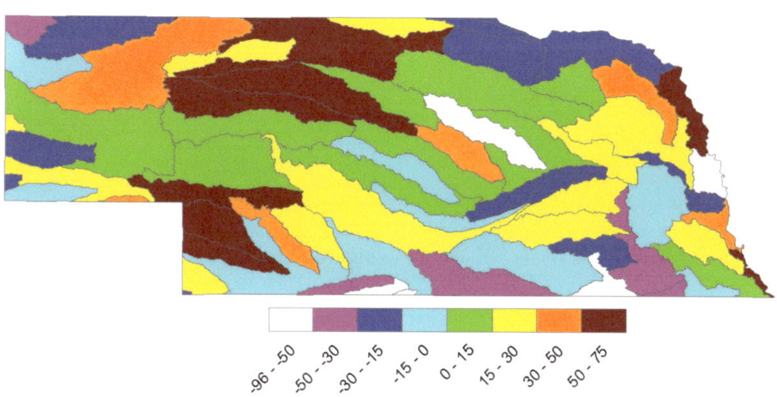

Figure 13. Distribution of the estimation error (mm) in the mean annual ET rates among the USGS HUC-8 watersheds within Nebraska, 2000-2009.

Figure 14 summarizes the ET verification results. It can be seen that in the vast majority of the USGS HUC-8 watersheds the estimated values are within 10% of the simplified water balance derived values. Five of the seven overestimates (above the upper intermittent line) of Figure 14, found between 400 and 500 mm, correspond to the large groundwater depletion area in and around the Republican River basin, displayed in Figure 9, so in those cases the CREMAP ET estimates may better represent reality than the simplified water balance derived values. The explained variance, R^2, has a value of 0.87, meaning that 87% of the spatial variance found in the HUC-8 water-balance derived ET values is explained by the CREMAP estimates. In summary, the annual and monthly ET maps are recommended for use in future regional-scale water-balance calculations with the resolution and accuracy of the estimates kept in mind. The maps are certainly not recommended for reading off individual cell values, because the exact cell coordinates maybe slightly off due to the geographically referenced data manipulations necessary to produce those maps. For example, the author found some problem with coordinate referencing when cells are extracted from a grid employing another grid with differing cell size. The maps are best suited for studies of spatial scale larger than one km.

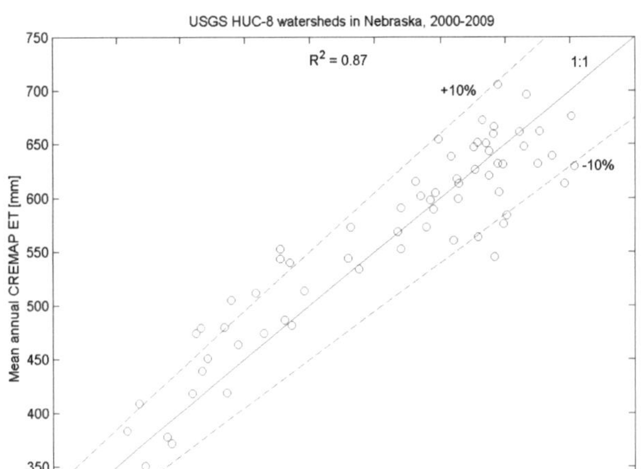

Figure 14. 14. Regression plot of the water-balance derived and CREMAP-estimated mean annual ET rates (mm) among the USGS HUC-8 catchments. R^2 is the portion of the spatial variance of the water balance ET rates that is explained by the CREMAP estimates. The upper and lower envelope lines designate the P – Ro value plus or minus 10%.

Acknowledgements

This work has been supported by the Hungarian Scientific Research Fund (OTKA, #83376) and the Agricultural Research Division of the University of Nebraska. This work is connected to the scientific program of the "Development of quality-oriented and harmonized R+D+I strategy and functional model at BME" project. This project is supported by the New Szechenyi Plan (Project ID: TAMOP-4.2.1/B-09/1/KMR-2010-0002).

Disclaimer: The views, conclusions, and opinions expressed in this study are solely those of the writer and not the University of Nebraska, state of Nebraska, or any political subdivision thereof.

Author details

Jozsef Szilagyi[1,2*]

Address all correspondence to: jszilagyi1@unl.edu

1 Department of Hydraulic and Water Resources Engineering, Budapest University of Technology and Economics, Budapest, Hungary

2 School of Natural Resources, University of Nebraska-Lincoln, Lincoln, Nebraska, USA

References

[1] Szilagyi, J., Kovacs, A., & Jozsa, J. (2011). A calibration-free evapotranspiration mapping (CREMAP) technique. In: Labedzki L. (ed.) Evapotranspiration. Rijeka: InTech Available from http://www.intechopen.com/books/show/title/evapotranspiration (accessed 28 June 2012)., 257-274.

[2] GEWEX Continental Scale International Project (2012). Surface Radiation Budget. GCIP-SRB. http://metosrv2.umd.edu/~srb/gcip/cgi-bin/historic.cgiaccessed 28 June).

[3] United States Geological Survey, USGS. (2012). Computed runoff. http://waterwatch.usgs.gov/new/accessed 28 June).

[4] Bouchet, R. J. (1963). Evapotranspiration reelle, evapotranspiration potentielle, et production agricole. *Annales Agronomique*, 14, 543-824.

[5] Brutsaert, W., & Stricker, H. (1979). An Advection-Aridity approach to estimate actual regional evapotranspiration. *Water Resources Research*, 15, 443-449.

[6] Morton, F., Ricard, F., & Fogarasi, S. (1985). Operational estimates of areal evapotranspiration and lake evaporation- Program WREVAP. *National Hydrological Research Institute (Ottawa, Canada) Paper*, 24.

[7] National Aeronautics and Space Administration. (2012). Moderate Resolution Imaging Spectroradiometer. MODIS data. http://modis.gsfc.nasa.govaccessed 28 June).

[8] Szilagyi, J., & Jozsa, J. (2009). Estimating spatially distributed monthly evapotranspiration rates by linear transformations of MODIS daytime land surface temperature data. *Hydrology and Earth System Science*, 13(5), 629-637.

[9] Priestley, C., & Taylor, R. (1972). On the assessment of surface heat flux and evaporation using large-scale parameters. *Monthly Weather Review*, 100, 81-92.

[10] Bastiaanssen, W., Menenti, M., Feddes, R., & Holtslag, A. (1998). A remote sensing surface energy balance algorithm for land (SEBAL): 1. Formulation. *Journal of Hydrology*, 212, 198-212.

[11] Allen, R., Tasumi, M., & Trezza, R. (2007). Satellite-based energy balance for mapping evapotranspiration with internalized calibration (METRIC)-model. *Journal of Irrigation and Drainage Engineering*, 133(4), 380-394.

[12] Daly, C., Neilson, R. P., & Phillips, D. L. (1994). A statistical-topographic model for mapping climatological precipitation over mountainous terrain. Journal of Applied Meteorology Available from www.prism.oregonstate.eduaccessed 28 June 2012)., 33, 140-158.

[13] Brutsaert, W. (2005). Hydrology: An Introduction. *Cambridge University Press, Cambridge, United Kingdom.*

[14] Huntington, J., Szilagyi, J., Tyler, S., & Pohll, G. (2011). Evaluating the Complementary Relationship for estimating evapotranspiration from arid shrublands. *Water Resources Research*, 47, W05533.

[15] Szilagyi, J., Hobbins, M., & Jozsa, J. (2009). A modified Advection-Aridity model of evapotranspiration. *Journal of Hydrologic Engineering*, 14(6), 569-574.

[16] Szilagyi, J., Zlotnik, V., Gates, J., & Jozsa, J. (2011). Mapping mean annual groundwater recharge in the Nebraska Sand Hills, USA. *Hydrogeology Journal*, 19(8), 1503-1513.

[17] Korus, J., Burbach, M., & Howard, L. (2012). Groundwater-level changes in Nebraska- Spring 2000 to Spring 2010. Conservation and Survey Division, School of Natural Resources, University of Nebraska-Lincoln. http://snr.unl.edu/csdaccessed 26 June).

[18] Dappen, P., Ratcliffe, I., Robbins, C., & Merchant, J. (2012). Map of 2005 Center Pivots of Nebraska. University of Nebraska- Lincoln http://www.calmit.unl.edu/2005landusestatewide.shtml (accessed 12 June).

[19] Dappen, P., Ratcliffe, I., Robbins, C., & Merchant, J. (2012). Map of 2005 Land Use of Nebraska. University of Nebraska- Lincoln http://www.calmit.unl.edu/2005landuse/statewide.shtml (accessed 12 June).

[20] Sophocleous, M. (2009). Aquifer storage and recovery and the Lower Republican River valley, Kansas. *KGS open file report-18, Lawrence, Kansas.*

Chapter 3

Quantifying the Evapotranspiration Component of the Water Balance of Atlantis Sand Plain Fynbos (South Africa)

Nebo Jovanovic, Richard Bugan and Sumaya Israel

Additional information is available at the end of the chapter

http://dx.doi.org/10.5772/53405

1. Introduction

The Cape Floral Kingdom, which experiences the Mediterranean climate of the Western Cape (South Africa), is home to about 9,000 species of the fynbos and succulent karoo biomes, 68% of which are endemic [1]. Vegetation types or veld types of the Cape Floral Kingdom are commonly classified as Mountain Fynbos, Coastal Fynbos, Strandveld and Coastal Rhenosterbosveld. The fynbos biome includes three large taxonomic groups: i) proteoids (tall, deep-rooted shrubs), ii) ericoids (fine leaves, shallow-rooted shrubs), and iii) restioids (graminoids) [2]. Future climate predictions indicate that the Western Cape region will become warmer, drier and subject to more extreme droughts [3,4], with potential risks of species extinctions and range shifts [5]. It is therefore imperative to consider the adaptation mechanisms of these species to drought and their contribution to the water balance as part of a sensitive ecosystem.

The physiological and morphological adaptation of fynbos to drought has been studied in the past. Differences in plant-water relations of two species of Protea (*Protea susannae* and *Protea compacta*) have been previously studied [6]. These two species exhibited different water use adaptation strategies, indicating that habitat specialization plays an important role in their distributions across landscapes. A similar plant-water behavior was observed in different species occurring in riparian zones and hillslopes, as they extract water from deeper soil layers through a well-developed root system [7]. Plant-water relations of several dominant fynbos species were investigated [8], where it was demonstrated that deep-rooted and isohydric species of fynbos tolerate drought better than shallow-rooted and anisohidric species. Rhenosterbos (*Elytropappus Rhinocerotis*) and its impacts on regulating the groundwater table through water uptake via a deep root system were also investigated [9].

The quantification of water resources and the water cycle are of utmost importance in water resources planning and management. This is particularly important in arid regions which experience water stress. In these regions, evapotranspiration (ET) is likely to be the dominant component of the water balance and potential evaporation is much higher than rainfall. It is therefore imperative that ET be accurately quantified. Under such climatic conditions, potential ET rates seldom occur as plants are subject to water shortages and stress due to limited soil water supply. As a physiological adaptation mechanism, plants close stomata resulting in the actual ET rates being below potential rates. This concept of atmospheric demand-soil water supply limited ET was described in detail in [10].

Knowledge of the water use of vegetation could have enormous implications for water resources management. The impact of alien invasive species on streamflow in the Kogelberg area of the Western Cape has been well documented [11]. The results of this study suggested that controlling the spread of alien species in this natural habitat of Mountain Fynbos could reduce water losses by up to 30%, thus increasing the water supply potential of the stream. Groundwater may also be an integral part of the hydrological cycle [12] and shallow groundwater can be a critical resource to natural vegetation [13]. However, the contribution of capillary rise from shallow groundwater to ET is difficult to measure. This study aims to highlight the importance of accurately quantifying ET and also aims to illustrate its effect on other components of the water balance, particularly groundwater recharge. The study was conducted in an area dominated by endemic fynbos vegetation. A lack of data exists, in terms of estimates of ET fluxes from fynbos vegetation, in particular the effects of shallow groundwater on root water uptake. The aim of this study was to quantify all components of the water balance in an area vegetated by Sand Plain Fynbos and characterized by shallow groundwater table. This included weather, vegetation, soil and groundwater. Monitoring from May 2007 to September 2011 served to generate a time series sufficiently long to account for rainfall and weather variability, and for model calibration.

2. Application area

The field trial was established in the Riverlands Nature Reserve, managed by Cape Nature Conservation, located about 10 km South of Malmesbury (Western Cape, South Africa; Figure 1). The reserve is in a predominantly flat area (slope <1%), the soils are deep, well-leached, generally acidic and coarse sandy of marine and aeolian origin (Luvic Cambisol; [14]). The reserve is situated on Cenozoic deposits with Cape granite outcrops occurring in the surroundings. A complete description of the topography, soil physical and chemical properties, and geology was given in [15]. The climate is Mediterranean with the mean annual rainfall being about 450 mm, occurring mainly from May to October. Mean potential annual evaporation is about 2,150 mm and daily evaporation exceeds rainfall for about 70% of the time.

http://dx.doi.org/10.5772/53405

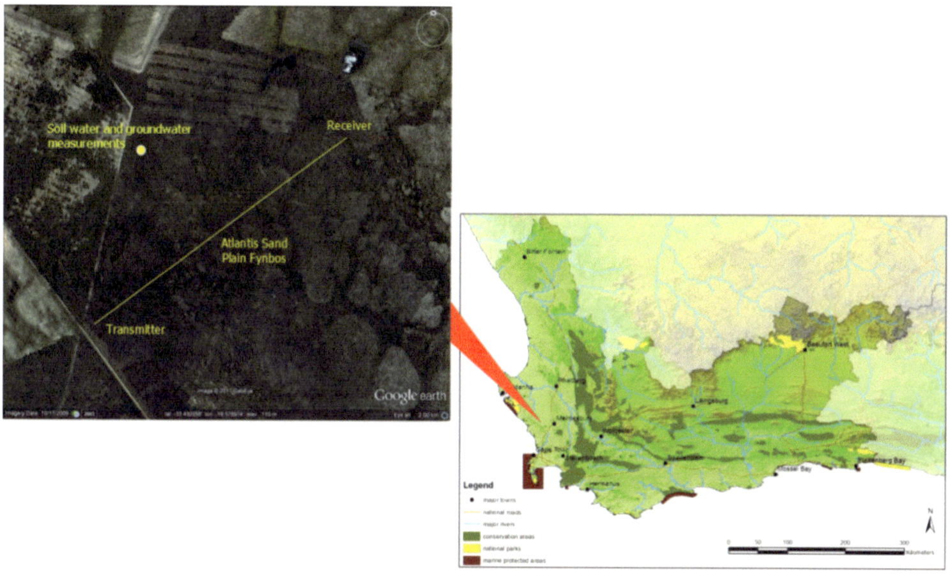

Figure 1. Location of the Riverlands Nature Reserve on the Western Cape map of conservation areas (right). The locations of the scintillometer measurement transect and soil water measurements (yellow circle) are shown in the Google Earth map (left)

The background information for the vegetation description was sourced primarily from [16], and supplemented from [17]. Botanical terminology follows [1] and [16]. The dominant vegetation type of the reserve is Atlantis Sand Plain Fynbos (FFd4, [16]), one of the 11 forms of Sand Plain Fynbos that occurs on the coastal plains of the western and southern coast of the Western Cape. Figure 2 depicts Atlantis Sand Plain Fynbos showing the restio dominated community of the lower-lying areas in the foreground and the taller shrubs of the higher-lying community in the background. The vegetation type is classified as vulnerable with only about 6% conserved, mainly at Pella, Riverlands Nature Reserve (1,111 ha) and Paardeberg. About 40% of the vegetation type has been transformed for agriculture, urban and industrial development, and plantations of eucalypts (for firewood and windbreaks) and pines (windbreaks). Large areas have been invaded by *Acacia saligna* and *A. cyclops* which were used to control drift sands from the mid-1800s up to the 1950s, often in areas that were denuded of vegetation by grazing and excessive burning. The reserve has at least 400 plant species, a number of which are only known from the area.

Figure 2. A view of the Atlantis Sand Plain Fynbos in the Riverlands Nature Reserve

The vegetation is dominated by 1-1.5 m tall emergent shrubs with a dense mid-storey of other shrubs and Restionaceae, a ground layer of recumbent shrubs, herbaceous species, geophytes and grasses with occasional succulents. The Atlantis Sand Fynbos at Riverlands is characterized by a relatively high cover of shrubs of the Proteaceae, Ericaceae and Rutaceae. Shrubs of *Euclea racemosa* and *Diospyros glabra* are also reasonably prominent. The vegetation has two different communities that seem to be controlled by the micro-topography and groundwater depth (Figure 2). Slightly higher-lying areas are dominated by *Protea scolymocephala, Leucadendron salignum, Leucadendron cinereum* and *Leucospermum calligerum* with *Erica mammosa, Erica* species, *Euclea, Diospyros, Phylica cephalantha, Staavia radiata* and shrubs in the Rutaceae. In the lower-lying areas, the dominant species were from the Restionaceae – *Chondropetalum tectorum, Willdenowia incurvata, Staberoha distachyos, Thamnochortus spicigerus* - with *Diastella proteoides, Berzelia abrotanoides, Serruria decipiens* and *S. fasciflora*. The prostate, spreading shrub *Leucospermum hypophyllocarpodendron* (subspecies *canaliculatum*) occurred in both communities, but was more common in the higher-lying areas. The ground layer included a wide variety of geophytic species in the Liliaceae and Iridaceae, seasonal herbs and a few grass species.

http://dx.doi.org/10.5772/53405

3. Method used

The principle of monitoring the entire hydrological system (weather, vegetation, soil and groundwater) was used in this study. Daily weather records for the study period were available from the South African Weather Services (Malmesbury station) and from the Agricultural Research Council (Langgewens holdings of the Western Cape Department of Agriculture). Daily rainfall data were collected with a manual rain gauge at the Riverlands Nature Reserve.

Total evaporation (ET) can be defined as the algebraic sum of all processes of water movement into the atmosphere. Soil evaporation (E) and transpiration (T) occur simultaneously and are determined by the atmospheric evaporative demand (mainly the available energy and the vapour pressure deficit of the air), soil water availability and canopy characteristics (canopy resistances) [18]. Total evaporation is also referred to as ET [19]. In this study, total evaporation represents ET and refers to the sum of evaporation from the soil surface, transpiration by vegetation, and evaporation of water intercepted by vegetation, as estimated with large aperture scintillometers [20]. The energy balance theory and methods for measurement of ET were extensively discussed in [20] and [21].

A Scintec boundary layer large aperture scintillometer system (BLS900, Scintec AG, Germany) was used to estimate fynbos ET in the period 14-27 October 2010. This window period for ET measurements was chosen to be at season change in spring, at a time when both sunny days with high atmospheric evaporative demand and overcast days with low ET occurred.

The BLS900 system measures the path-averaged structure parameter of the refractive index of air (C_N^2) over a horizontal path. The BLS900 system determines C_N^2 and ET over distances of 500 m to 5 km. Estimates of total evaporation are spatially averaged over the area between the transmitter and receiver sensor with a larger proportion of the flux emanating from the middle of the transect. Measurements of C_N^2 together with standard meteorological observations (air temperature, wind speed, air pressure and vertical temperature gradients) collected with an automatic weather station were used to derive the sensible heat flux density (H). The net irradiance was measured using a North – facing net radiometer (CNR1, Kipp & Zonen, Delft, The Netherlands) installed in the middle of the transect over representative vegetation, while the soil heat flux was measured at three different locations within the scintillometer transect using pairs of soil heat flux plates (Campbell Scientific. Ltd, USA) installed at depths of 0.03 and 0.08 m. The latent heat flux (LE) was subsequently calculated as a residual of the simplified surface energy balance equation, from measurements of net irradiance, soil heat flux and H (estimated with the large aperture scintillometer [18]). The assumptions were closure of the surface energy balance and that the energy used for processes like photosynthesis was negligible.

Figure 3 shows the equipment, including the transmitter and receiver of the scintillometer, and the weather station. All data were collected and stored with CR23X data loggers (Campbell Scientific Ltd, USA) for the weather, available energy data and soil heat flux, and in the Signal Processing Unit of the scintillometer for H. The components of the energy balance were measured every half hour. The calculated LE values in W m^{-2} (energy used to evaporate water) were converted into the equivalent water depth units cumulated over the day in mm d^{-1} (ET).

Figure 3. Scintillometer set-up: transmitter (bottom) and receiver (top left) of the scintillometer; and weather station and energy balance system (top right)

The measurements with the BLS900 system were done over a 1,160 m transect indicated in Figure 1. The coordinates at the transmitter were 33.49665 °S, 18.57265 °E and the altitude 114 mamsl. The receiver was located at 33.50103 °S, 18.58454 °E and 111 mamsl. Most of the reserve is relatively young following fires in 2004 and 2005, but the area of ET measurements was situated in a 11-15 years old stand. This compared well with an estimated age of 12-13 years based on counts of shoot growth increments on *Protea scolymocephala* shrubs.

Canopy cover was measured with an AccuPar light sensor in the range of photosynthetically active radiation (Decagon Devices Inc., USA). Readings with the AccuPar were taken in October 2010 at 10 locations along the ET measurement transect (average ratio of 10 readings below the canopy and 10 readings above the canopy). A comparison between readings obtained from manual rain gauges below and clear of vegetation canopy was done in order to estimate canopy interception for rain events in the period between 12/06/2007 and 18/02/2008. Root density was determined in soil samples collected at different depths. A composite soil sample (about 5 kg) was taken from each soil horizon displaying different characteristics down to 0.8 m. The samples were sieved and washed to separate roots from the mineral particles. The roots were then dried in the oven at 40ºC for at least two days. Root density was expressed as mg of dry roots per kg of soil.

Soil water contents were measured at different depths in the profile (0.1, 0.4 and 0.8 m), both adjacent to bushes and in open space areas (Figure 1). Continuous hourly records were collected with Echo-TE sensors and logged with Echo-loggers (Decagon Devices Inc., USA). The location of soil water content measurements was considered to be representative of the scintillometer transect (Figure 1), as vegetation, canopy cover and sandy soil were relatively uniform across the field. The groundwater level was monitored during the study period at shallow monitoring wells equipped with Leveloggers (Leveloggers model 3001; Solinst Ltd., Georgetown, Canada). Surface water, with the exception of occasional ponding in the low-lying areas, did not occur due to the sandy nature of the soil and high infiltration rates.

4. Research course

Well-drained, alluvial, sandy soils are a typical example of a system where vertical water fluxes dominate. Rain water infiltrates in the unsaturated zone, it generates a wetting front and it refills soil layers from the surface towards the bottom of the soil profile. Infiltrating rain water is available for ET. Excess water is drained into deeper soil layers and eventually recharges the unconfined aquifer. The amount of direct groundwater recharge is therefore dependent on initial soil water content, rainfall amounts and distribution, and ET. The main purpose of the experiment at Riverlands was to quantify the various components of the one-dimensional soil water balance (rainfall, soil water storage, evapotranspiration and groundwater recharge/capillary rise). Measurements of rainfall, soil water storage and ET allowed for the quantification of direct groundwater recharge/capillary rise as the residual component of the soil water balance.

A coupled atmospheric-unsaturated zone model was developed to determine the soil water balance of Atlantis Sand Plain Fynbos. The first step in the coupling of models was to apply an atmospheric model to calculate potential evapotranspiration (PET) of Atlantis Sand Plain Fynbos. For this purpose, grass reference evapotranspiration (ETo) was first calculated from weather data with the Penman-Monteith formula [22] and used to determine PET with the following equation:

$$PET = Kc_{max}\ ETo$$

where Kc_{max} is a coefficient dependent on vegetation (i.e. height, morphology) and environmental conditions (i.e. weather variables), and PET represents the evapotranspiration immediately after a rainfall event [22]. Daily PET was then used as input in the unsaturated zone model HYDRUS-2D. Caution should be exercised in the use of this approach for natural vegetation that is usually heterogeneous.

HYDRUS-2D is computer software that can be used to simulate one- and two-dimensional water flow, heat transport and movement of solutes in unsaturated, partially saturated and fully saturated porous media [23]. It uses Richards' equation for variably-saturated water flow and the convection-dispersion equations for heat and solute transport, which is based on Fick's Law. The water flow equation accounts for water uptake by plant roots through a sink term. The heat transport equation considers transport due to conduction and convection with flowing water, whilst the solute transport equation considers convective-dispersive transport in the liquid phase, as well as diffusion in gaseous phase. The solute flux equations account for non-linear, non-equilibrium reactions between the solid and liquid phases, linear equilibrium reactions between the liquid and gaseous phases, zero-order production and two first-order degradation reactions, the one independent of other solutes, the other providing sequential first-order decay reactions. A dual-porosity system can be set up for partitioning of the liquid phase into mobile and immobile regions and for physical non-equilibrium solute transport. A database of soil hydraulic properties is included in the model. The HYDRUS-2D model does not account for the effect of air phase on water flow. Numerical instabilities may develop for convection-dominated transport problems when no stabilizing options are used, and the programme may crash when extremely non-linear flow and transport conditions occur.

The HYDRUS-2D model allows the user to set up the geometry of the system. The water flow region can be of more or less irregular shape and having non-uniform soil with a prescribed degree of anisotropy. Water flow and solute transport can occur in the vertical plane, horizontal plane or radially on both sides of a vertical axis of symmetry. The boundaries of the system can be set at constant or variable heads or fluxes, driven by atmospheric conditions, free drainage, deep drainage (governed by a prescribed water table depth) and seepage. The HYDRUS-2D version includes a CAD programme for drawing up general geometries and the MESHGEN-2D mesh generator that automatically generates a finite element unstructured mesh fitting the designed geometry.

The HYDRUS-2D model was used to calculate the soil water balance, in particular soil water fluxes towards the groundwater table (i.e. groundwater recharge) and from the shallow groundwater table upwards (i.e. capillary rise). Input data used in the simulations are summarized in Table 1. The main processes simulated were water flow and root water uptake. A vertical plane in rectangular geometry was simulated with a homogeneous profile. The initial condition in water pressure head was established by setting pressure head = 0 at the bottom nodes with equilibrium from the bottom nodes upwards. The hydraulic properties model was van Genuchten-Mualem with no hysteresis. The hydraulic parameters (water flow parameters) were obtained from textural analyses, soil water retention properties and an average bulk density (1.53 g cm^{-3}) [15].

The vertical rectangular dimension of the simulated geometry was 1.5 m, which corresponded approximately to the depth of water table at the beginning of the simulation. The boundary conditions were:

i. Atmospheric top boundary flux (rainfall, potential evapotranspiration).

ii. Constant head = 0 at the bottom nodes to simulate a shallow groundwater table.

iii. No flux at all other boundaries.

Parameters and variables	**Inputs**
Main processes	Water flow, root water uptake
Length units	cm
Type of flow	Vertical plane
Geometry	Rectangular
Number of materials and layers in the soil profile	1
Time units	Days
Initial time	0 (1 May 2007)
Final time	1602 (19 September 2011)
Initial time step	0.05 (default)
Minimum time step	1e-006 (default)
Maximum time step	0.5 (default)
Number of time-variable boundary records	1602
Maximum number of iterations	20 (default)
Water content tolerance	0.0005 (default)
Pressure head tolerance	0.05 (default)
Lower optimal iteration range	3 (default)
Upper optimal iteration range	7 (default)
Lower time step multiplication factor	1.3 (default)

Parameters and variables	Inputs
Upper time step multiplication factor	0.3 (default)
Lower limit of the tension interval	1e-006 (default)
Upper limit of the tension interval	10000 (default)
Initial condition	In the pressure head
Hydraulic model	Van Genuchten-Mualem
Hysteresis	No
Residual water content (Qr)	0.02
Saturation water content (Qs)	0.35
α of the soil water retention function	0.036
n of the soil water retention function	1.56
Saturated hydraulic conductivity (cm d^{-1})	47.85
l of the soil water retention function	0.5
Water uptake reduction model	Feddes
Potential evaporation and transpiration	Daily values calculated from weather data and vegetation characteristics [22]
Horizontal rectangular dimension (cm)	1
Vertical rectangular dimension (cm)	150
Slope of the base	0
Number of vertical columns	2
Number of horizontal columns	150
Mesh	Generated with MeshGen
Root distribution	Uniform down to 1.5 m (bottom of geometry)
Atmospheric boundary condition	Top nodes
Constant boundary condition	Pressure head = 0 at bottom nodes
Initial pressure head	Pressure head = 0 at bottom nodes with equilibrium from the bottom nodes
Depth of observation nodes (cm)	5 and 40 cm

Table 1. Summary of inputs used in the simulation with HYDRUS-2D

http://dx.doi.org/10.5772/53405

The HYDRUS-2D model calculates actual evapotranspiration from PET and applies the method of Feddes to predict reduced transpiration due to water stress. The Feddes' water uptake reduction model incorporated in HYDRUS-2D was used with no solute stress. Actual evaporation from the soil surface was calculated from soil water fluxes at the atmospheric boundary. Observation nodes were set at 0.05 and 0.4 m soil depth to write records of simulated soil water contents. These were also depths of installation of soil water sensors.

5. Results

The canopy cover measured with an AccuPAR in the range of photosynthetically active radiation was between 29.0 and 48.9% (average of 39.2%). The LAI calculated with the AccuPAR varied between 1.12 and 1.54 (average of 1.30). The average rainfall intercepted by the canopy was 0.06%. Intercepted water may have been lost through evaporation or may have reached the soil through leaves and stem flow paths. There were rain events when more water collected in rain gauges below the canopy than in those clear of vegetation. Additionally, the readings obtained under the fynbos canopy were not consistent, indicating that the nature of rainfall and wind conditions may determine the amount of rainfall intercepted by this bushy type of vegetation. The results of the root density measurements are shown in Figure 4. High readings of root density were recorded at about 0.8 m soil depth. This may be indicative of the phreatophytic behaviour of fynbos, with enhanced root development close to the water table. These results, however, need to be confirmed through the analysis of additional samples, given the high spatial variability of plant rooting systems. Given the results and uncertainties of the root density measurements, root distribution was set uniform down to the water table in HYDRUS-2D (Table 1).

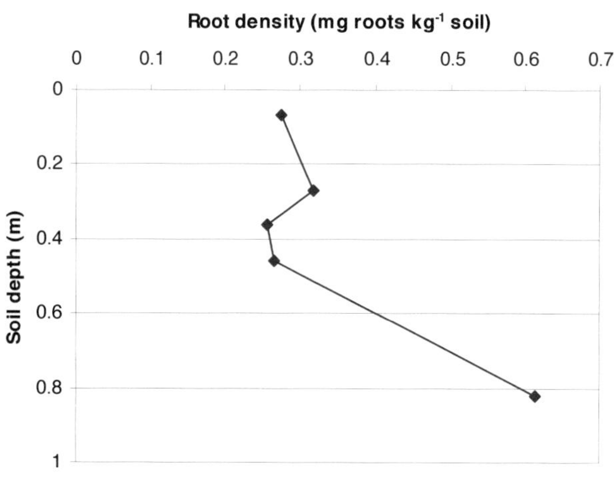

Figure 4. Root density distribution of fynbos

Total evaporation values measured with the scintillometer are shown in Figure 5 for the period 14-27 October 2012. These ET values represent the actual evapotranspiration from an area dominantly vegetated by Atlantis Sand Plain Fynbos. It should however be noted that different vegetation and bare patches also occur at the site. The ET values ranged between 0.8 mm d^{-1} on 21 October 2010 (rainy day) and 5.3 mm d^{-1} on 26 October 2010 (sunny day). High ET values were measured as a considerable amount of water is stored in the soil for ET at the end of the rainy season (14-27 October 2010), a shallow water table occurs (~1.5 m depth on average) and well-established fynbos species have root systems deeper than 0.8 m. Additionally, water stress conditions are interpreted to occur seldom as a result of the shallow groundwater table. The ET values measured in this study could therefore approximate PET of this vegetation. For comparative purposes, ETo calculated with the Penman-Monteith equation [22] was also displayed in Figure 5. The ETo ranged between 2.6 mm d^{-1} (21 October 2010) and 6.8 mm d^{-1} (27 October 2010). The average ratio of ET/ETo for the measurement period was 0.69 with a standard deviation of 0.18.

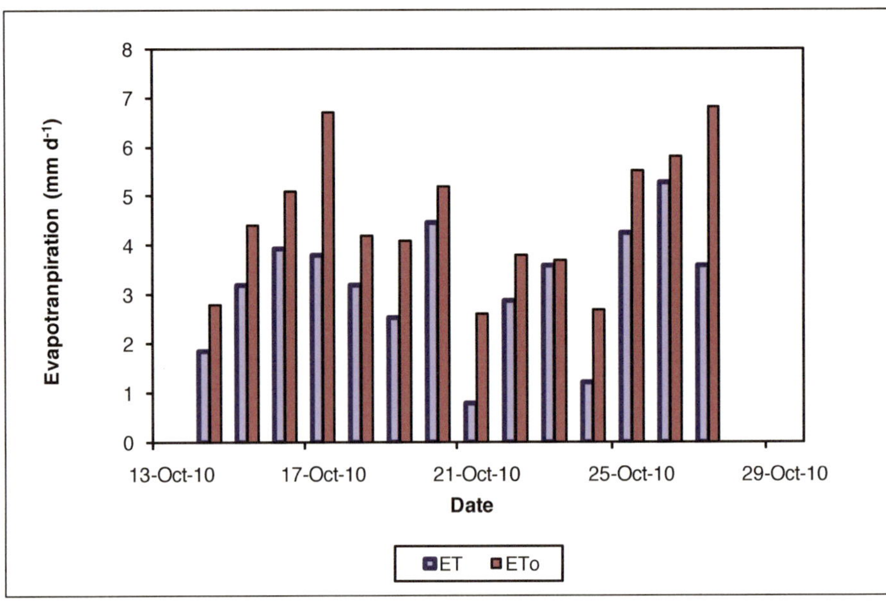

Figure 5. Evapotranspiration (ET) measured with the scintillometer and reference evapotranspiration (ETo) calculated with the Penman-Monteith equation on 14-27 October 2010 at Riverlands

The HYDRUS-2D model was used to simulate direct groundwater recharge and capillary rise as the residual components of the water balance at the study site. Figure 6 shows the daily rainfall data recorded at Riverlands with a manual rain gauge and the cumulative rainfall flux produced by HYDRUS-2D at the atmospheric boundary (input). Total rainfall for the period of simulation from 1 May 2007 to 19 September 2011 was 2,778 mm. The flux units on the Y-axis of the HYDRUS-2D graph represent cm of rainfall. The flux is negative because the water is entering the system.

http://dx.doi.org/10.5772/53405

Figure 7 represents measured and simulated volumetric soil water contents at 0.05 and 0.40 m soil depth. Missing measurements in the first half of 2010 (Figure 7, top graph) were due to logger malfunction. It can be visually noted that the trends and ranges of values obtained with HYDRUS-2D replicated observed measurements very well. As an indication of model performance, this comparison gave confidence in the soil water balance estimation with the model.

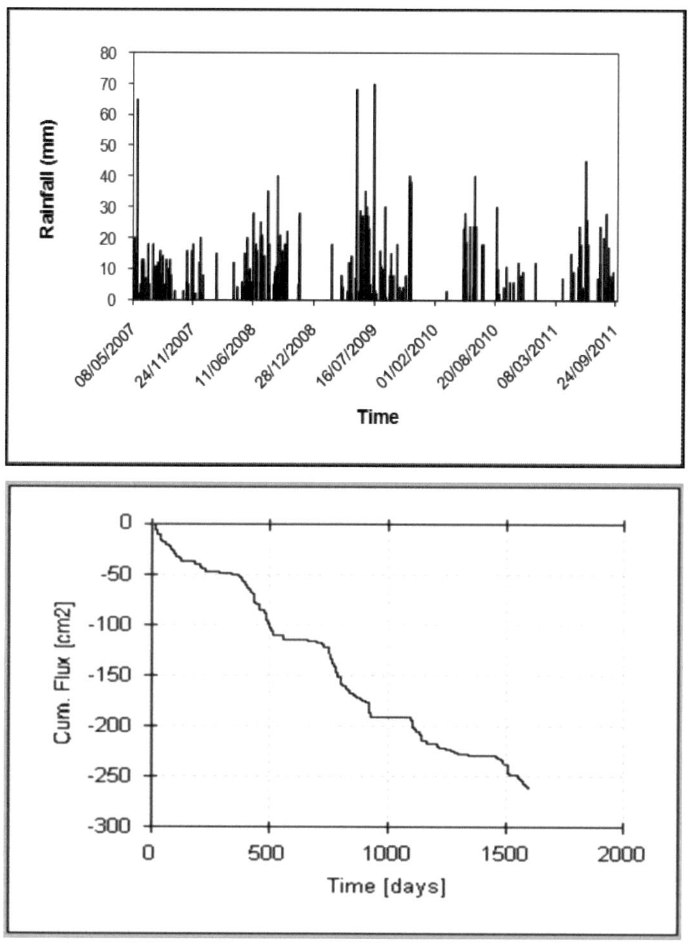

Figure 6. Daily rainfall data recorded at Riverlands with a manual rain gauge (top graph) and cumulative rainfall flux produced by HYDRUS-2D at the atmospheric boundary (bottom graph, screen printout)

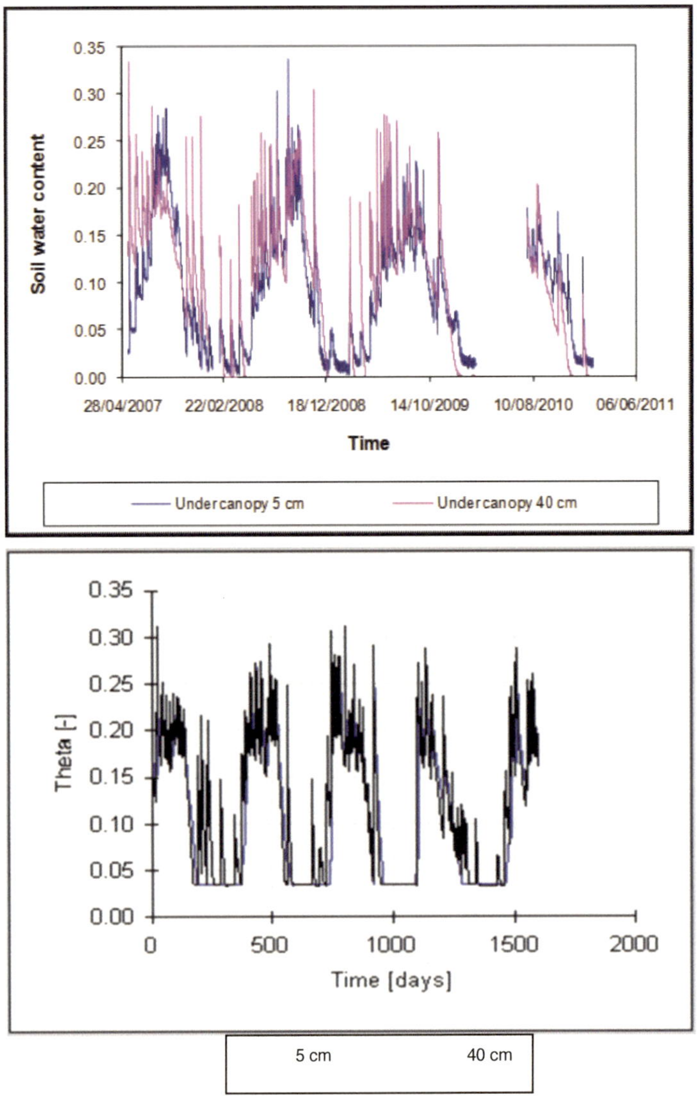

Figure 7. Hourly measurements of volumetric soil water content with Echo sensors (Decagon Inc., USA) (top graph) and volumetric soil water contents (Theta) simulated with HYDRUS-2D (bottom graph, screen printout) at 5 and 40 cm soil depth at Riverlands

The HYDRUS-2D model calculates daily actual ET from PET values by applying the method of Feddes to predict reduced ET due to water stress. The model set-up allowed for root water uptake from the shallow water table by setting a constant pressure of 0 (groundwater table) at the bottom of the soil profile (1.5 m depth). Figure 8 represents HYDRUS-2D output graphs of cumulative potential root water uptake (input) and actual root water uptake calculated with Feddes' model. The fluxes are positive because water is leaving the system. The units on the

http://dx.doi.org/10.5772/53405

Y-axes correspond to cm of root water uptake. Cumulative potential root water uptake for the simulation period was 6,118 mm and actual root water uptake was 4,183 mm (68%). In the two-weeks of scintillometer measurements done in October 2010, the ratio of actual to grass reference evapotranspiration was found to be 69%.

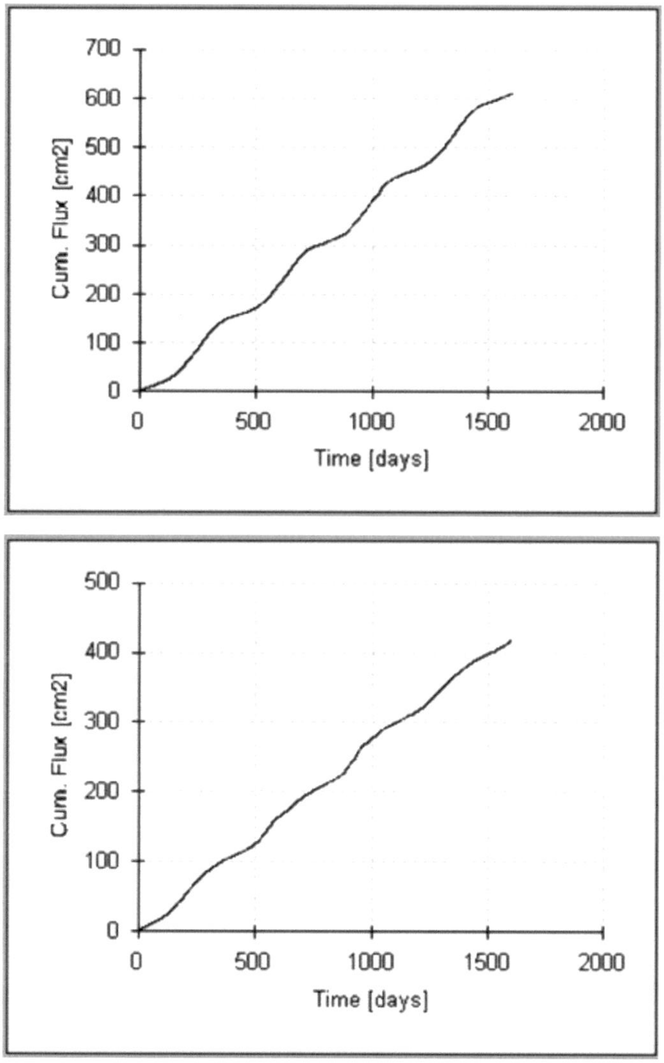

Figure 8. HYDRUS-2D simulations of cumulative potential root water uptake (top graph, input data screen printout) and actual root water uptake calculated with the Feddes' model (bottom graph, screen printout) for fynbos at Riverlands

Figure 9 represents the cumulative fluxes at the bottom boundary (groundwater table). Positive fluxes represent water leaving the system (groundwater recharge) and negative fluxes represent water entering the system (capillary rise from shallow groundwater). The units of the Y-axis correspond to cm. It can be noted that the cumulative flux increased on five occasions (five rainy winter seasons) and it decreased four times (four summers). The increases in flux corresponded to groundwater recharge that occurred during five rainy seasons. Net recharge was largely negative (1,566 mm) because capillary rise from shallow groundwater was much larger than downward water fluxes. Lateral groundwater sources and sinks were not considered in the one-dimensional simulation.

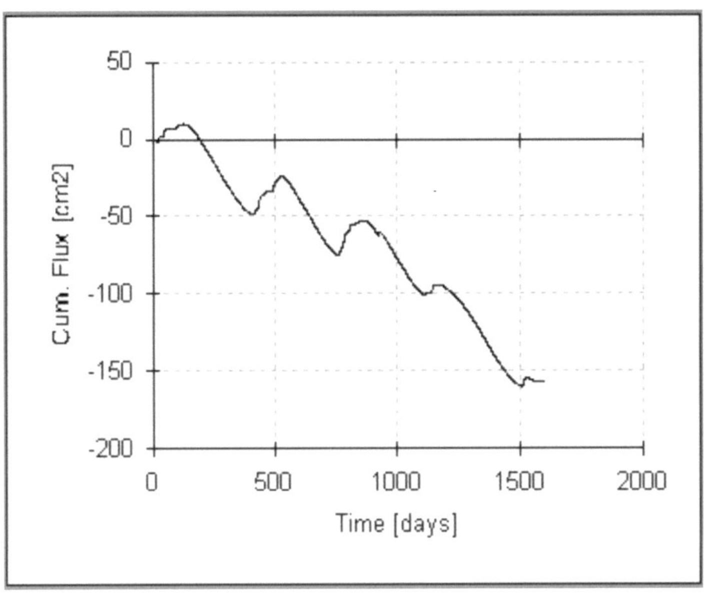

Figure 9. Cumulative bottom boundary flux simulated with HYDRUS-2D (screen printout)

Table 2 presents annual rainfall, groundwater recharge (increases in cumulative flux in Figure 9) and the groundwater recharge as a percentage of rainfall ($R^2 = 0.76$ between annual recharge and rainfall). Comparatively, groundwater recharge in the same catchment was estimated to be 81 mm a^{-1} [24], and 25% of annual rainfall using a mass balance approach at Atlantis (20 km South of Riverlands) [25]. Other estimates of groundwater recharge for the same catchment included 15.4% of mean annual rainfall using a Cl mass balance approach [26], 5% using a GIS-based groundwater recharge algorithm [26] and 13% in the vicinity of Riverlands [27].

Year	Rainfall (mm)	Groundwater recharge (mm)	Groundwater recharge (% of rainfall)
2007	509	98	19
2008	718	253	35
2009	804	227	28
2010	390	65	17
2011	357	55	15
Total	2778	698	25

Table 2. Annual rainfall and groundwater recharge at Riverlands

6. Conclusions and further research

The monitoring of weather, soil water content, vegetation and groundwater was very beneficial in terms of model calibration, to gain understanding of natural systems and ultimately in quantifying water balance processes accurately. This highlights the importance of long term monitored hydrological and hydrogeological data. Scintillometer measurements carried out in October 2010 represented the first direct estimates of ET from Atlantis Sand Plain Fynbos. These measurements were invaluable in gaining understanding of the water use and water balance dynamics of the study area.

The HYDRUS-2D model made good predictions of seasonal variations (trends and ranges) in soil water content in Riverlands. This gave confidence in the soil water balance estimation with the model. Total rainfall for the period of simulation from May 2007 to September 2011 was 2,778 mm. Evapotranspiration rates depended on weather conditions, vegetation (root distribution and canopy cover) and the soil water storage capacity. Cumulative potential root water uptake for the simulation period was 6,118 mm and actual root water uptake was 4,183 mm (68% of the potential and which is approximately equal to the ratio of ET/ETo). The model results also indicated that groundwater contributed considerably to ET, as a result of the fynbos having a well-established root system. The capillary rise from shallow groundwater was simulated to be 1,566 mm more than the downward water fluxes (groundwater recharge). A well-developed fynbos root system at Riverlands allowed the vegetation to tap the shallow groundwater and which resulted in high ET rates. The ratio of simulated actual and potential root water uptake was similar to the ratio of actual (determined using scintillometer measurements) and grass reference evapotranspiration (ET/ETo) during a two-week window period in October 2010.

Uncertainties in the estimates of the water balance components depend on the accuracy of measured input data into the model (e.g. scintillometer measurements, weather data etc.), but especially on temporal variabilities (e.g. rainfall) and spatial variabilities (e.g. rainfall, vegetation, groundwater levels, soil hydraulic properties etc.). For example, the average of 25% of annual rainfall derived for annual direct groundwater recharge (Table 2) should be used with caution as large variations in annual rainfall may result in large variations of recharge (15 to 35% in the five-year time series). It is therefore imperative to account for the seasonality and temporal distribution of rainfall and the other water balance components. Vegetation is spatially variable in terms of canopy cover, structure and speciation. This may have effects on the relation between ETo and PET, root depth and root water uptake and, ultimately, on the water balance. The need to provide an accurate description of the spatial variability of environmental variables (e.g. variability in vegetation) is highlighted. This has implications not only on the estimation of hydrological processes, but also on water management, decision-making and risk associated with the water resource.

A number of recommendations emanated from this research. The study highlighted the need for an accurate conceptualization of the system. The concept of atmospheric demand-soil water supply should be employed in the quantification of actual evapotranspiration. A daily time step is recommended in the calculation of water balance variables to account for daily actual evapotranspiration and rainfall distribution. In some instances, the high temporal resolution of the daily time step can be traded off for speed of calculation (e.g. in numerical models like HYDRUS-2D) and the monthly time step can be adopted to account at least for the seasonality of rainfall and other water balance components. An accurate spatial description of environmental variables is essential (e.g. vegetation, groundwater levels, soil properties etc.). Continuous long-term monitoring of all environmental components (weather, soil water content, vegetation and groundwater) is invaluable for understanding natural systems and calibrating models. Calibration of models should be seen as an on-going process as new data become available. Model sensitivity analyses are essential in order to identify which model parameters are sensitive and which thus need accurate input data. The sensitive parameters in HYDRUS-2D were found to be root distribution, soil properties and potential evapotranspiration. Process models are generally suitable in terms of quantifying the water balance, in particular because computers are able to handle more and more detailed information. However, assumptions, limitations and potential sources of error need to be well-defined (e.g. complexity of systems, interpolation of spatial data, lack and patchiness in input data etc.).

Acknowledgements

The authors wish to acknowledge the Water Research Commission (Pretoria, South Africa) for funding this research, and Cape Nature Conservation for providing the pilot study sites at the Riverlands Nature Reserve and for supplying records of daily rainfall. The South African Weather Services (SAWS) and the Agricultural Research Council are thanked for supplying weather data for Malmesbury and Langgewens, respectively. The authors also thank Dr D. Le Maitre (CSIR, Natural Resources and Environment, South Africa) for the botanical description of fynbos, Dr S. Dzikiti (CSIR, Natural Resources and Environment, South Africa), Dr C. Jarmain and Dr C. Everson (University of KwaZulu-Natal, South Africa) for collecting evapotranspiration data with scintillometry.

Author details

Nebo Jovanovic*, Richard Bugan and Sumaya Israel

*Address all correspondence to: njovanovic@csir.co.za

CSIR, Natural Resources and Environment, Stellenbosch, South Africa

References

[1] Goldblatt P, Manning JC. Plant Diversity of the Cape Region of Southern Africa. Annals of the Missouri Botanical Garden 2002; 89 281–302.

[2] Stock WD, Van der Heyden F, Lewis OAM. Plant Structure and Function. In: Cowling R.M, (ed.) The Ecology of Fynbos: Nutrients, Fire and Diversity. Cape Town, South Africa: Oxford University Press; 1992. p226–240.

[3] Hewitson B, Tadross M, Jack C. Scenarios from the University of Cape Town. In: Schulze R.E. (ed.) Climate Change and Water Resources in Southern Africa: Studies on Scenarios, Impacts, Vulnerabilities and Adaptation. Pretoria, South Africa: Water Research Commission; 2005. p39–56.

[4] IPCC. Pachauri RK, Reisinger A (eds.) Climate Change 2007: Synthesis Report. Geneva, Switzerland: IPCC; 2007.

[5] Hannah L, Midgley G, Hughes G, Bomhard B. The View from the Cape. Extinction Risk, Protected Areas, and Climate Change. BioScience 2005; 55 231–242.

[6] Richards MB, Stock WD, Cowling RM. Water Relations of Seedlings and Adults of Two Fynbos Protea Species in Relation to their Distribution Patterns. Functional Ecology 1995; 9 575-583.

[7] Richardson DM, Kruger FJ. Water Relations and Photosynthetic Characteristics of Selected Trees and Shrubs of Riparian and Hillslope in the South-Western Cape Province, South Africa. South African Journal of Botany 1990; 56(2) 214-255.

[8] West AG, Dawson TE, February EC, Midgley GF, Bond WJ, Aston TL. Diverse Functional Responses to Drought in a Mediterranean-Type Shrubland in South Africa. New Phytologist 2012. doi: 10.1111/j.1469-8137.2012.04170.x

[9] Vermeulen T. Plant Water Relations of *Elytropappus Rhinocerotis* with Specific Reference to Soil Restrictions on Growth. MSc thesis. University of Stellenbosch; 2010.

[10] Jovanovic N, Israel S. Critical Review of Methods for the Estimation of Actual Evapotranspiration in Hydrological Models. In: Irmak A. (ed.) Evapotranspiration – Remote Sensing and Modelling. Rijeka: InTech; 2012. p329-350.

[11] Le Maitre DC, Van Wilgen BW, Chapman RA, McKelly DH. Invasive Plants and Water Resources in the Western Cape Province, South Africa: Modelling the Consequences of a Lack of Management. Journal of Applied Ecology 1996; 33 161-172.

[12] Alley WM, Healy RW, Labaugh JW, Reilly TE. Flow and Storage in Groundwater Systems. Science 2002; 296 1985-1990.

[13] Clarke R, Lawrence A, Foster S. Groundwater: A Threatened Resource. Environment Library No. 15. Nairobi, Kenya: United Nations Environment Programme; 1996.

[14] FAO. World Reference Base for Soil Resources. Rome, Italy: United Nations Food and Agricultural Organization; 1998.

[15] Jovanovic NZ, Hon A, Israel S, Le Maitre D, Rusinga F, Soltau L, Tredoux G, Fey MV, Rozanov A, Van der Merwe N. Nitrate Leaching from Soils Cleared of Alien Vegetation. Report No. 1696/09. Pretoria, South Africa: Water Research Commission; 2009.

[16] Rebelo AG, Boucher C, Helme N, Mucina L, Rutherford MC. Fynbos Biome. In: Mucina L, Rutherford MC. (eds.) The Vegetation of South Africa, Lesotho and Swaziland. Strelitzia 19. Pretoria, South Africa: South African National Biodiversity Institute; 2006. p53-219.

[17] Yelenik SG, Stock WD, Richardson DM. Ecosystem Level Impacts of Invasive *Acacia saligna* in the South African Fynbos. Restoration Ecology 2004; 12(1) 44-51.

[18] Rosenberg NJ, Blad BL, Verma SB. Microclimate: The Biological Environment. 2nd edition. New York, USA: Wiley; 1983. p495.

[19] Kite G, Droogers P. Comparing Estimates of Actual Evapotranspiration from Satellites, Hydrological Models, and Field Data: A Case Study from Western Turkey. Research Report no. 42. Colombo, Sri Lanka: International Water Management Institute; 2000. p32.

[20] Jarmain C, Everson CS, Savage MJ, Mengistu MG, Clulow AD, Walker S, Gush MB. Refining Tools for Evaporation Monitoring in Support of Water Resources Management. Report No. K5/1567/1/08. Pretoria, South Africa: Water Research Commission; 2009.

[21] Savage MJ, Everson CS, Odhiambo GO, Mengistu MJ, Jarmain C. Theory and Practice of Evaporation Measurement, with Special Focus on SLS as an Operational Tool for the Estimation of Spatially-Averaged Evaporation. Report No. 1335/1/04. Pretoria, South Africa: Water Research Commission; 2004.

[22] Allen RG, Pereira LS, Raes D, Smith M. Crop Evapotranspiration: Guidelines for Computing Crop Water Requirements. Irrigation and Drainage Paper 56. Rome, Italy: United Nations Food and Agriculture Organization; 1998. p300.

[23] Simunek J, Sejna M, Van Genuchten MTh. The HYDRUS-2D Software Package for Simulating Two-Dimensional Movement of Water, Heat, and Multiple Solutes in

Variably-Saturated Media, Version 2.0. Riverside, California: U.S. Salinity Laboratory, USDA, ARS; 1999.

[24] Vegter JR. An Explanation of a Set of National Groundwater Maps. Report No. TT74/95. Pretoria, South Africa: Water Research Commission; 1995.

[25] Bredenkamp DB, Vandoolaeghe MAC. Die Ontingbare Grondwaterpotensiaal van die Atlantisgebied. Technical report No. Gh 3227. Pretoria, South Africa: Department of Water Affairs and Forestry, Geohydrology Division; 1982.

[26] DWAF. Groundwater Resource Assessment II: Recharge Literature Review Report 3aA. Project No. 2003-150. Pretoria, South Africa: Department of Water Affairs and Forestry; 2006.

[27] Woodford AC. Preliminary Assessment of Supplying Eskom's Ankerlig Power Station with Water from Local Groundwater Resources. Report No. 374624. Cape Town, South Africa: SRK Consulting; 2007.

Chapter 4

Seasonal and Regional Variability of the Complementary Relationship Between Actual and Potential Evaporations in the Asian Monsoon Region

Hanbo Yang and Dawen Yang

Additional information is available at the end of the chapter

http://dx.doi.org/10.5772/53067

1. Introduction

The concept of the complementary relationship (CR) between actual and potential evaporations was first proposed by Bouchet [1]. The underlying argument of the CR can be developed as follows. For one reason independent of energy considerations, actual evaporation LE decrease below wet environment evaporation LE_w, a certain amount of energy not consumed in evaporation becomes sensible heat fluxΔH, which can be expressed as

$$LE_w - LE = \Delta H \tag{1}$$

At the regional scale, this residual energy ΔH affects temperature, humidity, and other variables of air near the ground surface, which lead to an increase in potential evaporation LE_p, and one will have

$$LE_p = LE_w + \Delta H \tag{2}$$

Combination of equations (1) and (2) yields the complementary relationship

$$LE + LE_p = 2LE_w \tag{3}$$

Theoretically, the CR has been heuristically proven based on a series of restrictive assumptions [2, 3]. Also, it has been proved in many applications, such as interpreting the evaporation paradox and estimating actual evaporation. The evaporation paradox was referred as that an increase in actual evaporation estimated by water balance methods over large areas, and a decrease in pan evaporation from measurements in many regions have been recently reported [4, 5], which can be interpreted based on the CR [6]. Using the CR, Brutsaert [7] estimated actual evaporation increase at about 0.44 mm/a2, according to typical values of global trends of net radiation, temperature and pan evaporation. Direct measurement of actual evaporation over large areas is still difficult [8]. Consequently, the CR in which the feedback of potential evaporation with actual evaporation is considered suggests an attractive method for estimating *LE* over a large region, without knowing underlying surface conditions such as soil moisture. This has been widely applied for actual evaporation estimation over different time scales, such as monthly [9-13], daily [14, 15], and hourly [16].

Nevertheless, it was found that the Bouchet hypothesis (Equation 3) was only partially fulfilled [17,18]. In fact, Bouchet [1] documented that Equation (3) was generally modified with consideration of changes to water vapor and energy exchanges of the system with its surroundings, so that$LE + L\ E_p \leq 2L\ E_w$. Whereupon the expression was modified [19, 20] as $LE + L\ E_p = mL\ E_w$, where m is a constant of proportionality. Based on 192 data pairs from 25 basins over the United States, Ramirez et al. [19] determined a mean m of 1.97, but with high observed variability.

In the CR, wet environment evaporation (LE_w) was suggested [14] to be given by the Priestley-Taylor equation [21]:

$$LE_w = \alpha \frac{\Delta}{\Delta + \gamma}\left(R_n - G\right) \tag{4}$$

where α is a parameter,.Δ. (kPa/oC) is the slope of saturated vapor pressure at the air temperature, γ(kPa/oC) is a psychometric constant, R_n(mm/day) is net radiation, and G (mm/day) is soil heat flux. Central to wet environment evaporation is the concept of equilibrium evaporation. According to a theory for surface energy exchange in partly open systems, embracing a fully open system and fully closed system as limits, Raupach [22] asserted that a steady state with a steady-state LE_w could be attained; the time to reach steady state (a steady proportion of available energy transforming into latent heat$\alpha \frac{\Delta}{\Delta + \gamma}$) was 1–10 hours for a shallow convective boundary layer. Because of water vapor and energy exchanges between the system and surroundings, the proportion of available energy transforming into latent heat is usually modified. Raupach [23] parameterized the effect of air exchange between system and surroundings on equilibrium evaporation, and suggested conservation equations for entropy and water vapor in an open system. This revealed that advection was likely to modify air temperature and entropy at the system reference height, causing change in the proportion$\alpha \frac{\Delta}{\Delta + \gamma}$.

On calculating LE_w in the CR, Brutsaert and Stricker [14] suggested an average α on the order of 1.26–1.28. The value $\alpha = 1.32$ was predicted by Morton [11]. Hobbins et al. [9] obtained a value of $\alpha = 1.3177$ using data from 92 basins across the conterminous United States. Xu and Singh [24] determined α values in the advection-aridity (AA) model of Brutsaert and Stricker [14] for three study regions at 1.18, 1.04, and 1.00. Yang et al. [25] furnished an average $\alpha = 1.17$ with range 0.87–1.48 from 108 catchments in the Yellow River and Hai River basins of China, whereas Gao et al. [26] suggested an α of 1–1.23 for nine sub-basins of the Hai River basin. Using data from flux measurement stations #40 and #944 from the First International Land Surface Climatology Field Experiment (FIFE) but not in the same period, Pettijohn and Salvucci [27] and Szilagyi [20] obtained different values ofα, 1.10 and 1.18 (or 1.15), respectively. According to data from Weishan flux measurement station, Yang et al. [28] indicated an α range of 1–1.5 for a daytime hourly average. These variable values of the Priestley-Taylor parameter α may imply the variability of the CR.

Under the condition without water limitation, *LE* equals LE_p, and thus Equation (3) transforms into

$$LE = LE_w \tag{5}$$

This provides a simplified condition to study CR variability. According to analysis of saturated surface evaporation, Priestley and Taylor [21] gave an α range from 1.08 to 1.34, and took 1.26 as the average. Numerous papers report an average α of 1.26 [29-32]. Nevertheless, some details about α in these studies are noteworthy. Means in June, July and September were 1.27, 1.20 and 1.31, respectively [29], and α was less than 1.26 when *LE* was large, maybe in June or July [31]. Additionally, data in these studies were obtained only in particular months of the year, such as September and October [30], June to September [29], July [31], and June, July and September [32]. Using observations from April to October over a large, shallow lake in the Netherlands, [33] found α had a seasonal variation from 1.20 in August to 1.50 in April. Seasonal variation of α in the Priestley-Taylor equation for calculating LE_w can be considered an indicator of CR variability.

This chapter tried to examine quantitatively the seasonal and regional variability of the CR on the basis of observation data from 6 flux experiment sites and 108 catchments in the Asian monsoon region, and then to find an explanation for CR seasonal and regional variability.

2. Study area and data available

2.1. Flux experiment sites

A flux observation data set was collected from six flux experiment sites (the information was shown in Table 1 and Figure 1). These sites covered a wide range of climate and vegetation conditions from low latitude to high latitude in Asia. Therein five sites belong to the GEWEX (Global Energy and Water cycle Experiment) Asian Monsoon Experiment (GAME), including the Kogma site in the Thailand, Tibet_MS3637 site (renamed as Tibet site in this paper) and Hefei site in China, Yakutsk site and Tiksi site in Russia. Another site (Weishan experiment site) was located at the downstream of the Yellow River, China, which was set up by Tsinghua University. This data set includes meteorological elements (air temperature, relative humidity, wind direction and speed, air pressure), radiation (longwave and shortwave radiation, net radiation), soil temperature, precipitation, soil moisture, skin temperature, sensible heat flux, latent heat flux, and soil heat flux. Data were recorded as hourly averages. The energy balance closure problem was solved before data release at five of the six sites except Weishan site. At Weishan site, closure of the energy balance of approximately 0.8 was evaluated, according to data from 2005 to 2006.

The Kogma watershed is covered by a hilly evergreen forest in which only a few species lose their leaves, and canopy top is about 30 m [34, 35]. The Kogma site, part of the GEWEX (Global Energy and Water Cycle Experiment) Asian Monsoon Experiment (GAME), is located in the Kogma watershed of northern Thailand, with a 50 m observation tower. Air temperature and humidity were measured at 43.4 m using a psychrometer (HMP45D, Vaisala). Wind speed was measured at 43.4 m using an anemometer (AC750, Makino Ohyosokki). Downward and upward solar radiations were measured with pyranometers (MS-801 and MS-42, Eiko Seiki Co.) at 50.5 and 43.4 m, respectively. Downward and upward long-wave radiations were measured with an infrared radiometer (MS-200, Eiko Seiki Co.) at 50.5 and 43.4 m. Sensible and latent heat flexes were measured using the eddy correlation system, and the sonic anemometer-thermometer (DA-600, Kaijo) was installed at 41.5 m. Soil heat flux was measured using a probe (MF81, Eiko Seiki Co.)

The Tibet site was setup on May 1998 in the wet grassland between Amdo and Naqu, in the GAME-Tibet region, which was closed in September 1998. Air temperature and humidity were measured using sensors (50Y, Vaisala) at 7.8 and 2.3 m. Wind speed was measured at 9.8 m using an anemometer (R.M. Young Prop-Vane). Downward and upward solar radiations were measured with two pyranometers (CM21, Kipp-Zonen) respectively. Downward and upward long-wave radiations were measured with two pyrgeometers (PSP, Eppley). Sensible and latent heat flexes were measured using the eddy correlation system with a sonic anemometer- thermometer (R3A, Gill). Soil heat flux was measured with a probe (HFT-3.1, REBS).

The Hefei site is set up in the Shouxian Meteorological Observatory, Anhui province for surface flux observation in the Huaihe River Basin. The vegetation of surrounding area consists of mostly rice paddy and partly farmland. Shouxian is located in the middle of intensified observation area of GAME-HUBEX. Air temperature and humidity were measured using sensors (50Y, Vaisala). Wind speed was measured at 9.8 m using an anemometer (09101, R.M. Young Prop-Vane). Downward and upward solar radiations were measured with two pyranometers (Kipp-Zonenn), respectively. Downward and upward long-wave radiations were measured with two pyrgeometers (PIR, Eppley). Sensible and latent heat flexes were measured using the eddy correlation system with a sonic anemometer- thermometer (Gill). Soil heat flux was measured with a probe (HFT-3.1, REBS).

The Yakutsk site is located in the middle reaches of the Lena and is in a region of continuous permafrost, the Sakha Republic of Russian, where the climate exhibits a strong continentality [36]. Air temperature and humidity were measured using sensors (HMP-35D, Vaisala) at 17.2 and 13.4 m. Wind speed was measured at 9.8 m using an anemometer (AC-750, Makino). Downward and upward solar radiations were measured with pyranometers (CM-6F, Kipp-Zonen) at 18.2 and 15.9, respectively. Downward and upward long-wave radiations were measured with pyrgeometers (MS-201F, EKO). Sensible and latent heat flexes were measured using the eddy correlation system with a sonic anemometer- thermometer (DA-600, KAIJO) and an open pathe H2O gas analyser (AH-300, KAIJO).

The Tiksi site is performed in the Siberian tundraregion near Tiksi, Sakha Republic, Russian Federation. Air temperature and humidity were measured using sensors (HMP-45D, Vaisala) at 10 m. Wind speed was measured at 10 m using an anemometer (AC860, Makino). Downward solar radiation was measured at 1.5 m with a pyranometer (MS-802F, EKO). Downward and upward long-wave radiations were measured at 1.5 m with pyrgeometers (MS-802F, EKO). Sensible and latent heat flexes were measured using the eddy correlation system with a sonic anemometer- thermometer. Soil heat flux was measured with a probe (MF-81, EKO) at 0.01 and 0.08 m. More details about the five sites are provided at the GAME website (http://aan.suiri.tsukuba.ac.jp/).

Site	**Location**	Altitude (m)	Vegetation type	**Data period**
Kogma	18°48.8'N, 98°54.0'E	1268	Evergreen forest	Feb. – Dec., 1998
Tibet	31°01.0'N, 91°39.4'E	4820	Grass	May – Sep., 1998
Hefei	32°34.8'N, 116°46.2'E	23	Rice paddy	Aug., 1998 , Apr., Nov., and Dec., 1999
Weishan	36°38.9'N, 116°03.3'E	30	Wheat, corn	May 18, 2005 – Dec. 31, 2006
Yakutsk	62°15.3'N, 129°37.1'E	210	Larch forest	Apr. – Aug., 1998 Apr. – Jul., 1999
Tiksi	71°35.2'N, 128°46.5'E	40	Tundra	Jun. – Sep., 2000

Table 1. Descriptions of the flux experiment sites

The Weishan site is located in a downstream reach of the Yellow River. Most of this region is farmland, with flat topography. Winter wheat and maize are the two major crops, rotationally cultivated [37]. Winter wheat planting season is in early October, and the growing period is from March to mid-June. The experimental field is near the center of the irrigation district, and is a 400 m by 500 m rectangular field. Typical meteorological instruments are installed atop a 10 m tall tower, along with a radiometer and an eddy correlation system for sensible and latent heat fluxes. Air temperature and humidity were measured using sensors (HMP-45C, Vaisala) at 10 m. Wind speed was measured at 10 m using an anemometer (05103, Young Co.). Downward and upward solar and long-wave radiations were measured at 3.5 m with pyranometers (CNR-1, Kipp-Zonen). Sensible and latent heat flexes were measured using the eddy correlation system at 3.7 m with a sonic anemometer-thermometer (CSAT3, Campbell). Soil heat flux was measured with a probe (HFP01SC, Hukseflux). Observations were recorded as 30-minute averages.

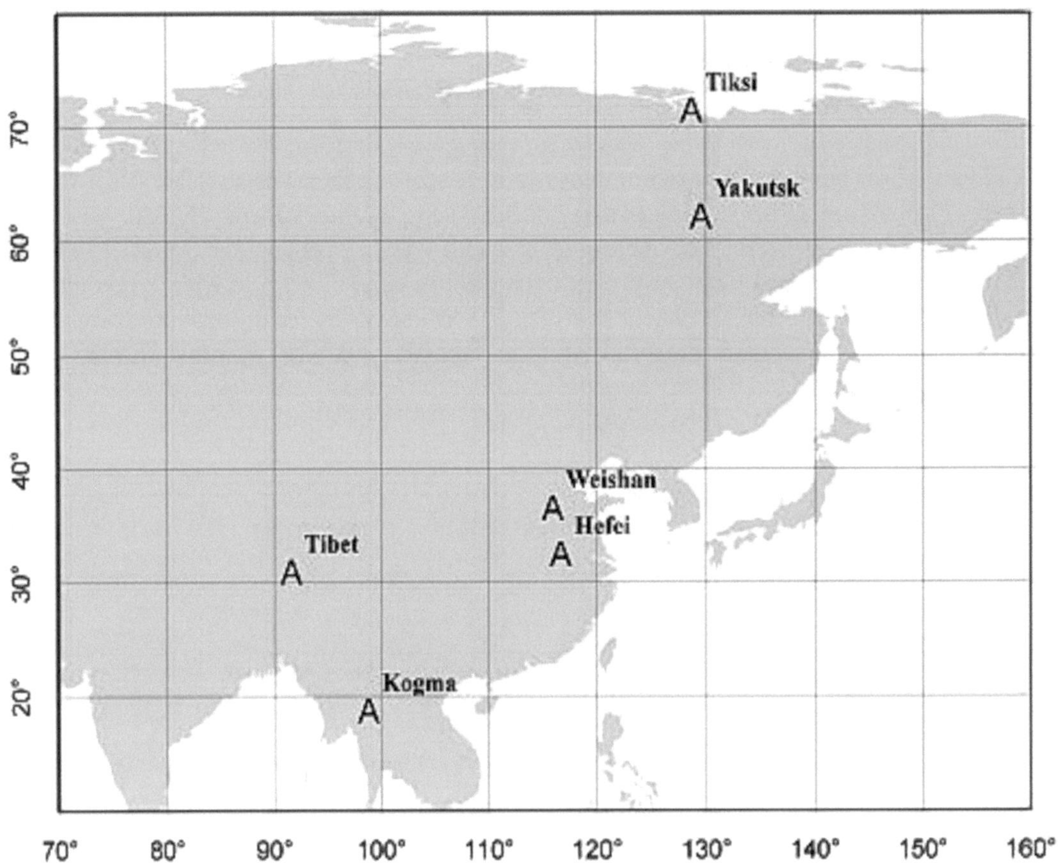

Figure 1. Location of the two flux experiment sites

2.2. Study catchments

To examine the regional variability, 108 catchments, locating in the Yellow River basin, the Hai River basin, and the Inland Rivers basin in the non-humid region of China were chose. Their drainage areas cover a range 272–94,800 km^2. Climate arid index covers from 1–7, and the runoff coefficient ranges 0.02–0.32. The hydrologic and meteorological data were collected from each catchment from 1953–1998. Figure 2 presents the distribution of hydrologic and meteorological stations in the study region. Furthermore, more information on the 108 catchments was given by [38].

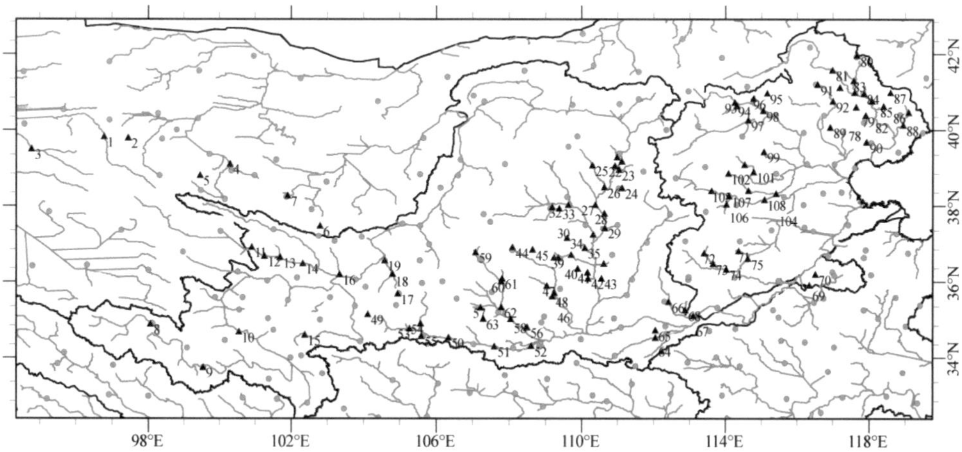

Figure 2. Distribution of the hydrologic and meteorological stations in the 108 catchments in the non-humid regions of China (the solid triangle represents a hydrologic station at the outlet of catchments and the grey solid circle represents a meteorological station)

3. Method

In a closed system without advection, the CR can be expressed as equation (3) in which LE_w is estimated using equation (4) with $\alpha = 1.26$. However, in a real open environment the horizontal advection can't be neglected. Therefore, taking the effect of the horizontal advection into account, we have two methods, and the one is

$$A = LE + LE_p - 2 \times 1.26 \frac{\Delta}{\Delta + \gamma} \left(R_n - G \right), \tag{6}$$

i.e., the CR is modified as $LE + LE_p = 2LE_w + A$. The value of A indicated the effect of the horizontal advection of both energy and water vapor. Morton [10] suggested a similar equation $LE + LE_p = 2 \times 1.26 \frac{\Delta}{\Delta + \gamma}(R_n - G + A_m)$, where A_m was an empirical correction factor for advection. If there is a seasonal and regional variability in the CR, A should have a seasonal and regional variation. The other one is focusing on the Priestley-Taylor parameter α to reveal seasonal and regional variability of the CR. According to Equations (3) and (4), α can be calculated as

$$\alpha = \frac{\gamma + \Delta}{2\Delta} \cdot \frac{LE + LE_p}{R_n - G} \tag{7}$$

Similar to A, α should have a seasonal and regional variation if there is a seasonal and regional variability in the CR.

The wet environment evaporation LE_w was estimated by the Priestley-Taylor equation. To calculate potential evaporation LE_p, the Penman equation [39] has been suggested by [10, 14, 40]

$$LE_p = \frac{\Delta}{\Delta + \gamma}(R_n - G) + \frac{\gamma}{\Delta + \gamma} E_A, \tag{8}$$

where E_A is the drying power of the air. This can be estimated by

$$E_A = f(u)\left(e^* - e\right), \tag{9}$$

where e^* (kPa) and e (kPa) are the saturated and actual vapor pressures at the same air temperature, respectively. The wind function $f(u)$ can be estimated as

$$f(u) = 0.26\left(1 + 0.54u\right), \tag{10}$$

where u (m/s) is mean wind speed at 2 m height.

Actual evaporation LE was observed using the eddy correlation technique at the flux experiment sites, and was calculated from the annual water balance by ignoring the inter-annual change of water storage in these catchments.

4. Results

4.1. Seasonal and regional variability observed in the flux experiment sites

The horizontal advection term A in the modified CR and the Priestley-Taylor parameter α at the six sites were calculated using equations (6) and (7) respectively based on the daily data. Then the daily values were averaged on each ten-day from the starting date, and the results were shown in Figures 3 and 4. With regard to the value of A, Figure 3 shows different variance ranges at different sites. However, a seasonal variation is observable, i.e. the value of A reaches the minimum in July or August, then rises up to the maximum in March or April, and decreases at last until July or August. Similarly, Figure 4 shows the Priestley-Taylor's parameter α has a seasonal variation with a maximum in winter and a minimum in summer. In general, we discern seasonal variation ofα, although the points are scattered in winter.

In particular, at Kogma, αhas a maximum of 1.7 approximately in February, then falls to a minimum of 1.1 approximately in summer; it increases thereafter, until winter, which ranges from 1.08–1.40 between April and October. At Weishan, αhas a mean of 1.18 during the summer monsoon period, and about 1.92 during the winter monsoon. Monthly α varies from 1.02–1.40 between May and October.

Figure 3 also shows that the Priestley-Taylor's parameter α increases with latitude increasing, which is the largest at the Yakutsk site in Siberia of Russia and the smallest at the Kogma site in Thailand in the same season.

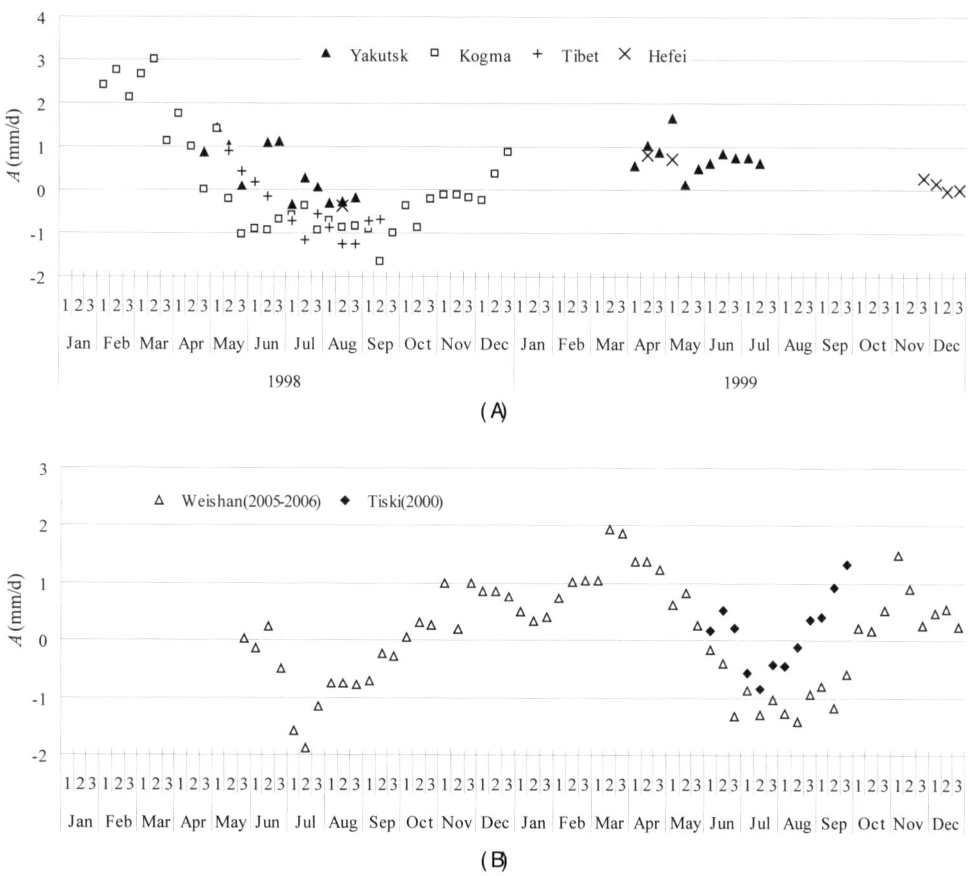

Figure 3. Seasonal and regional variability of the horizontal advection.

4.2. Regional variability observed in the 108 catchments

The parameter α for the 108 catchments has a large variance, which ranges from 0.87 to 1.48 (with an average of 1.17). Nevertheless, the relation of α with latitude can be revealed, as shown in Figure 3. The parameter α increases with the latitude increasing over the region ranging 33–40 °N, while it decreases with the latitude increasing over the region ranging 40–42 °N. Also, the relation of the parameter α with the longitude was plotted in Figure 4. It can be found that the catchment with larger longitude is approximately closer to the ocean. Therefore, Figure 4 shows that α increases with the distance from ocean decreasing.

5. Discussion

5.1. Seasonal variability

The value of A reaches the minimum in July or August, then rises up to the maximum in March or April, and decreases at last until July or August. As shown in Figure 4, the Priestley-Taylor's parameter α has a seasonal variation with a maximum in winter and a minimum in summer [41], which is similar to that A has. Also, most studies did not introduce an advection item, but instead adjusted the Priestley-Taylor parameter for advection. Therefore, we focus on the variation of α in this chapter.

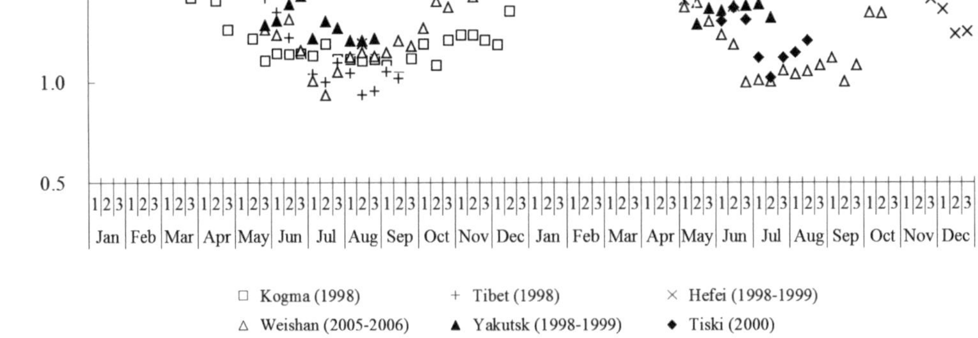

Figure 4. Seasonal and regional variability of the Priestley-Taylor's parameter α

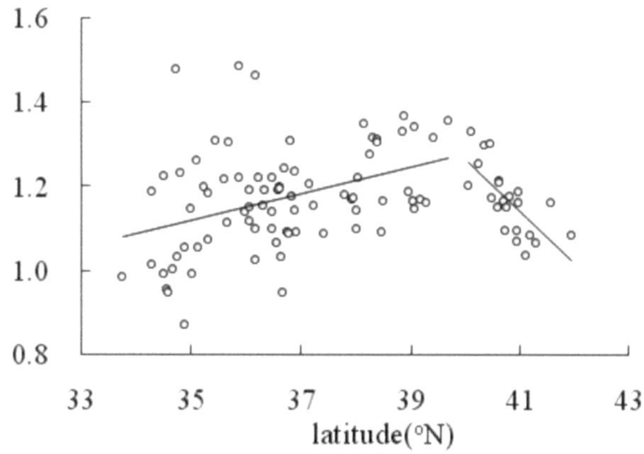

Figure 5. Relation of parameter *α* with latitude

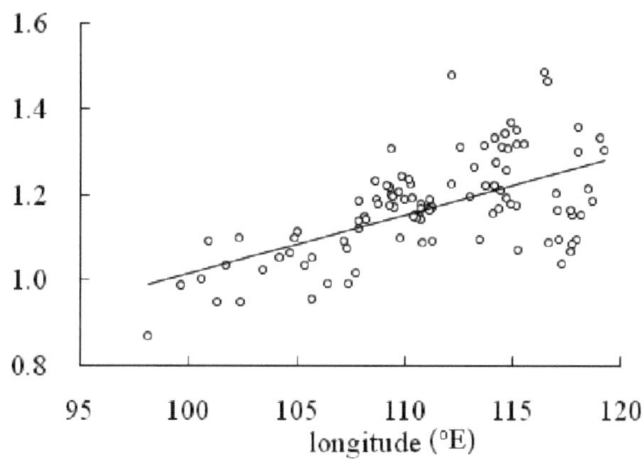

Figure 6. Relation of parameter *α* with longitude (larger longitude indicating the catchment being nearer the ocean)

DeBruin and Keijman [33] reported that α differed slightly from 1.26 in May–September, but was about 1.50 in April and October. Similarly, the α value at Kogma was between 1.07–1.26 in May–September and was 1.40 in April. At the Weishan site, αhad a similar seasonal variation but larger values, up to 1.60 in April and 1.38 in October. At the same time, another phenomenon, that the Asian monsoon is from ocean to continent in June–October and from continent to ocean in October–June [42], should be noticing. As a result, the monsoon leads to air temperature decreasing and humid increasing in June–October, while both air temperature and humidity increasing in October–June above the continent. It was therefore speculated that there is a certain relation between the seasonal variation in α and the monsoon.

The energy balance near the ground surface can be expressed as

$$R_n - G = LE + H \tag{11}$$

Advection impacts the CR by modifying air temperature, water vapor pressure, and others. As a result, the partition of available energy into latent and sensible heats will change, and the presence of advection causes $LE > R_n - G$ [43-45] when the direction of sensible heat H is downward. Differences in thermodynamic properties between land and ocean produce generally higher temperatures and less water vapor over continents than over oceans in summer, and lower temperatures and less water vapor over continents than oceans in winter. Consequently over continents, atmospheric circulation between land and ocean decreases temperature and increases vapor during summer, and increases both temperature and vapor in winter. It seems paradoxical that the winter monsoon increases temperature over continents. In fact, we find that the distribution of isotherms is not completely latitudinal; temperature has an inverse relationship with distance from the ocean in identical latitude continental regions. This indicates heat transport from ocean to continent by advection. We speculate that the circulation increases temperature over land, and the increase weakens with distance from the ocean, as a result of sensible heat transport.

Advection possibly affects the major assumption of the CR, that energy release from a decrease in actual evaporation compensates the increase in potential evaporation [46]. The monsoon transports water vapor and sensible heat between ocean and continent, which causes additional seasonal changes to air humidity and temperature. The effects of these changes on the two sides of Equation (3) are asymmetric. On the left side, the terms LE_p and LE can be determined by climate variables (such as air temperature and vapor pressure), which include the effect of horizontal advection. On the right side, the effect of horizontal advection on LE_w is parameterized as only the change of air temperature (if the effect of radiation is neglected), not including changes of wind speed and humidity.

We assume a system without horizontal advection, where Equation (3) is satisfied. Since the summer monsoon imports a large amount of water vapor and reduces latent heat, the drying power of the air E_A decreases, and increases the ratio $H/(R_n - G)$ (i.e., LE decreases). This reduces $(LE + LE_p)$ but causes less change in LE_w. This translates into a smaller α in Equation (3). The winter monsoon increases E_A and $LE/(R_n - G)$, which produces an increase in $(LE + LE_p)$ but less change in LE_w, resulting in a larger in Equation (3). Following the same reasoning, we can explain the seasonal variation in α revealed by [33]. According to the CR, with an unlimited water supply above a lake, the evaporation LE equals the potential evaporation LE_p. In summer, horizontal advection reduces $(LE + LE_p)$, resulting in a small α value, but a large α in winter.

5.2. Regional variability

In addition, the effect of horizontal advection also has a regional variation. Since energy is transported by atmospheric and oceanic circulations from low to high latitudes, and water vapor transported from the lower atmospheric layer over the ocean to that over land. This therefore leads to a regional variability in the CR.

From the Figure 3, it can be seen that the magnitude of the horizontal advection effect is the largest at the Yakutsk site and the smallest at the Kogma site in the same season. This indicates that the magnitude of the horizontal advection effect increases with latitude increasing. Figure 4 shows that the Priestley-Taylor's parameter α increases with latitude increasing, which is the largest at the Yakutsk site and the smallest at the Kogma site in the same season. This is also consistent with the results given by Xu and Singh [23], in which α =1.0, 1.04 and 1.18 at the catchments in Eastern China (29°15′N, 121°10′E), Northwestern Cyprus (about 35°N), and Central Sweden (59°53′N, 17°35′E) respectively for a long-term mean.

Across the 108 catchments, the parameter α increases with the latitude increasing over the region ranging 33–40 °N but decreases with the latitude increasing over the region ranging 40–42 °N. As shown in Figure 2, the catchments ranging 40–42 °N belong to the Hai River basin, which are adjacent to the Bohai Sea. Those catchments have an increasing distance from the sea with their latitude increasing. The possible cause is that the change in distance from the sea has a larger effect on the horizontal advection than increasing latitude has. This was revealed by Figure 6, i.e. the catchments farther from the sea have larger parameterα. In addition, Figure 6 shows larger dispersion in the relation between α and the longitude, the possible cause for which was that the flexuous coastline results in the catchments with same longitude having different distance from the ocean.

6. Conclusion

The complementary relationship (CR) between actual evaporation and potential evaporation has been widely used to explain the evaporation paradox, as well as to estimate regional evaporation. The theoretical foundation of the CR is the Bouchet hypothesis, including the constraint that exchanges of water vapor and energy between the considered system and its exterior are constant. In reality, the atmosphere does not always satisfy the constraint. In the Asian monsoon region, atmospheric motions have a significant seasonal variation, accompanied by transport of water vapor and energy. Through analyzing seasonal variation in parameter α of the Priestley-Taylor equation for calculating wet environment evaporation, this chapter analyzed effects of horizontal advection on the CR. αhas a significant seasonal variation, which is larger in winter than in summer. The possible cause is that the summer monsoon increases water vapor content and decreases air temperature, whereas the winter monsoon increases both water vapor and air temperature. The parameter α increases with latitude, as a result of the annual transport of the energy and vapor from low latitudes to high latitudes through the atmospheric and oceanic flows. Atmospheric circulation between continent and ocean transports vapor from the oceans to the land, so α decreases with distance from ocean.

Acknowledgements

Data were from the Global Energy and Water Cycle Experiment (GEWEX) Asian Monsoon Experiment in a Tropical region (GAME-T), and Weishan flux observation was supported by the National Natural Science Foundation of China (grant nos. 50909051, 50939004, and 51025931). This research was also supported by the Ministry of Science and Technology of China (2011IM011000).

Author details

Hanbo Yang* and Dawen Yang

*Address all correspondence to: yanghanbo@tsinghua.edu.cn

State Key Laboratory of Hydro-Science and Engineering, Department of Hydraulic Engineering, Tsinghua University, Beijing, China

References

[1] Bouchet, R. (1963), Evapotranspiration reelle evapotranspiration potentielle, signification climatique, Int Assoc Sci Hydrol Proc, 62, 134-142.

[2] Morton, F. I. (1971), Catchment evaporation and potential evaporation: further development of a climatologic relationship, J Hydrol, 12, 81-99.

[3] Szilagyi, J. (2001), On Bouchet's complementary hypothesis, J Hydrol, 246(1-4), 155-158.

[4] Milly, P. C., and K. A. Dunne (2001), Trends in evaporation and surface cooling in the Mississippi River basin, Geophys Res Lett, 28(7), 1219-1222.

[5] Walter, M. T., D. S. Wilks, J. Y. Parlange, and R. L. Schneider (2004), Increasing evapotranspiration from the conterminous United States, J Hydrometeorol, 5(3), 405-408.

[6] Brutsaert, W., and M. B. Parlange (1998), Hydrologic cycle explains the evaporation paradox, Nature, 396(6706), 30.

[7] Brutsaert, W. (2006), Indications of increasing land surface evaporation during the second half of the 20th century, Geophys Res Lett, 33, L20403, doi: 10.1029/2006GL027532.

[8] Kahler, D. M., and W. Brutsaert (2006), Complementary relationship between daily evaporation in the environment and pan evaporation, Water Resour Res, 42(5), doi: 10.1029/2005WR004541.

[9] Hobbins, M. T., J. A. Ramirez, and T. C. Brown (2001), The complementary relationship in estimation of regional evapotranspiration: An enhanced Advection-Aridity model, Water Resour Res, 37(5), 1389-1403.

[10] Morton, F. I. (1975), Estimating evaporation and transpiration from climatological observations, Journal of Applied Moteorology, 14, 488-497.

[11] Morton, F. I. (1983), Operational estimates of areal evapotranspiration and their significance to the science and practice of hydrology, J Hydrol, 66, 1-76.

[12] Szilagyi, J., and J. Jozsa (2008), New findings about the complementary relationship-based evaporation estimation methods, J Hydrol, 354(1-4), 171-186.

[13] Szilagyi, J., M. T. Hobbins, and J. Jozsa (2009), Modified Advection-Aridity Model of Evapotranspiration, JOURNAL OF HYDROLOGIC ENGINEERING, 14(6), 569-574.

[14] Brutsaert, W., and H. Stricker (1979), An advection-aridity approach to estimate actual regional evapotranspiration, Water Resour Res, 15(2), 443-450.

[15] Han, S. J., H. P. Hu, D. W. Yang, and F. Q. Tian (2011), A complementary relationship evaporation model referring to the Granger model and the advection-aridity model, Hydrol Process, 25(13), 2094-2101.

[16] Parlange, M. B., and G. G. Katul (1992), An advection-aridity evaporation model, Water Resour Res, 28, 127-132.

[17] Kim, C. P., and D. Entekhabi (1998), Feedbacks in the land-surface and mixed-layer energy budgets, Bound-Lay Meteorol, 88(1), 1-21.

[18] Sugita, M., J. Usui, I. Tamagawa, and I. Kaihotsu (2001), Complementary relationship with a convective boundary layer model to estimate regional evaporation, Water Resour Res, 37(2), 353-365.

[19] Ramirez, J. A., M. T. Hobbins, and T. C. Brown (2005), Observational evidence of the complementary relationship in regional evaporation lends strong support for Bouchet's hypothesis, Geophys Res Lett, 32, L15401, doi:10.1029/2005GL023549.

[20] Szilagyi, J. (2007), On the inherent asymmetric nature of the complementary relationship of evaporation, Geophys Res Lett, 34, L02405, doi:10.1029/2006GL028708.

[21] Priestley, C. H. B., and R. J. Taylor (1972), On the assessment of surface heat flux and evaporation using large-scale parameters, Mon Weather Rev, 100(2), 81-92.

[22] Raupach, M. R. (2000), Equilibrium evaporation and the convective boundary layer, Bound-Lay Meteorol, 96(1-2), 107-141.

[23] Raupach, M. R. (2001), Combination theory and equilibrium evaporation, Q J Roy Meteor Soc, 127(574), 1149-1181.

[24] Xu, C. Y., and V. P. Singh (2005), Evaluation of three complementary relationship evapotranspiration models by water balance approach to estimate actual regional evapotranspiration in different climatic regions, J Hydrol, 308(1-4), 105-121.

[25] Yang, H.B., Yang, D.W., Lei, Z.D., Sun, F.B. and Cong, Z.T., 2008. Reginal varibility of the complementary relationship between actual and potential evapotranspirations (in Chinese). Journal of Tsinghua University (Science and Technique), 48(9): 1413-1416.

[26] Gao, G., C. Xu, D. Chen, and V. P. Singh (2012), Spatial and temporal characteristics of actual evapotranspiration over Haihe River basin in China, Stoch Env Res Risk A, 26(5): 655-669.

[27] Pettijohn, J. C., and G. D. Salvucci (2006), Impact of an unstressed canopy conductance on the Bouchet-Morton complementary relationship, Water Resour Res, 42(9), W09418, doi:10.1029/2005WR004385.

[28] Yang, H. B., D. W. Yang, Z. D. Lei, F. B. Sun, and Z. T. Cong (2009), Variability of complementary relationship and its mechanism on different time scales F-5023-2011 G-4441-2010, Sci China Ser E, 52(4), 1059-1067.

[29] Davies, J. A., and C. D. Allen (1973), Equilibrium, potential, and actual evaporation from cropped surfaces in southern Ontario, J Appl Meteorol, 12, 649-657.

[30] Eichinger, W. E., M. B. Parlange, and H. Stricker (1996), On the concept of equilibrium evaporation and the value of the Priestley-Taylor coefficient, Water Resour Res, 32(1), 161-164.

[31] Stewart, R. B., and W. R. Rouse (1976), A simple method for determining the evaporation from shallow lakes and ponds, Water Resour Res, 12, 623-628.

[32] Stewart, R. B., and W. R. Rouse (1977), Substantiation of the Priestley-Taylor parameter α= 1.26 for potential evaporation in high latitudes, J Appl Meteorol, 16, 649-650.

[33] DeBruin, H. A. R., and J. Q. Keijman (1979), The Priestley-Taylor evaporation model applied to a large shallow lake in the Netherland, J Appl Meteorol, 18, 898-903.

[34] Komatsu, H., N. Yoshida, H. Takizawa, I. Kosaka, C. Tantasirin, and M. Suzuki (2003), Seasonal trend in the occurrence of nocturnal drainage flow on a forested slope under a tropical monsoon climate, Bound-Lay Meteorol, 106(3), 573-592.

[35] Kume, T., H. Takizawa, N. Yoshifuji, K. Tanaka, C. Tantasirin, N. Tanaka, and M. Suzuki (2007), Impact of soil drought on sap flow and water status of evergreen trees in a tropical monsoon forest in northern Thailand, Forest Ecol Manag, 238(1-3), 220-230.

[36] Dolman, A. J., T. C. Maximov, E. J. Moors, A. P. Maximov, J. A. Elbers, A. V. Kononov, M. J. Waterloo, and M. K. van der Molen (2004), Net ecosystem exchange of carbon dioxide and water of far eastern Siberian Larch (Larix cajanderii) on permafrost, BIOGEOSCIENCES, 1(2), 133-146.

[37] Lei, H. M., and D. W. Yang (2010), Interannual and seasonal variability in evapotranspiration and energy partitioning over an irrigated cropland in the North China Plain, Agr Forest Meteorol, 150(4), 581-589.

[38] Yang, D. W., F. B. Sun, Z. Y. Liu, Z. T. Cong, G. H. Ni, and Z. D. Lei (2007), Analyzing spatial and temporal variability of annual water-energy balance in nonhumid regions of China using the Budyko hypothesis, Water Resour Res, 43, W04426, doi: 10.1029/2006WR005224.

[39] Penman, H. L. (1948), Natural evaporation from open water, bare soil and grass, Proceeding of the Royal Society of London. Series A, Mathematical and Physical Sciences, 193(1032), 120-145.

[40] Morton, F. I. (1976), Climatological estimates of evapotranspiration, J. Hydraul. Div. Proc. ASCE, 102, 275-291.

[41] Yang, H. B., D. W. Yang, and Z. D. Lei (2012), Seasonal variability of the complementary relationship in the Asian monsoon region, Hydrol Process, DOI: 10.1002/hyp. 9400.

[42] Ye, D. Z., Tao, S. Y., and Li, M. C., 1958. The abrupt change of circulafion over the Northern Hemisphere during June and October (in Chinese). Acta Meteorologica Sinica, 29, 249-263.

[43] Rijks, D. A. (1971), Water Use by Irrigated Cotton in Sudan. III. Bowen Ratios and Advective Energy, The Journal of Applied Ecology, 8, 643-663.

[44] Rosenberg, N. J., and S. B. Verma (1976), Extreme Evapotranspiration by Irrigated Alfalfa: A Consequence of the 1976 Midwestern Drought, J Appl Meteorol, 17, 934-941.

[45] Wright, J. L., and M. E. Jensen (1972), Peak water requirements of crops in southern Idaho, Journal of Irrigation and Drainage Division, 96, 193-201.

[46] Lhomme, J. P., and L. Guilioni (2006), Comments on some articles about the complementary relationship, J Hydrol, 323(1-4), 1-3.

Chapter 5

Satellite-Based Energy Balance Approach to Assess Riparian Water Use

Baburao Kamble, Ayse Irmak, Derrel L. Martin, Kenneth G. Hubbard, Ian Ratcliffe, Gary Hergert, Sunil Narumalani and Robert J. Oglesby

Additional information is available at the end of the chapter

http://dx.doi.org/10.5772/52929

1. Introduction

Previous studies across the High Plains and the Arid West of the United States have produced widely varying impacts of riparian evapotranspiration (ET) on surface and ground water. Many producers as well as various state agencies have advocated removing all trees along the river basins as a method of riparian control for water reclamation. Although eradication of trees might be an effective method for water reclamation in the short-term, it has not been yet proven whether such water savings are possible on a stream level. Mean water use of riparian trees has been reported in relatively few studies, and most of the previous studies have been of short duration. The water use for saltcedar (Tamarix spp.) was estimated at 15.9 L d^{-1} for 10 cm^2 sap wood area (swa) (Smith et al. 1998), 56.8 L d^{-1} for 33 cm^2 swa (Nagler et al. 2003), and 29.9 L d^{-1} for 100 cm^2 swa (Owens and Moore, 2007). The water use for Fremont cottonwood (Populus fremontii S. Wats.) varied from 57.6 L d^{-1} for 33 cm^2 swa (Nagler et al. 2003) to as high as 499.7 L d^{-1} for 833 cm^2 swa (Schaeffer et al. 2000). Riparian plant communities are complex ecosystems that, through an intimate relationship with the fluvial dynamics of river systems, are as much described by their continual cycle of disturbance and succession as by the vegetation that makes up their multi-storied habitats. Currently, there is uncertainty in the water use of riparian systems due to the narrow and sparse vegetation commonly associated with them. Local, state and federal water management regulatory agencies need good quality water use estimates on unmanaged riparian systems. High frequency micrometeorological flux measurements such as Eddy Correlation System (ECS) have been used to estimate water use by balancing fluxes of sensible and latent heat with total energy incident on a riparian area. However, the technique is most effective when applied to an agricultural land where the plant canopy is relatively homogeneous both in composition and in height, and the fetch is relatively large. Estimating riparian ET in semi-arid river basins is difficult because of the complicated geometry of a typical riparian zone (Goodrich et al., 2000). Riparian forests are typically characterized by long narrow strips of vegetation directly adjacent to stream channels. These strips of forest are often relatively high (~5-20m), not more than 20 m wide on either side of a watercourse and may consist of several different species and size classes of trees (Goodrich et al., 2000). This geometry precludes the use of classical meteorological flux measurements as the required fetch requirements usually are not satisfied. Without the required fetch, the total flux measured using this system will not representf the riparian zone. Water use varies spatially even in the same riparian species because of variation in tree age, height, density, and surroundings. Therefore, the estimation based on in-situ measurements is at stand scale and adds uncertainty when applied to riparian corridors of larger scale. In addition, application of high frequency meteorological flux measurements to quantify ET along stream channels in a basin is limited due to the number of measurement sites needed and the operational expense of such a dense network. The tree sap flow measurements capture variations in transpiration demand as a function of atmospheric demand and water availability. However, there is some uncertainty associated with estimation of stand-level transpiration from individual plant sap flow measurements.

Remote sensing measurements can provide information with a broad spatial coverage and a repeat temporal coverage and avoid the need to rely on field databases. Calculation of water consumption by remote sensing has benefited from significant research efforts over the last 30 years, especially the dedicated energy balance models like the TSEB (Norman et al., 1995), SEBAL (Bastiaanssen, 1998a; 1998b), SEBI (Menenti, 2000), and METRIC^{tm}(Allen et al. 2007b). In this chapter, the method used to estimate ET in riparian areas employs the model known as Mapping Evapotranspiration at high Resolution with Internalized Calibration (METRIC^{tm}) by Allen (2007b). The METRIC^{tm} model (Allen et al. 2007b) requires parameterization of the energy balance and estimates surface energy fluxes based on spectral satellite measurements. The model has auto-calibration capabilities for each satellite image using ground-based calculation of alfalfa reference ET based on hourly weather data. Modifications to the energy balance algorithms for narrow riparian regions that may experience advection and different turbulence characteristics than shorter and more homogeneous surfaces is accomplished by running the METRIC^{tm} model with airborne data collected from June through the end of October 2009 using AISA hyperspectral system hyperspectral system at Center for Advanced Land Management Information's (CALMIT's), onboard Piper Saratoga aircraft. Furthermore, frequent and intensive aerial images help to investigate patterns of surface temperature at various spacings inside riparian structures and the variation with wind speed and wind direction. This data is also used to understand the partitioning of the available energy between understory and overstory in the riparian system. There is some uncertainty in estimating ET from riparian systems with SRS based energy balance due to the narrow and sparse vegetation commonly associated with them. However, the high resolution of Landsat images is extremely useful for assessing ET patterns and might be the most suitable method (Irmak and Kamble, 2009;, Kamble and Irmak, 2011;2008; Irmak et al., 2011). The literature also shows that the thermal signatures of the riparian systems indicate a pretty complete picture of the amount of evaporative cooling within boundaries of the ecosystem, including evaporation from wet soil or under structure. Therefore, the METRIC^{tm} or related remote sensing based energy balance approach is one of the better means to make the ET estimates and to monitor before and after vegetation and land-use modification.

The information presented here is taken from the results of projects conducted at the University of Nebraska. Our primary goal in this chapter is to illustrate spatiotemporal estimation of ET using satellite-and aerial derived spectral radiances in conjunction with a land surface energy balance model and flux measurements to evaluate water use and water distribution within and between riparian systems, including invasive species, along the North Platte River Basin (NPRB) in Nebraska (NE). Specific objectives were to quantify daily and seasonal distributions of ET in riparian systems within the NPRB, quantify surface energy balance flux components for riparian systems, and compare water use among riparian species, both native and invasive, by utilizing ET maps together with riparian species distribution map.

2. Material and methods

2.1. Study area

The Nebraska Panhandle is an area in western Nebraska bounded by the mountainous (Rocky Mountains) states of Wyoming and Colorado on the west and the city of North Platte,Nebraska on the east, wherein the main stem of the North Platte River flows from west to east. The Nebraska panhandle roughly encompasses the area in Nebraska between 102° and 104°W longitude and 41° and 43°N latitude. The elevation ranges from 3,000-5,000 ft and the growing season is characterized by hot days and cool nights. The Panhandle additionally has a high desert-type semi-arid climate receiving 14-16 inches rainfall per year. Flows in the North Platte River are derived mainly from snowmelt runoff from the Rocky Mountains and surface runoff and ground-water discharge (Bentall and Shaffer, 1979; Gutentag et. al., 1984). The eastern and central parts of Nebraska predominantly are cultivated-agricultural land with increasing amounts of rangeland to the west (Center for Advanced Land Management Information Technologies, 2000; U.S. Geological Survey, 1999-2000). Riparian vegetation is restricted to lowlands located along the North Platte River and its tributaries. Forests and grasslands (including wet meadows) in the riparian zone are the predominant communities, but other categories are present including herbaceous, shrub, and emergent wetlands (Currier et. al., 1985).

2.2. Satellite image processing

A total of 8 Landsat5 and Landsat7 satellite images (Path 33, Row 31) from 2005 were used for this study (table 2). Each Landsat scene size is approximately 170 km X 185 km with a repeat cycle of 16 days. The Thematic Mapper (TM) sensor on-board Landsat 5 has seven spectral bands with 30 m spatial resolution in reflective bands and 120 m resolution in thermal bands. The Enhanced Thematic Mapper (ETM) on-board Landsat 7 has eight spectral bands including panchromatic band (not used in this study). UTM zone 13 and NAD 1983 was the projection and datum used. For Landsat 7 imagery, the high gain on the thermal band was used. The Landsat 7 thermal band is acquired in both low and high gain. The low gain provides an expanded dynamic range generating less saturation at high values, but lower radiometric resolution (sensitivity).

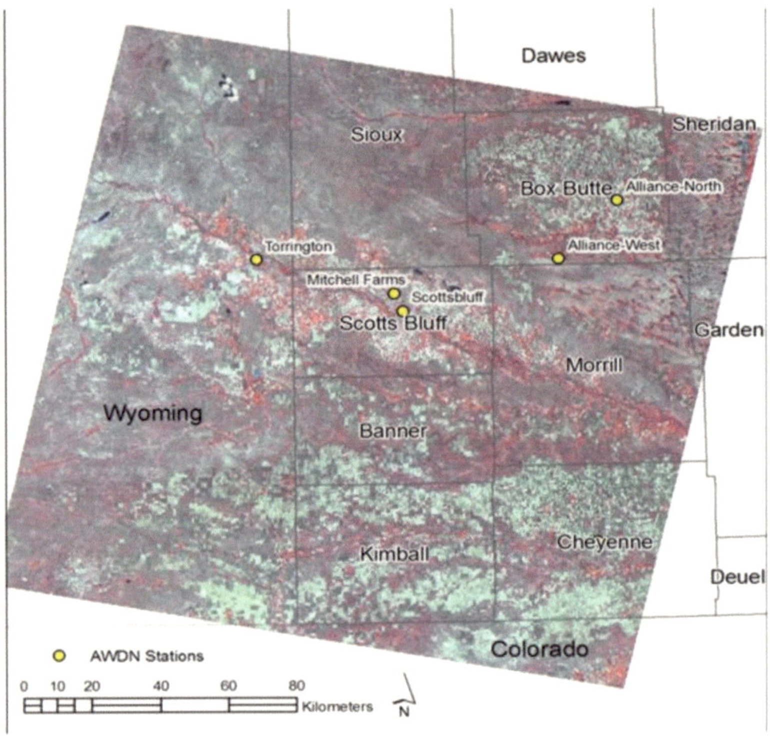

Figure 1. Geographic footprint of Landsat path 33, row 31. Images cover parts of the Nebraska Panhandle, Wyoming, and Colorado.

The thermal high gain band has higher radiometric resolution but provides a less dynamic range.The spatial resolution of ETM in reflective bands is 30 m and 60 m in the thermal bands. Landsat7 ETM images have missing data in the form of a wedge due to failure of Scan Line Corrector (SLC) on May 31, 2003 referred to as SLC off images. Processing of SLC-off images requires replacing the missing data. Gap filling was used utilizing same time images with spectral information taken from the neighbouring pixels. The convolution filtering algorithm with majority function was used to replace the missing data. The METRIC^{tm} model (Allen et al. 2007b) estimated energy fluxes using the remotely sensed data as input (A general overview of the METRIC^{tm} model is presented in the next section). The model maker tool of Erdas Imagine® image processing software (Leica Geosystems Geospatial Imaging, LLC) was used to code the METRIC^{tm} algorithms. An iterative procedure was followed for sensible heat flux estimation using hot and cold pixels. From each processed image, average of 9 (3 X 3) pixels cantered over the field measurement location was used for the comparison of model estimated fluxes with the field measurements.

2.3. Meteorological data

High quality hourly weather data consisting of air temperature, relative humidity, wind speed, incoming solar radiation, and precipitation are required for the operation of the METRIC^{tm} model. Hourly weather data were acquired from the High Plains Regional Climate Center's (HPRCC) Automated Weather Data Network (AWDN). Weather data were acquired for 2005 from the Scottsbluff (latitude: 41.22 N; longitude: 103.02 W; elevation=1208 m) AWDN station to calibrate METRIC^{tm} model. The weather data was quality controlled following the recommendations of Allen et al., 1996; 1998; 2005 and by ASCE-EWRI (2005) for all the weather stations in and out of Landsat path. Table 2 shows the list of AWDN stations used in this analysis. Hourly and daily observed solar radiation (R_s) values were compared with that of calculated clear sky solar radiation (R_{so}).

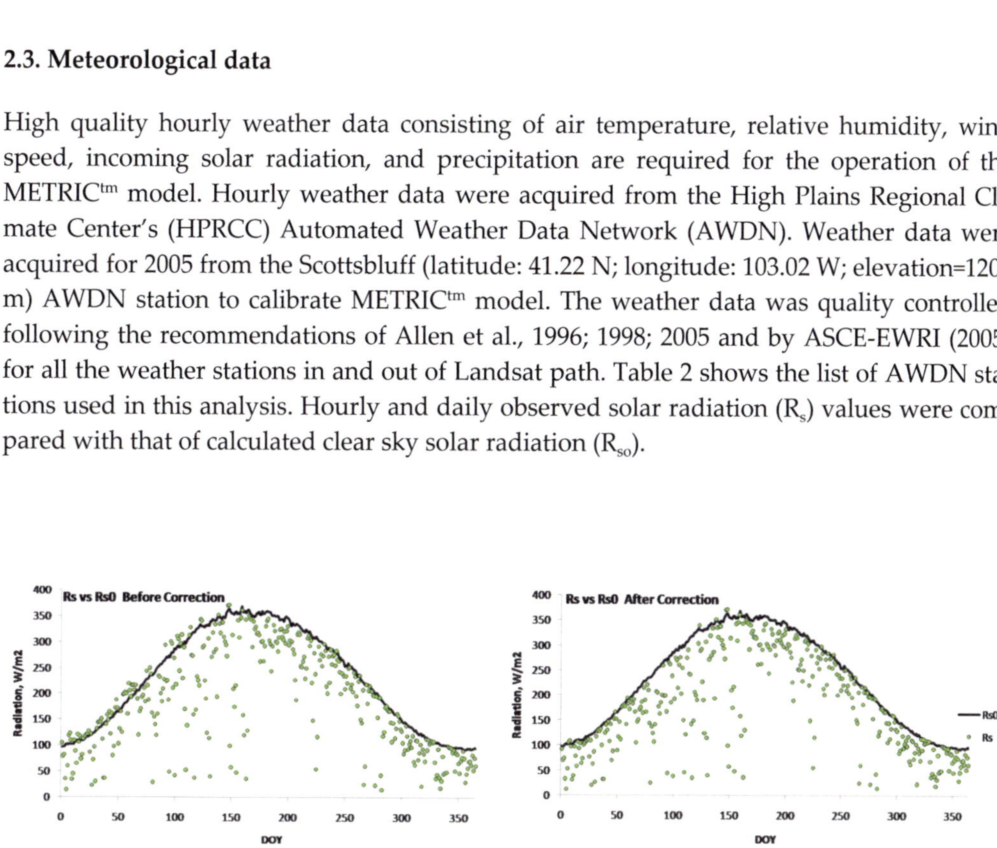

Figure 2. Scottsbluff 2005 corrected observed solar radiation (W/m^2) and calculated clear sky solar radiation (W/m^2). For DOY 1-123 (03 May) a 6% decrease in Rs values were applied.

R_{so} is the theoretical estimate of incoming solar radiation to the ground surface on a clear sky day with low atmospheric aerosol content (e.g. no haze, dust, smoke from fires, etc.) and is calculated based on atmospheric pressure, sun angle, and precipitable water in the atmosphere (i.e. figure 2). Corrections are only applied when the data exhibits systematic errors eg. R eg. Rs>Rso. Individual values are corrected. Reasons for errors in the solar radiation values can be due to misalignment or a nonrepresentative calibration of the sensor. reference evapotranspiration (ETr) values were calculated using the ASCE-EWRI (2005) standardized Penman-Monteith equation for alfalfa reference. These calculations were carried out using Ref-ET software (University of Idaho and Allen, 2003

Station	Latitude	Longitude	Elevation (m)
Alliance-North, NE	42.18	102.92	3980.10
Alliance-West, NE	42.02	103.13	3980.00
Arapahoe Prairie, NE	41.48	101.85	3600.00
Arthur, NE	41.65	101.52	3598.08
Gordon, NE	42.73	102.17	3638.45
Gudmundsen, NE	42.07	101.43	3441.60
Mitchell Farms, NE	41.93	103.70	3602.36
Scottsbluff, NE	41.88	103.67	3963.25
Sidney, NE	41.22	103.02	4320.87
Sterling, CO	40.47	103.02	3937.01
Torrington, WY	42.03	104.18	3989.50

Table 1. Lists of observation stations for calibration and validation

Figure 3 shows the riparian species as documented by the AISA system onboard CALMIT's Piper Saratoga aircraft on NE Panhandle. This map was integrated with seasonal ET maps obtained with METRIC^{tm} model. Hyperspectral remote sensing holds great promise for research on invasive species. Spectral information provided by hyperspectral sensors on AISA system can detect invaders at the species level across a range of community and ecosystem types. This gigh resolution classification provides a valuable description of spatial variation in riparian, serves as a baseline to measure ET in riparian, and is essential for prioritizing riparian restoration treatments in the basin.

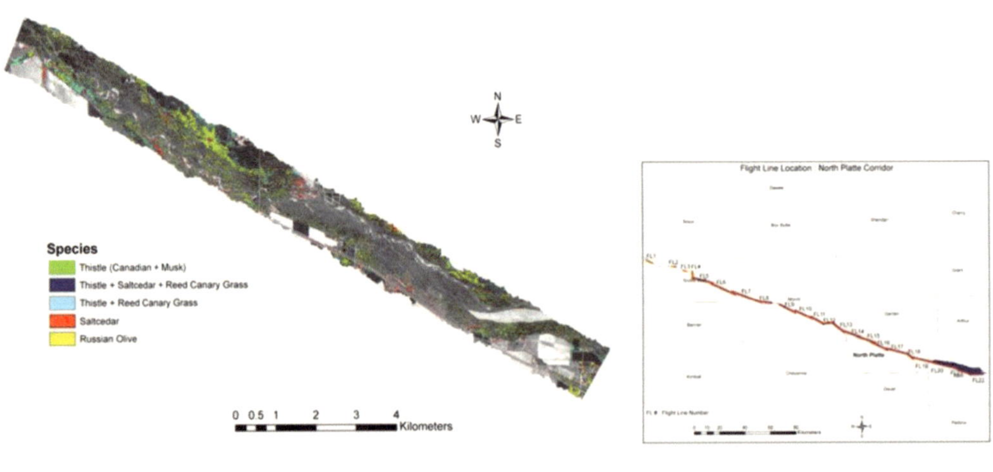

Figure 3. Invasive species distribution in 2005 for flight line #3 on Path 33, Row 31.

2.4. Land surface energy balance model

The landsat images for 2005 for Path32 Row 31 were processed using the algorithms in the METRIC^{tm} (2007a; 200b) model which requires parameterization of the energy balance and estimation of surface energy fluxes based on spectral satellite measurements (Allen et al., 2007). The model is originated from the SEBAL model and a "hybrid" energy balance model that uses thermal bands from Landsat imagery to compute ET. In particularly, it combines remotely-sensed energy balance (satellite) data and ground-based ETr (reference ET) data to determine ET. METRIC^{tm} computes LE as a residual of the energy balance as:

$$LE = R_n - G - H \tag{1}$$

where R_n is the net radiation, G is the soil heat flux, H is the sensible heat flux, and LE is the latent heat flux. The units for all the fluxes are in W m^{-2}. METRIC^{tm} calculates net radiation (R_n) as the difference between incoming radiation at all wavelengths and reflected short-wavelength (~ 0.3 – 3 μm) and both reflected and emitted long-wavelength (3~60 - μm) radiation (Allen et al., 2007a). The LE time integration was split into two steps. The first step was to convert the instantaneous value of LE into daily values of actual ET (ET_{24}) values by holding the reference ET fraction constant (Allen et al., 2007b). An instantaneous value of ET (ET_{inst}) in equivalent evaporation depth is the ratio of LE to the latent heat of vaporization. usually range from 0 to 1.05 and is defined as the ratio of instantaneous ET (ET_{inst}) for each pixel to the alfalfa-reference ET calculated using the standardized ASCE Penman-Monteith equation for alfalfa (ET_r) following the procedures given in ASCE-EWRI (2005):

$$\mathrm{ETrF} = \frac{\mathrm{ET_{inst}}}{\mathrm{ET_r}} \tag{2}$$

The procedures outlined in ASCE-EWRI (2005) were used to calculate parameters in the hourly ET_r equation. The daily ET at each pixel was estimated by consideringETrF and 24 hour ET_r as:

$$\mathrm{ET_{24}} = \mathrm{ET_rF\ xET_{r-24}} \tag{3}$$

where ET_{24} is the daily value of actual ET (mm day^{-1}), ET_{r-24} is 24 hour ET_r for the day of image and calculated by summing hourly ET_r values over the day of image. In order to produce monthly and seasonal ET maps, individual ET_rF maps from each image in the analysis were generated from METRIC and interpolated using a cubic spline model. The spline model is deterministic interpolation method which fits a mathematical function through data points to create a surface (Hartkamp, 1999). The spline surface was achieved through weights (λ_i) and number of points (N). A regularized spline was used because this method results in a smoother surface. Daily images were generated by interpolation used for monthly and seasonal ET calculation.

3. Result and discussion

A daily soil water balance model was applied for 2005 using precipitation and ETr from the for all weather stations (Table1). The water balance model estimates residual evaporation from bare soil for each of the Landsat image dates. The model is based on the two-stage daily soil evaporation model of the United Nations Food and Agriculture Organization's Irrigation and Drainage Paper 56 (Allen et al., 1998). The soil water balance is set up assuming a loam soil having a water content at field capacity and the wilting point of 0.3 cm^3/ cm^3 and 0.15 cm^3/ cm^3, respectively and having 10 mm of readily evaporable water in the upper 12.5 cm of soil. Figure 4 shows a simulation of evaporation from bare soil. The results from the soil water balance were used to determine reference evapotranspiration fraction (EtrF) for hot pixel selection, an internal calibration step for running METRIC^{tm}.

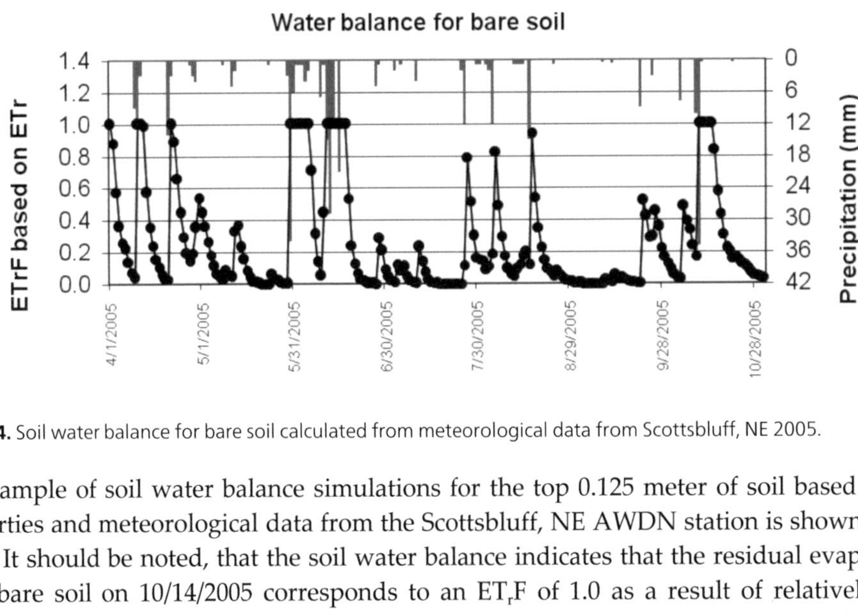

Figure 4. Soil water balance for bare soil calculated from meteorological data from Scottsbluff, NE 2005.

An example of soil water balance simulations for the top 0.125 meter of soil based on soil properties and meteorological data from the Scottsbluff, NE AWDN station is shown in Figure 4. It should be noted, that the soil water balance indicates that the residual evaporation from bare soil on 10/14/2005 corresponds to an ET_rF of 1.0 as a result of relatively large amounts of precipitation a few days prior to the image date. An ET_rF value for the hot pixel of this magnitude leaves a very small margin up to the ET_rF of 1.05 generally assigned to the cold pixel. For reasons discussed under the individual images below, the ET_rF for the hot pixel has been assigned the value 0.8 for this image.

We utilized both satellite and air-bone remote sensing data with an energy balance model to provide a better understanding and quantification of evapotrabspiration for selected invasive species. We utilized METRIC^{tm} to quantify spatial distribution and seasonal variation of actual ET over riparian zone in North Platte River during growing season for 2005. Next, we

http://dx.doi.org/10.5772/52929

integrated ET maps with invasive species map to estimate the mean and the range of water use for each riparian species. The invasive species map developed using hyperspectral aerial imagery (AISA) in 2005 at 1.5 meter resolution for the North Platte River Basin was used. The measured components of the water balance (precipitation and ETrF based on ETr) from Scottsbluff, NE were evaluated to determine ETrF for the hot pixel selection and to determine the net ground-water recharge that occurred during the study. Precipitation and ETrF were the dominant components of the water balance with ground-water storage being a comparatively minor term. The ETrF is highly variable over the landscape because of the variability in landuse, climate, soil properties, and management practices. Soil properties affect surface soil evaporation and energy balances, including soil heat flux and sensible flux; causing within-field and across field variability in ETrF. Much of this variability occurs at the field scale, making it nearly impossible to quantify ET spatially using more traditional and conventional methods. Figure 6 shows monthly ETrF for path 33, row 31 in 2005. The spline model requires two images each in the preceding and subsequent months for the month to be interpolated. Because only one image was available for the month of April, a new ETrF image was created for April 23rd from MODIS 250m NDVI data. The methods used were the same as the cloud filling method using MODIS 250m NDVI data.

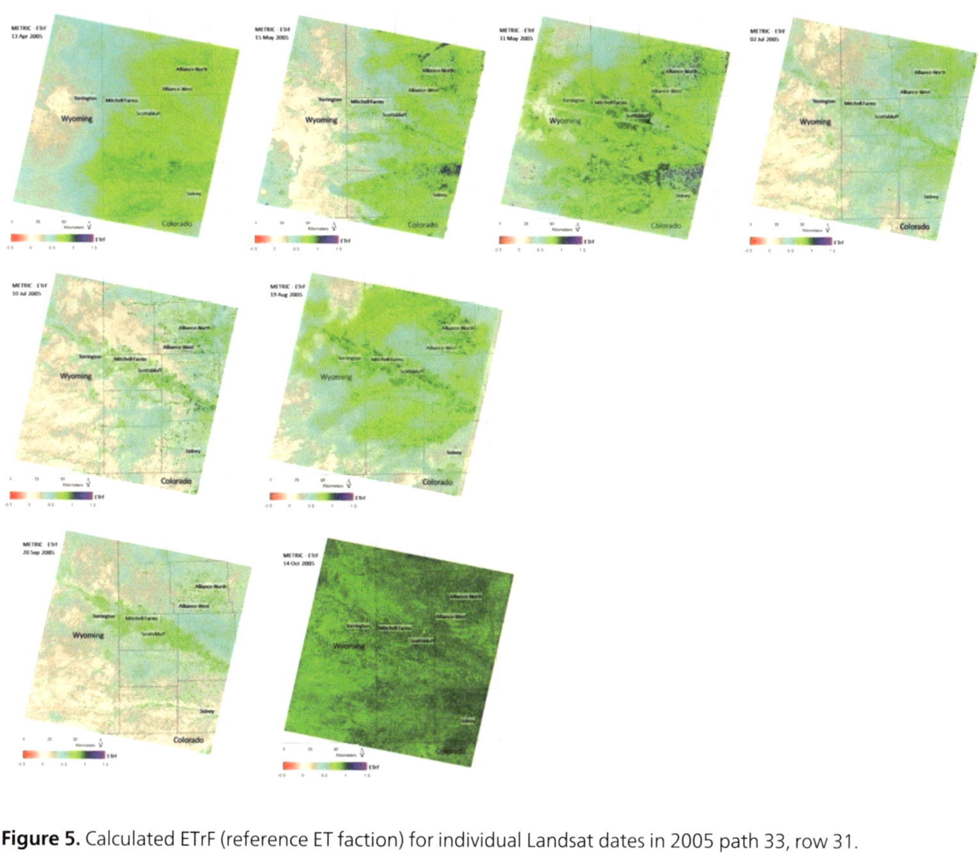

Figure 5. Calculated ETrF (reference ET faction) for individual Landsat dates in 2005 path 33, row 31.

Figure 6 shows the expected progression of ETrF during a growing season as surface conditions changed. The spatial distribution of daily ETrF estimations using the Landsat overpass on May 31, 2005 and August 19 2005 were highly variable ranging from 0.6 mm day^{-1} to as high as 0.9 mm day^{-1} across the images. Most of the variability was due to differences in land use and riparian species. The land use in the top part of the study area is mainly agricultural land that is devoid of standing crops in early May. The bottom part of the study area is mostly grazed rangeland or natural vegetation dominated by green vegetation in early spring, resulting in higher ET.

Monthly ET maps were summed to obtain total seasonal ET for the study area. Figure 6 shows the progression of ET from May through September in 2007 across the Panhandle derived using spline interpolation algorithm. The monthly ET maps generated by the METRICtm model showed a good progression of ET during the growing season as surface conditions continuously changed. Results showed that salt cedar water use was lowest compared to other invasive species. Russian olive also has substantial water use during growing season. For most species seasonal actual ET ranged from 20 to 35 inches. From figure 2 of hyperspectral image of invasive species distribution and figure 6 of monthly ET maps produced with METRICtm for Landsat path 33 row 31 in 2005 we have an indication of the water use during the growing season for individual invasive species. Comparison is difficult due to non-uniform distribution of species and differences in age. Water use varied considerable even in the same species due to plant density, plant distribution and plant height of individual species. ET roughly ranged from 12 in to as high as 43 in for all the invasive species for May 1st to September 31st. Average actual seasonal ET ranged from 27 inches to 30 inches for all the invasive species. Overall, the remote sensing based energy balance approach based on landsat image in conjunction with high resolution hyperspectral image was useful to obtain distribution of ET estimates from riparian systems.

The ET was lower early in the growing season and gradually increased as the riparian species increasingly transpire water towards the mid season. The METRIC^{tm} model was also able to estimate the decreasing evaporative losses towards the end of the season and after the harvest. However, in figures 6 subfigures July, August and September show visible distinctions in ET among the riparian species. To calculate species wise distribution one needs full knowledge of the study area land use with hyperspectral imagery classification. Since requirement of accurate riparian species type identification can increase costs of ET mapping at larger scales, this is an advantage of METRIC^{tm} because the model does not require information on soil and management practices.

July is usually the peak ET month with high incoming solar radiation, high temperatures, and large vapor pressure deficit all contributing to increased ET. The ET shows variation throughout the district as a function of different ET rates of various land covers, including riperian species, agriculture crops and natural vegetation, etc. With physiological maturity, leaf aging and senescence, ET starts to decrease gradually in September. At the start of fall, leaves of plants start falling in October, most of the ET in this month represents the soil evaporation component of ET. As shown in monthly ET maps, mapping ET on large scales can provide vital information on the progression of ET for various vegetation surfaces over

http://dx.doi.org/10.5772/52929

time. Information gained enables the prediction of the timing and the spatial extent of potential depletions or gains in both the short-term and in the long-term management of surface and ground water.

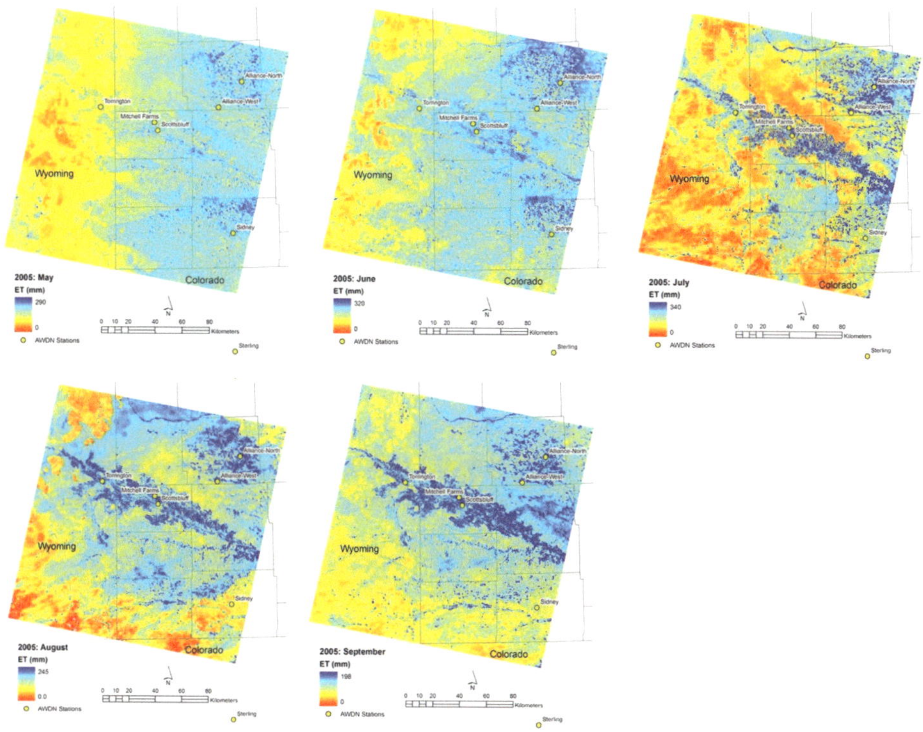

Figure 6. Monthly ET (mm) maps produced with METRIC^{tm} for Path 33 Row 31 in 2005.

The seasonal ET (mm) maps generated by the METRIC^{tm} model showed spatial and temporal distribution of relative ET during the 2005 season as land surface conditions continuously changed (Figure. 8). The information also allowed us to follow the seasonal trend in ET for major land use classes on the image. Water consumption by the riperian species is higher than the water consuption for other landuse in the Panhandle.

The frequency distribution and the basic statistics for seasonal ET including all species are presented in Figure 8 and and summarized in Table 2, respectively. Statistics provided in Table 2 are for the water use during the growing season for individual invasive species. Comparison is difficult due to the non-uniform distribution of species and difference in age. Water use varied considerable even in the same species due to plant density, plant distribution and plant height of individual species. ET roughly ranged from 12in to as high as 43 in for all the invasive species for May 1st to September 31st. Average actual seasonal ET ranged from 27 inches to 30 inches for all the invasive species.

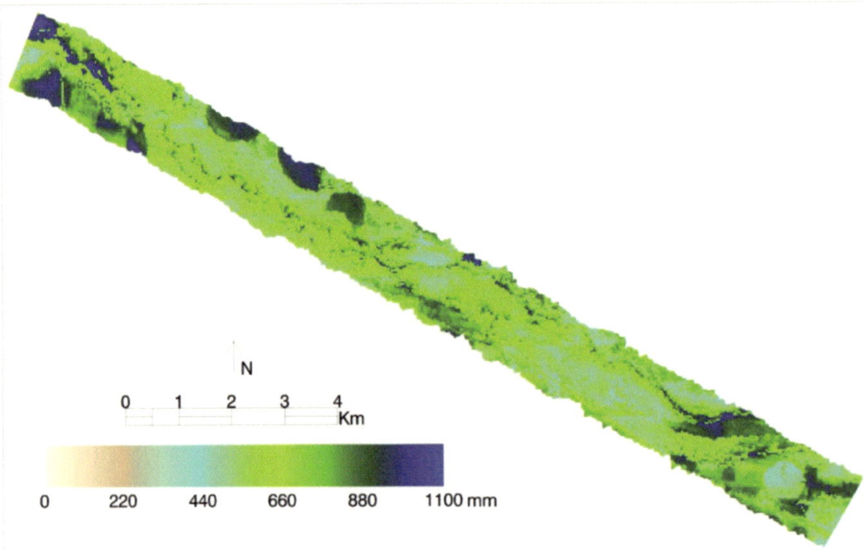

Figure 7. Seasonal ET (mm) map from May 01 through September 30 in 2005 for Path 33 Row 31.

Species	Minimum	Maximum	Average
Russian Olive	12	42	29
Salt Cedar	14	40	27
Thistle	12	42	30
Thistle+Salt Cedar+Reed Canary Grass	15	41	28
Thistle+Reed Canary Grass	16	43	30

Table 2. Comparison of seasonal water use in inches from May 1st to September 30th in 2005.

Figure 8 shows the histogram of seasonal actual ET for individual and combination of various riparian species. The y axis on the figure (histogram) shows the number of AISA pixels (1.5 m) for each riparian species while x axis shows corresponding actual ET values in inches. We observed that there is no single ET value for invasive species. Results showed that salt cedar water use was lowest compared to other invasive species. Russian olive also has substantial water use during growing season. For most species seasonal actual ET ranged from 20 to 35 inches. Overall, the remote sensing based energy balance approach based on the landsat image in conjunction with high resolution hyperspectral image was useful to obtain distribution of ET estimates from riparian systems.

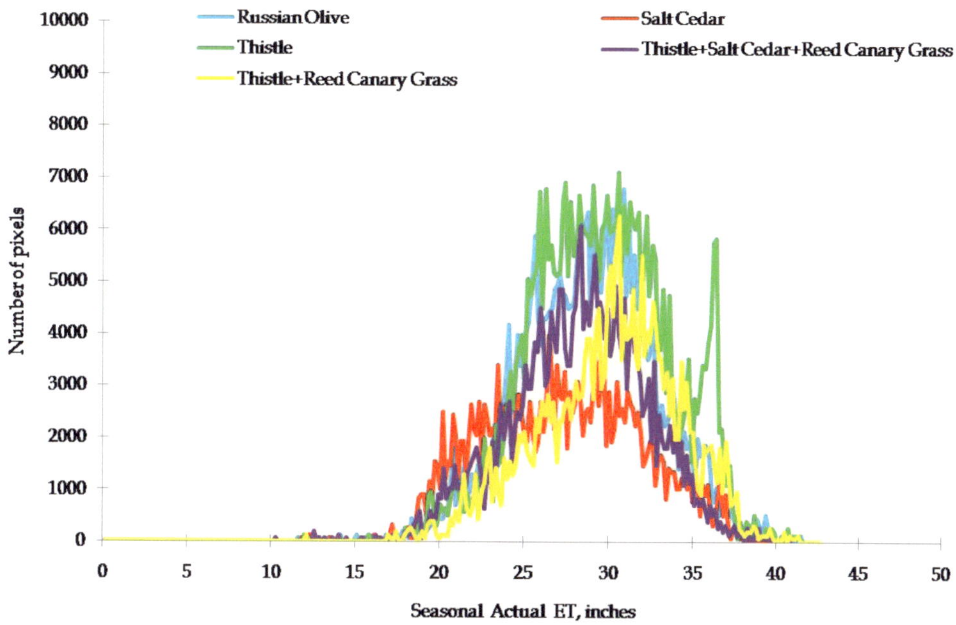

Figure 8. The comparison of water use (seasonal actual ET, inch) for individual and combination of various riparian species.

4. Conclusion

This study focused on the use of multi-temporal landsat data and hyperspectral remote sensing data to calculate daily and seasonal actual ET based upon satellite energy balance model such as METRIC^{tm}. The remotely sensed measurements using METRIC^{tm}.provide the estimation of spatial distribution of instantaneous ET, which can be integrated into daily and seasonal ET values. The seasonal spatial distribution maps help to explain the water consumption for the di ▪ erent riperean species season. Based on maps we have developed guidelines on riparian water use which will be of benefit to the state, particularly with regard to riparian control for water acclamation. Satellite-based measurements can provide such information and avoid the need to rely on field databases. The ET values for a given species varies due to the density and age of the plant species. By integrating the hyperspectral images with the METRIC^{TM} model, we obtain the following seasonal ET values (May 1 through September 1, 2005) for the following invasive species: Salt Cedar (Tamarisk) 26.8 inches (680 mm) or 2.23 acre-feet water, Russian Olive 26.7 inches (677 mm) or 2.23 acre-feet water, average Canada and Musk Thistle 27.9 inches (708 mm). Using developed figures, this vegetation transpires approximately 30,453 acre-feet of water per season. Since willows, cottonwood trees and some grasses transpire approximately the same amount of water, revegetation with these species would result in a net "no gain" of water conservation. Based on the analysis of images from AISA and Landsat, we show that ET varies even in the same species (i.e. salt cedar) because of variation in tree age, height, density, and surrounding area. Nevertheless, further studies are necessary to expand this method in conjuction with hyperspectral data to obtain species wise water use distribution. The METRIC^{tm} model should be tested for its suitability for other climate conditions found in Nebraska and an assessment made of the spatial variability of the calibration parameters is needed.

Acknowledgements

The. authors thank to the University of Nebraska Foundation, Anna H. Elliot Fund grants which supported this work and University of Idaho for METRICtm modeling support.

Author details

Baburao Kamble, Ayse Irmak, Derrel L. Martin, Kenneth G. Hubbard, Ian Ratcliffe, Gary Hergert, Sunil Narumalani and Robert J. Oglesby

University of Nebraska-Lincoln (UNL), Lincoln, USA

References

[1] Allen, R. G. (1996). Assessing Integrity of Weather Data for Use in Reference Evapotranspiration Estimation. J. Irrig. Drain. Eng ASCE, , 122, 97-106.

[2] Allen, R. G., Pereira, L., Raes, D., & Smith, M. (1998). Crop Evapotranspiration, Food and Agriculture Organization of the United Nations, Rome, It. 925-1-04219-530-0p.

[3] Allen, R. G., Tasumi, M., & Trezza, R. (2007a). Satellite-based energy balance for mapping evapotranspiration with internalized calibration (METRIC)- Model. ASCE J. Irrigation and Drainage Engineering , 133(4), 380-394.

[4] Allen, R. G., Tasumi, M., Morse, A. T., Trezza, R., Kramber, W., Lorite, I., & Robison, C. W. (2007b). Satellite-based energy balance for mapping evapotranspiration with internalized calibration (METRIC)- Applications. ASCE J. Irrigation and Drainage Engineering , 133(4), 395-406.

[5] -E, A. S. C. E., & , W. R. I. (2005). The ASCE Standardized reference evapotranspiration equation. ASCE-EWRI Standardization of Reference Evapotranspiration Task Comm. Report, ASCE Bookstore, 078440805pages.(40805)

[6] Bastiaanssen, W. G. M., Menenti, M., Feddes, R. A., & Holtslag, A. A. M. (1998a). A remote sensing surface energy balance algortithm for land (SEBAL). Part 1: Formulation. J. of Hydrology , 198-212.

[7] Bastiaanssen, W. G. M., Pelgrum, H., Wang, J., Moreno, Y. J. F., Roerink, G. J., Roebeling, R. A., & van der Wal, T. (1998b). A remote sensing surface energy balance algortithm for land (SEBAL). Part 2: Validation. J. of Hydrology 212-213: 213-229.

[8] Bentall, Ray., & Shaffer, F. B. (1979). Availability and use of water in Nebraska, 1975: Conservation and Survey Division, University of Nebraska-Lincoln, Nebraska Water Survey Paper 48, 121 p.

[9] Center for Advanced Land Management Information Technologies (CALMIT), 2000, Delineation of 1997 land use patterns for the Cooperative Hydrology Study in the central Platte River Basin: Lincoln, Nebraska, CALMIT, 73 p., accessed August 29, 2008, at http://www.calmit.unl.edu/cohyst/data/1997_finalreport.pdf

[10] Currier, P. J., Lingle, G. R., & Van Derwalker, J. G. (1985). Migratory bird habitat on the Platte and North Platte Rivers in Nebraska: Grand Island, Nebr., Platte River Whooping Crane Maintenance Trust, 177 p.

[11] Goodrich, D. C., Scott, R., Qi, J., Goff, B., Unkrich, C. L., Moran, M. S., Williams, D., Schaeffer, S., Snyder, K., Mac, R., Nish, T., Maddock, D., Pool, A., Chehbouni, D. I., Cooper, W. E., Eichinger, W. J., Shuttleworth, Y., Kerr, R., & Marsett, W. N. (2000). Seasonal estimates of riparian evapotranspiration using remote and in situ measurements: Agricultural and Forest Meteorology, , 105(1-3), 281-309.

[12] Gowda, P. H., Chávez, J. L., Colaizzi, P. D., Evett, S. R., Howell, T. A., & Tolk, J. A. (2008). ET mapping for agricultural water management: present status and challenges. Irrig. Sci. , 26, 223-237.

[13] Gutentag, E. D., Heimes, F. J., Krothe, N. C., Luckey, R. R., & Weeks, J. B. (1984). Geohydrology of the High Plains aquifer in parts of Colorado, Kansas, Nebraska, New Mexico, Oklahoma, South Dakota, Texas, and Wyoming: U.S. Geological Survey Professional Paper 1400B, 63 p.

[14] Hartkamp, A. D., De Beurs, K., Stein, A., & White, J. W. (1999). Interpolation Techniques for Climate Variables, NRG-GIS Series 9901CIMMYT: Mexico, DF, http://www.cimmyt.org/Research/nrg/pdf/NRGGIS%2099pdf [1 Sep 2004].

[15] High Plains Regional Climate Center,(2006). National Weather Service surface observations and automated weather data network data: Lincoln, Nebraska, University of Nebraska, digital data, accessed November 1, 2006, at http://www.hprcc.unl.edu/.

[16] Irmak, A., & Kamble, B. (2009). Evapotranspiration Data Assimilation with Genetic Algorithms and SWAP Model for On-demand Irrigation. Irrigation Science. 28:101-112 (DOIs00271-009-01939).

[17] Irmak, A., Ratcliffe, I., Ranade, P., Hubbard, K. G., Singh, R. K., Kamble, B., Allen, R. G., & Kjaersgaard, J. (2010). Estimation of land surface evapotranspiration: A Satellite Remote Sensing Procedure. Great Plains Research. 21(1):April 2011.

[18] Irmak, A., Rundquist, D., , S., Narumalani, G., Hergert, , & Stone, G. (2009). Satellite-Based Energy Balance to Assess Riparian Water Use. UNL. Lincoln, NE. USGS 104b final report. 11 p.

[19] Irmak, A., Ratcliffe, I., Ranade, P., Irmak, S., Allen, R. G., Kjaersgaard, J., Kamble, B., Choragudi, R., Hubbard, K. G., Singh, R., Mutiibwa, D., & Healey, N. (2010). Seasonal Evapotranspiration Mapping Using Landsat Visible and Thermal Data with an Energy Balance Approach in Central Nebraska. Remote Sensing and Hydrology. Special Issue IAHS Publ. 3XX, 2011

[20] Irmak, S., Howell, T. A., Allen, R. G., Payero, J. O., & Martin, D. L. (2005). Standardized ASCE Penman-Monteith: impact of sum-of-hourly vs. 24-hour time step computations at reference weather station sites. Trans. ASABE. 48

[21] Kamble, B., & Irmak, A. (2011). Remotely Sensed Evapotranspiration Data Assimilation for Crop Growth Modeling, Evapotranspiration, Leszek Labedzki (Ed.), 978-9-53307-251-7InTech.

[22] Kamble, B., & Irmak, A. (2008). Assimilating Remote Sensing-Based ET into SWAP Model for Improved Estimation of Hydrological Predictions," Geoscience and Remote Sensing Symposium, 2008. IGARSS 2008. IEEE International, no., pp.III-1036-III-1039, 7-11 July 2008 (DOI:10.1109/IGARSS.2008.4779530), 3

[23] Kjaersagaard, J., & Allen, R. G. (2010). Remote Sensing Technology to Produce Consumptive Water Use Maps for the Nebraska Panhandle. Final completion report submitted to the University of Nebraska. 60 pages.

[24] Landon, M. K., Rus, D. L., Dietsch, B. J., Johnson, M. R., & Eggemeyer, K. D. (2009). Evapotranspiration rates of riparian forests, Platte River, Nebraska, 2002-06: United States Geological Survey Scientific Investigations Report p., 2008-5228.

[25] Menenti, M. (2000). Evaporation. In: Schultz GA, Engman ET (eds.). Remote sensing in hydrology and water management. Springer Verlag, Berlin Heidelberg New York, , 157-188.

[26] Nagler, P. L., Glenn, E., & Thompson, T. L. (2003). Comparison of transpiration rates among saltcedar, cottonwood and willow trees by sap flow and canopy temperature models. Agricultural and Forest Meteorology , 116, 73-89.

[27] Norman, J. M., Divakarla, M., & Goel, N. S. (1995). Algorithms for extracting information from remote thermal-IR observations of the Earth's Surface. Remote Sensing of Environment, , 51, 157-168.

[28] Owens, K. M., & Moore, G. W. (2007). Saltcedar water use: realistic and unrealistic expectations. Range. Ecol. & Manage. , 60, 553-557.

[29] Ranade, P. (2010). Spatial Water Balance for Bare Soil. University of Nebraska. Report, 10 pp.

[30] Schaeffer, S. M., Williams, D. G., & Goodrich, D. C. (2000). Transpiration of cottonwood/willow forest estimated from sap flux. Agricultural and Forest Meteorology 105(1-3):257-270.

[31] Singh, R. K., Irmak, A., Irmak, S., & Martin, D. L. (2008). Application of SEBAL Model for Mapping Evapotranspiration and Estimating Surface Energy Fluxes in South-Central Nebraska. J. Irrigation and Drainage Engineering , 134(3), 273-285.

[32] Smith, S. D., Devitt, D. A., Sala, A., Cleverly, J. R., & Busch, D. E. (1998). Water relations of riparian plants from warm desert regions. Wetlands , 18, 687-696.

Chapter 6

Influence of Vegetation Cover on Regional Evapotranspiration in Semi-Arid Watersheds in Northwest China

Mir A. Matin and Charles P.-A. Bourque

Additional information is available at the end of the chapter

http://dx.doi.org/10.5772/52812

1. Introduction

Evapotranspiration (ET) is an important biospheric process whereby liquid water is vaporized from moist surfaces and from plant tissues [1]. It is one of the key processes in the hydrological cycle and the energy balance of watersheds. Level of ET from a surface is primarily dependent on the availability of moisture on the surface and the amount of energy available to evaporate that moisture. To understand and quantify ET, three conceptual definitions of ET exist in the scientific literature. Actual ET (AET) refers to the actual evaporation of liquid water from a surface under given atmospheric conditions. For the same atmospheric conditions, potential ET (PET) refers to the elevated evaporation of water when the amount of surface moisture is unlimited and vegetation conditions are ideal. Ideal conditions regarding vegetation is characterized by actively growing short vegetation covering a large surface area with unlimited supply of soil water [2, 3]. If the vegetation cover is standardized to grass or alfalfa, PET is considered as reference ET (i.e., ET_0) [1]. Understanding spatiotemporal trends in ET is critical to interpreting eco-hydrometeorological processes of complex landscapes [4]. Specific factors affecting ET are divided into three main categories, namely:

i. meteorological factors, including solar radiation, near-surface air temperature, wind velocity, and air humidity;

ii. surface factors, including surface water and soil water content; and

iii. plant factors, including rooting depth, leaf structure, and stomatal density and aperture [1, 5, 6].

Spatiotemporal variability in these factors determines the variability of ET across large areas. This variability increases as the complexity of the underlying terrain and vegetation cover intensifies.

Back and forth exchange of water vapor and liquid water from oases at the base of the Qilian Mountains (Northwest China) and from the Qilian Mountains to oases as surface and shallow subsurface flow has been previously shown to be a potentially significant mechanism in the long term maintenance of oases in westcentral Gansu, NW China [7]. In general, direct precipitation (rain + snow) to the foothills and oases at the base of the Qilian Mountains is inadequate to maintain vegetation in these areas. Maintenance of oases vegetation is shown to be dependent on the surface water flowing from the higher regions of the mountain range, where direct precipitation and snowmelt are greatest. Supply of atmospheric moisture that leads to the formation of precipitation in the mountains is sustained by seasonal ET at the base of the mountain range.

Influence of vegetation and landcover on local and regional climate is well documented in the scientific literature [7-9]. Changes in landcover result in changes in surface albedo, surface roughness, leaf area index (LAI), stomatal conductance, rooting depth, and soil texture and structure [10-12]. Changes in surface vegetation have been known to impact

i. the partitioning of net radiative energy into sensible and latent heat fluxes, and

ii. the formation of convective rainfall by affecting the state of the convective boundary layer [13].

Agricultural landuse affects climate by modifying the physiological attributes (e.g., canopy conductance) of the land [14].

2. AET-determination methods

2.1. Point measurements

Suitability of AET-measurement methods depend on the reasons for assessment and on the spatiotemporal dimensions of the problem. Measurement of AET is mainly done indirectly by measuring the effect of AET on the water and energy balance. *Rose and Sharma* [15] categorized AET-measurement methods into three broad categories, i.e.,

i. hydrological approaches, based on the residual of the water-budget equation;

ii. micrometeorological approaches, including the Bowen ratio [16], aerodynamic, and eddy covariance-based methods [17], and

iii. plant-physiological approaches at the scale of individual plants or groups of plants by collecting sap-flow or chamber measurements.

2.1.1. Hydrological methods

Hydrological methods include the soil-water balance and weighing-lysimeter methods [18]. AET-determination from the soil-water balance is possible with a rearrangement of the water-budget equation, such that

$$AET = P - R - D - \Delta S, \tag{1}$$

where P is the precipitation, R is the surface runoff, D is the soil drainage, and ΔS is the change in soil-water storage.

Each term in Eqn. (1) is given in unit catchment volume, in which proper delineation of catchment boundaries, distribution of precipitation and streamflow gauges, and accuracy of runoff are critical for the accurate assessment of AET [19].

Lysimeters are measuring devices with containers for soil and plants that estimate ET by measuring the difference in weight of water in individual containers [20]. Accuracy of AET with lysimeters rests on the precision of the weighing instrument and sampling frequency [18].

2.1.2. Micrometeorological methods

Micrometeorological methods estimate latent heat fluxes at potentially high temporal frequencies over homogeneous vegetation [18]. The Bowen ratio method estimates latent heat fluxes [i.e., λET (W m^{-2}), where λ is the latent heat of vaporization expressed in J kg^{-1}] from the ratio of surface energy to the Bowen ratio (non-dimensional), such that

$$\lambda ET = \frac{R_n - G}{1 + \beta}, \tag{2}$$

where R_n is the net surface radiation (W m^{-2}), G is the soil-heat flux (W m^{-2}), and β (Bowen ratio) is the ratio to sensible heat and latent heat fluxes estimated from air temperature and water vapor pressure gradients taken within the same sample of air [21], or

$$\beta = \gamma \frac{\Delta T_a}{\Delta e_a}, \tag{3}$$

where γ is the psychometric constant (0.67 hPa K^{-1}, at sea level) and ΔT_a (K) and Δe_a (hPa) are differences in air temperature and water vapor pressure taken at two separate levels [16].

The aerodynamic resistance method uses the atmospheric boundary-layer resistance to calculate the transfer of sensible heat (H; W m^{-2}) from the surface to the air [22] and is formally expressed as

$$H = \varsigma C_p \frac{T_s - T_a}{r_a}, \tag{4}$$

where r_a is the aerodynamic resistance (s m^{-1}), ς and C_p are the density and specific heat of the airmass involved (kg m^{-3} and J kg^{-1} K^{-1}, respectively), and T_s and T_a are the surface and air temperature (K). The aerodynamic resistance in Eqn. (4) is the most important and most difficult to define. Various methods have been proposed to estimate this resistance, many of which are summarized in [23]. Once H is determined, AET (i.e., AET = $\lambda ET \cdot \lambda^{-1}$) can be determined from

$$\lambda ET = R_n - G - H, \tag{5}$$

given corresponding measurements of R_n and G.

Eddy covariance facilitates determination of vertical fluxes of atmospheric gases (e.g., water vapor, carbon dioxide, and trace gases) and heat within the atmospheric boundary layer. Historical advancement of eddy-covariance concepts is summarized in [17]. Assuming that the vertical wind velocity is responsible for vertical fluxes of atmospheric gases and heat, there should be a high positive correlation between vertical wind velocities and individual fluxes [24]. The method monitors high-frequency fluctuations in vertical wind velocity (m s^{-1}) and water vapor mixing ratios (kg kg^{-1}), in estimating AET from the covariance of the two, i.e.,

$$\lambda ET = \lambda \overline{w'q'}, \tag{6}$$

where w' and q' are the instantaneous fluctuation (deviation from the interval mean) in vertical wind velocity and water vapor mixing ratio in a parcel of air.

The physiological approach determines AET for individual plants or groups of plants. Two of the more broadly used physiological methods, includes

1. the sap-flow method, and
2. the chamber method.

In the sap-flow method, sap flow in plants is assumed to be closely linked to plant transpiration and is quantified by applying heat pulses to the stem of plants and analyzing corresponding heat balances [18]. In a chamber setting, transpiring plants are enclosed in a transparent chamber and changes in within-chamber water vapor concentrations are quantified [25].

While AET-determination methods based on principles of micrometeorology and plant physiology provide suitable overall accuracy at point scales (within a few tens of metres), spatial interpolation of their results across entire landscapes is wholly inappropriate because of the inherent complexity of natural landscapes, particularly with respect to topography and landcover.

2.2. Regional estimates of AET from remote sensing-based data

For expansive landscapes, remote sensing (RS)-based methods have been gaining popularity during the past few decades with regard to estimating regional AET at daily, monthly, and annual time scales. Characterizing land-surface conditions with RS-based methods provides an important way of overcoming the difficulty of interpolating AET for complex landscapes [26]. Most methods using earth-observation data in approximating regional AET can be categorized into three main groups, i.e., methods based on:

i. the surface-energy balance equation [27-29];

ii. Penman-Monteith and Priestley-Taylor equations [30-32]; and

iii. the complementary relationship [33, 34].

While all methods require an assessment of available net energy at the surface (i.e., R_n-G) as primary input, they differ in the way they partition the energy into H and λET.

2.2.1. *Surface energy balance*

The surface-energy balance is based on the assumption that R_n is equal to the sum of H, λET, and G; the energy required for photosynthesis is insignificant here as it accounts for < 1% of incoming solar radiation [35]. Based on a rewriting of the surface-energy balance, the expression of latent heat flux becomes:

$$\lambda ET = R_n - G - H, \tag{7}$$

where

$$R_n = R_s \downarrow - R_s \uparrow + R_L \downarrow - R_L \uparrow . \tag{8}$$

In Eqn. (8), $R_s \downarrow$ and $R_s \uparrow$ represent incoming and outgoing shortwave radiation and $R_L \downarrow$ and $R_L \uparrow$, incoming and outgoing longwave radiation emitted by the atmosphere (including clouds, if present) and earth surface. Typically, R_n and G (generally expressed as a fraction of $R_s \downarrow$) are determined from sun-earth geometric relations and illumination angles imposed by variable terrain. Most of the functions and information for the determination of R_n and G are available in present day geographic information systems (GIS) and digital terrain models [36-39]. Reflected and emitted radiative components are estimated from surface albedo, land surface and air temperatures, and emissivities derived from RS-data, e.g., as those identified in Table 1.

Most popular among surface-energy balance models are the surface-energy balance system (SEBS; [29]) and surface-energy balance algorithm for land (SEBAL; [27]). A comprehensive review of various surface-energy balance procedures appearing in the scientitific literature can be found in [40]. The SEBS model uses Monin-Obukhov similarity theory for the atmospheric surface layer to derive land surface physical parameters and roughness lengths in determining the evaporative fraction at the limiting ends of accessible water, i.e., dry and wet limits [29]. At the dry end, evaporation (ET_{dry}) is assumed to be zero due to the limitation in soil moisture and, as a result, sensible heat (H_{dry}) is maximum, equalling net available energy (i.e., $H_{dry} = R_n - G$). At the wet end, evaporation (ET_{wet}) is assumed to be at its potential rate (i.e. $\lambda ET_{wet} = \lambda PET$) and sensible heat ($H_{wet}$) is minimum. Relative evaporation fraction is defined as:

$$EF_r = 1 - \frac{H - H_{wet}}{H_{dry} - H_{wet}}, \tag{9}$$

where H is the actual sensible heat flux and determined by solving a system of non-linear equations proposed by *Brutsaert* [29, 41]. AET is subsequently calculated with:

$$AET = \frac{1}{\lambda} EF_r \times \lambda PET, \tag{10}$$

where

$$\lambda PET = R_n - G - H_{wet}. \tag{11}$$

H_{wet} is calculated using an equation proposed by *Meneti* [42] based on the Penman-Monteith combination equation [43].

The SEBAL model [27] was designed to estimate regional AET from RS-data consisting of surface temperature, surface reflectance, Normalized Difference Vegetation Index (NDVI), and their corresponding relations. The SEBAL model calculates H from a linear relationship between T_s and T_a derived from plotted distributions of wet and dry pixels.

2.2.2. *Penman-Monteith equation*

To estimate regional ET using MODIS (or Moderate Resolution Imaging Spectroradiometer) data, *Cleugh et al.* [44] proposed the use of the Penman-Monteith equation, i.e.,

$$\lambda ET = \frac{\Delta(R_n - G) + \varsigma C_p \frac{e_s - e_a}{r_a}}{\Delta + \gamma\left(1 + \frac{r_s}{r_a}\right)}, \tag{12}$$

where Δ is the slope of the saturation-vapor-pressure-to-temperature-curve, r_a and r_s are the aerodynamic (atmospheric) and surface resistances (s m^{-1}) to the transfer of surface water vapor to the atmosphere, e_s is saturation vapor pressure (hpa), and e_a is actual vapor pressure (hPa). In their model, input to Eqn. (12) are based on MODIS-derived vegetation data and daily surface meteorological data. Surface conductance is estimated in the model as a function of MODIS LAI and NDVI.

The model was later modified by *Mu et al.* [5] by adding vapor pressure deficit (VPD) and minimum air temperature as two important factors influencing stomatal conductance. In this modification, EVI (enhanced vegetation index; [45]) was used in replacing NDVI in the calculation of surface-vegetation fraction. Also, treatment of soil-water evaporation was incorporated. The model was further modified by *Mu et al.* [46] to include calculation of daytime and nighttime ET, soil-heat flux, and canopy conductance for both dry and wet foliar conditions. This variant of the model along with climate-station data worldwide were used in the production of a global product of AET (i.e., MOD16; [46]) that was subsequently validated against flux data acquired from 46 AmeriFlux sites distributed throughout the Americas.

2.2.3. Complementary-based methods

Complementary-based methods were first introduced by *Bouchet* [33], considering that whenever a well-watered surface dries, the decrease in AET is coupled with a corresponding increase in PET and, as a result

$$AET + PET = 2E_w, \tag{13}$$

where E_w is defined as the value of potential evaporation, when AET and PET are equal for an unlimited moist surface [47].

The complementary relationship is based on the assumption that the energy required for evaporation is permanently available [48]. In such instance, drying of a wet surface would result in a reduction in AET and a corresponding reduction in energy consumption. The energy saved would lead to an increase in PET [49]. The complementary relationship was extended by *Granger* [34] for non-saturated surfaces and non-equal changes in AET and PET. He expressed these changes as

$$\frac{\partial AET}{\partial PET} = \frac{-\gamma}{\Delta}. \tag{14}$$

Granger [34] defined PET, AET, and E_w using a Dalton-type mass-transfer equation and introduced an alternative equation for the complementary relationship, i.e.,

$$AET + PET\frac{\gamma}{\Delta} = E_w\frac{\Delta + \gamma}{\Delta}. \tag{15}$$

For non-saturated surfaces, *Granger and Gray* [50] proposed a relative evaporation fraction with the notion that wind velocity (u) has an equal impact on AET and PET, i.e.,

$$EF_r = \frac{AET}{PET} = \frac{f(u)(e_a^s - e_a)}{f(u)(e_s^s - e_a)} = \frac{e_a^s - e_a}{e_s^s - e_a}, \tag{16}$$

where e_s^s and e_a^s are the saturated and actual water vapor pressure at the surface (hPa), e_a is the actual water vapor pressure of the air (hPa), and f(u) is a function of wind velocity.

Two of the most widely used models based on the complementary relationship are

i. the advection-aridity (AA) model of *Brutsaert and Stricker* [51] and

ii. the complementary relationship aerial evaporation (CRAE) model of *Morton* [52].

The AA-model combines

1. the Priestley-Taylor equation [32],
2. the Penman equation [30],
3. an empirical wind-velocity function for relative evaporation [31], and
4. the complementary relationship, giving rise to

$$\lambda AET = (2\alpha - 1)\frac{\Delta}{\Delta + \gamma}(R_n - G) - \frac{\gamma}{\Delta + \gamma}0.35\,(0.5 + 0.54u_2)(e_s - e_a), \tag{17}$$

where α is the Priestley-Taylor constant (i.e., 1.26) and u_2 is the wind velocity at 2-m above the ground or canopy surface (m s^{-1}).

The CRAE model estimates AET from PET and E_w, i.e.,

$$AET = 2E_w - PET. \tag{18}$$

PET is addressed by decomposing the Penman equation into its two main components: one component to address the energy balance at the surface and another, the transfer of water vapor from a moist surface, or

$$\lambda PET = R_n - f_P f_T (T_P - T_a), \tag{19}$$

and

$$\lambda PET = f_T(e_s^p - e_a), \tag{20}$$

where e_s^p is saturation vapor pressure at T_p and T_p is the equilibrium temperature. The equilibrium temperature is defined as the temperature at which the energy balance [Eqn. (19)] and mass transfer [Eqn. (20)] give the same result for PET, when solved numerically. Parameter f_T is a vapor-transfer coefficient considered constant for a specific pressure level and independent of wind velocity. Parameter f_P is a function of atmospheric pressure (Ψ; hPa) and T_p, such that

$$f_p = \gamma\Psi + 4\varepsilon_s\sigma(T_p + 273)^3 / f_T. \tag{21}$$

E_w is calculated with a regression equation,

$$\lambda_E = b_1 + b_2\left(1 + \frac{\gamma\Psi}{\Delta_{TP}}\right)^{-1} R_{nTP}, \tag{22}$$

where b_1 and b_2 are coefficients determined by regression, Δ_{TP} is the slope of the saturation-vapor-pressure-curve at T_p, and R_{nTP} is the net surface energy at T_p, ε_s is the surface emissivity (non-dimensional), and σ is the Stefan-Boltzmann constant (i.e., 5.67×10^{-8} W m^{-2} K^{-4}).

The two models were validated by *Hobbins et al* [53] using long term water-balance estimates from 120 watersheds distributed throughout the United States. Outcomes of this validation have revealed that the AA-model underestimated the annual AET by 10.6% of annual precipitation and the CRAE-model overestimated the same variable by 2.5% of annual precipitation. For highly arid watersheds, both models tended to overestimate AET. Validation of complementary-based models in other regions of the world, as well as comparing their results with direct observations of AET and PET strongly support the prevalence of the complementary relationship. *Ramirez et al* [54] found that complementary-based models tended to work well for temperate, humid regions of the world and less well for arid regions [55].

One important constraint of conventional complementary-based methods is the requirement of wind velocity in the calculation of relative evaporation, which is not always available or reliable. To overcome this limitation, *Venturini et al* [56] modified the relative evaporation fraction (EF_r) of Eqn. (16) by replacing the expression of water vapor pressure with the corresponding temperature used in calculating it. In the case of actual surface water vapor pressure, a hypothetical temperature is used. This hypothetical temperature (T_u) is defined as the temperature at which the surface becomes saturated without changing water vapor pressure; this is equivalent to the definition of dew point temperature for air (T_d; K). Using these temperatures, *Venturini et al.* [56] modified Eqn. (16) to the following:

$$EF_r = \frac{AET}{PET} = \frac{T_u - T_d}{T_s - T_d}. \tag{23}$$

The hypothetical temperature (T_u) is derived using the slope of the water vapor pressure curve at T_d (i.e., Δ_d) and T_s (Δ_s) and the relationship between water vapor pressures and corresponding temperatures, such that

$$T_u = \frac{e_s^s - e_a - \Delta_s T_s + \Delta_d T_d}{\Delta_d - \Delta_s}. \tag{24}$$

Combining Eqn.'s (15) and (23) and the Priestley-Taylor equation [32] with the definition of E_w, the equation of AET then becomes

$$\lambda AET = \alpha \frac{EF_r \Delta}{EF_r \Delta + \gamma} (R_n - G). \tag{25}$$

These changes render the method less error prone and simpler regarding its data requirements. The method is numerically robust, producing errors < 15% of daily AET-measurements for heterogeneous landscapes of the Southern Great Plains of the United States [56]. *Kalma et al.* [57] summarized results of validation studies of 30 RS-based AET methods, where it was found that *Venturini et al.*'s [56] method produced the highest accuracy among the methods using MODIS-derived data. In the current study (Section 5.2), we use the *Venturini et al.* [56] method to calculate regional AET for the extremely complex landscape of westcentral Gansu, NW China.

3. Data requirements and accessibility

Data requirements in the calculation of AET depend on the method or model used in the calculation procedure. In general, input data for all methods can be grouped into four main categories:

i. surface meteorological data, including near surface air temperature, wind velocity, and in-air water vapor pressure and humidity;

ii. radiative energy fluxes, including incoming shortwave ($R_s \downarrow$) and longwave radiation ($R_L \downarrow$) and their outgoing counterparts (i.e., $R_s \uparrow$ and $R_L \uparrow$);

iii. surface attributes, including temperature, emissivity, albedo, and soil water content; and

iv. vegetation attributes, including LAI, vegetation-cover density and extent, and stomatal density and aperture.

http://dx.doi.org/10.5772/52812

These data can be

i. acquired by direct measurement in the field,

ii. derived from other related variables by means of regression, or

iii. derived from RS optical reflectance or thermal emission data.

While optical and thermal RS has been providing spatially-distributed values for the assessment of landscape AET, it poses several important challenges. RS-images from LANDSAT, ASTER, and SPOT-systems have high spatial resolution, but are acquired infrequently (i.e., every 16 days for LANDSAT and ASTER-systems and every 26 days for SPOT). In contrast, high temporal-acquisitions with geostationary satellites provide coarse spatial resolutions that are largely inadequate for regional-estimate of AET.

The MODIS-instrument is crucial to the production of important global image-products for land- and ocean-surface monitoring [58]. MODIS-instruments were launched to space by NASA (National Aeronautics and Space Administration) as part of the earth-observing system (EOS). MODIS Terra and Aqua image-products are available since February 24, 2000 and June 24, 2002, respectively [59]. Raw image-products have a temporal-acquisition interval of one or two days, covering 36 spectral bands between 0.405 and 14.385 μm [59, 60]. Besides daily image scenes, several composite-products are also available at multiple-day intervals at different processing levels. Level-1 products provide the instrument data at full resolution for individual MODIS scenes, which can be exploited after converting to radiance values, surface reflectance, or brightness temperature. Level-2, 2G, and 3 images contain derived geo-biophysical parameters, while Level-4 images contain model-generated products. Level-2G and -3 image-products consist of aggregations of daily images to give 8 and 16-day cloud-free images. Table 1 lists a number of MODIS image-products valuable in estimating regional AET.

Variables	**MODIS image-products**	**Spatiotemporal Resolution**	
		m	**days**
Land surface temperature (T_s) Land surface emissivity (ε_s)	MODIS land surface temperature and emissivity data (**MOD11A2**)	1000	8
Normalized difference vegetation index (NDVI) Enhanced vegetation index (EVI)	MODIS vegetation indices (**MOD13Q1**)	250	16
Land surface albedo (A_s)	MODIS products combined with **BRDF**-albedo products	1000	16
Air temperature (T_a) Dew point temperature (T_d)	MODIS atmospheric profile data (**MOD07**)	5000	1

Table 1. Data needed in the calculation of regional AET and corresponding MODIS image-product sources and spatiotemporal resolutions.

3.1. Land surface temperature

MODIS land surface temperatures (T_s; MOD11A2) are estimated from MODIS thermal infrared (TIR) data collected in near-cloud-free conditions with the application of a split-window algorithm to avoid misidentification with cloud-top conditions [61]. The algorithm uses MODIS-emissivity bands 31 and 32 in the calculation of T_s [62]. Details of the calculation-procedure can be found in [63].

3.2. Vegetation indices

MODIS vegetation-products (MOD13Q1) provide two vegetation-indices (VI) both at 250-m and 16-day resolution [64]. NDVI is calculated as the normalized ratio of near infrared (NIR) and red bands. EVI (introduced earlier) was developed to improve the sensitivity of NDVI for high biomass regions [45]. NDVI and EVI are calculated from

$$NDVI = \frac{\rho_{NIR} - \rho_{red}}{\rho_{NIR} + \rho_{red}}, \tag{26}$$

and

$$EVI = B\frac{\rho_{NIR} - \rho_{red}}{\rho_{NIR} + C_1\rho_{red} - C_2\rho_{blue} + L}, \tag{27}$$

where ρ's are wavelength-specific atmospherically-corrected or partially atmosphere-corrected (for Rayleigh and ozone absorption) surface reflectances; B is the gain factor; L is the canopy background adjustment which addresses nonlinear, differential NIR and red-radiant transfer through vegetative canopies; and C_1, C_2 are the coefficients of aerosol resistance, which uses the blue band to correct for aerosol influences on the red band.

Coefficients adopted in the EVI-algorithm are L=1, C_1=6, C_2 = 7.5, and G= 2.5 [45]. While NDVI shows some problems with dense forests [45], both vegetation indices provide similar accuracy when used for cropland delineation [65].

3.3. Land surface albedo

Land surface albedo is the fraction of shortwave radiation reflected in all directions and is a critical parameter in estimating surface net shortwave radiation [66]. MODIS-BRDF (bidirectional reflectance distribution function)-based albedo image-products combine atmospherically-corrected surface reflectance from multiple dates and sensors in creating 16-day, 1-km resolution images [67]. The MODIS product, MCD43B3, includes both black and white sky albedo acquired with the Terra and Aqua satellites [68], covering all seven spectral bands and three broad bands [60].

3.4. Air and dew point temperature

Air and dew point temperatures provided by the MODIS-MOD07 product (at 5-km resolution) is currently the best possible estimate of the two variables derived from RS methods [69]. MOD07 atmospheric profile information includes temperature and atmospheric humidity at 20 different vertical pressure levels between 5 to 1000 hPa [70].

4. Study area description

The study area consists of the Shiyang and Hei River watersheds in westcentral Gansu, NW China (Fig. 1). The Shiyang River originates from the Qilian Mountains and flows northwestward before terminating in the Minqin-lake district [71]. The total basin area is approximately 49,500 km^2. Elevation in the Shiyang River basin varies from 1,284 to 5,161 m above mean sea level (AMSL), with an average elevation of 1,871 m AMSL. The Hei River watershed, with a land surface area of approximately 128,000 km^2, is the second largest inland river basin in NW China [72]. The Hei River watershed includes the Zhangye watershed, with a total land area of about 31,100 km^2. Elevation in the Zhangye watershed varies from 1,287 to 5,045 m AMSL, with an average elevation of 2,679 m AMSL (Fig. 1).

The natural landscape of the study area comprises of mountains, oases, and deserts, all interacting with each other [73]. The oases play an important role in the sustainability of the region's overall ecological and socio-economic integrity [74]. The area overlaps four distinct ecoregions [75]. The northern part, noted for its arid to semi-arid conditions, includes portion of the Badain Jaran and Tengger deserts and oases in the southwest portions of the Alashan Plateau. Liangzhou, at the south, and Minqin, at the north, are two important oases in the Shiyang River watershed [71]. Zhangye is the main oasis in the Zhangye watershed. Spring wheat is the main food crop grown in the oases, which is usually supported by irrigation [76]. In the deserts, salt-tolerant, xerophytic shrub species, i.e., saxaul (*Haloxylon ammodendron*) and *Reaumuria soongolica* [77] are common.

Locally-generated rainfall in the oases is normally insufficient (< 170 mm yr^{-1}) to support agriculture in the oases [7]. The main source of water to the oases is the runoff from the Qilian Mountains generated from snowmelt and in-mountain precipitation transported by the Shiyang and Hei River systems [78, 79]. Glacial meltwater contributes to about 3.8% and 8.3% of total runoff in the Shiyang and Hei Rivers, respectively [78]. The meltwater usually flows during the spring-summer period due to the warming of the mountain glaciers and previous snow-season's snow cover [74].

A primary source of water in the rivers during the summer is orographic precipitation [80] formed in the Qilian Mountains [81]. Annual precipitation in the watersheds is < 80 mm in the desert-portion of the watershed and > 800 mm in the Qilian Mountains, proper. Annual

PET in the deserts ranges between 2,000 to 2,600 mm and 700 to 1,200 mm in the mountains [82]. Most of the precipitation occurs during June-August. About 94% of the water delivered from the mountains is through surface runoff. Average annual runoff delivered by the Shiyang River is about 15.8 ×10^8 m^3 and about 37.7 × 10^8 m^3 by the Hei River [82].

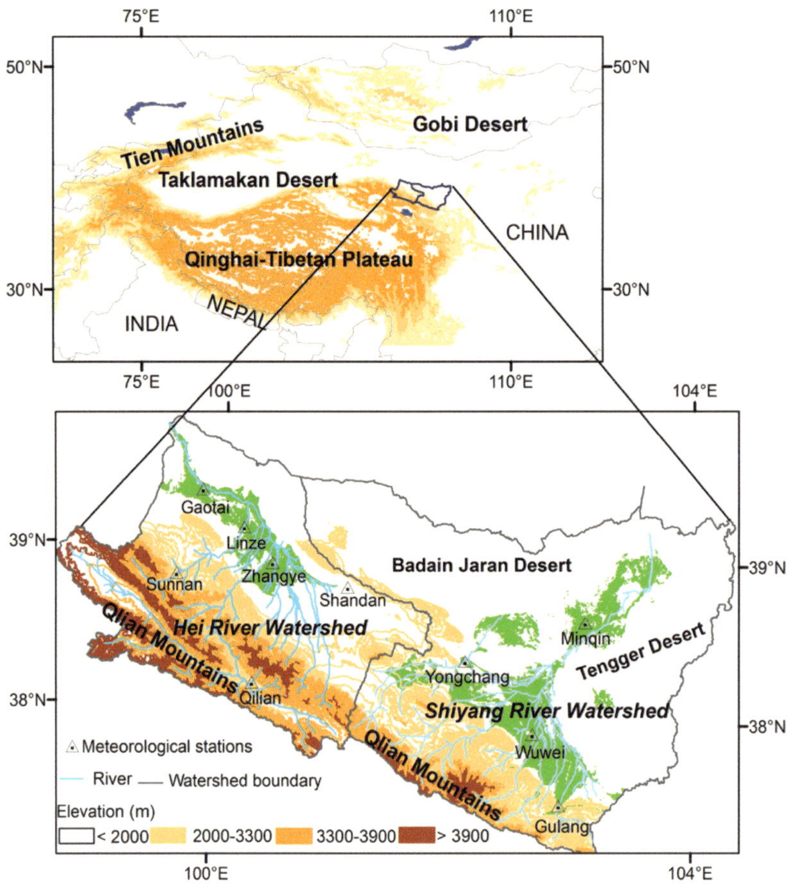

Figure 1. (a) Location of the study area along the northeast flank of the Qinghai-Tibetan Plateau. Yellow and orange areas correspond to mountain ranges and plateau, respectively. Map (b) gives the physical boundaries of the Shiyang and Hei River watersheds, respectively. Green areas in (b) illustrate the geographic extent of major oases at the base of the Qilian Mountains. Meteorological data relevant to the study were collected at individual climate stations identified in (b).

Presence of the Qinghai-Tibetan Plateau at the south of the study area blocks the northward passage of southwest monsoonal precipitation and the westerly airflow on the northern side of the plateau interferes with southerly airflow from reaching the region [83]. Dry northwesterly winds during summer generated from the Azores high pressure system and cold dry northerly winds during the winter generated from the Siberian high pressure system limit the outside contribution of moisture to this area [84]. Cyclic exchange of atmospheric and surface and shallow subsurface water in the region forms a nearly perfect close system [74]. ET in the series of oases at the base of the Qilian Mountains is shown to play an important role in the recycling of water in the region and seasonal evolution of snow cover in the Qilian Mountains [7, 85].

5. Seasonal and annual variation of AET for different landcover types

5.1. Landcover classification

The MODIS annual global landcover map currently available (as of 2012) is produced from seven spectral maps, BRDF-adjusted reflectance, T_s, EVI, and an application of supervised classification using ground data from 1860 field sites [86]. Assessments of the product have shown that this map is not entirely realistic for zones of transition or for mountainous regions [87]. Improved landcover definition at regional or local scales with supervised classification usually involves much greater amounts of ground data that are normally available for most regions. Recently, decision tree-based classification has been applied to RS-data and has been shown to produce better results than other classification systems based on maximum likelihood or unsupervised clustering and labeling [88]. One advantage of decision tree-based classification is that it is able to use local knowledge of vegetation characteristics together with other pertinent data, such as terrain characteristics. In the current study, we use chronological-sequences of MODIS-based EVI and digital terrain information (e.g., aspect, elevation) to classify landcover with decision trees.

Vegetation distribution in the study area has a unique preferential association with elevation, slope, and slope direction [89]. North-facing slopes of the Qilian Mountains support alpine meadow at elevations between 2,500 to 3,300 m AMSL. At elevations above 3,300 m AMSL, deciduous shrubs represent the most dominant vegetation type. Isolated patches of conifer forests in the Qilian Mountains are classified as a separate ecoregion [90] found at elevations between 2,500 m to 3,300 m AMSL. Vegetation density and seasonal vegetation growth vary as a function of vegetation type and, consequently, landcover.

Based on vegetation site preferences, the study area is subdivided into four main elevation zones defined by elevations:

i. < 2,500;

ii. between 2,500 to 3,300;

iii. between 3,300 to 3,900; and

iv. > 3,900 m AMSL.

Different landcover types in these elevation zones were then identified based on EVI and terrain attributes, in particular slope orientation (aspect). Landcover types and their discrimination are summarized in Table 2.

Elevation zone	Landcover	Discrimination criteria
Zone 1	Desert	Mean growing-season EVI < 0.1130
(< 2,500 m AMSL)	Crop	Maximum growing-season EVI > 0.27 and mean growing-season EVI < 0.1130
	Dense grass	Maximum growing-season EVI > 0.27, but different from cropland
	Sparse grass and/or shrub	Mean growing-season EVI between 0.113-0.27
Zone 2	Alpine meadow	Maximum growing-season EVI > 0.27 on north-facing slopes
(2,500-3,300 m AMSL)	Coniferous forest	Maximum growing-season EVI > 0.27 on other than north-facing slopes
	Sparse grass and/or shrub	Mean growing-season EVI between 0.113 – 0.27
	Bare land	Mean growing-season EVI < 0.113
Zone 3	Deciduous shrub	Maximum growing-season EVI > 0.27
(3,300-3,900 m AMSL)	Bare land	Mean growing-season EVI < 0.113
Zone 4	Sparse shrub	Mean growing-season EVI between 0.113-0.27
(> 3,900 m AMSL)	Snow and/or ice	Mean growing-season EVI < 0.113

Table 2. Landcover definition as a function of elevation zone, EVI, and slope orientation (aspect).

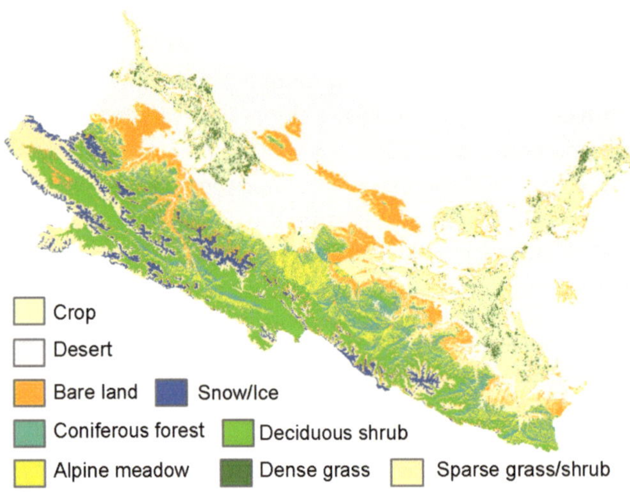

Figure 2. Study-area distribution of dominant landcover types.

Ten landcover maps were generated for 2000-2009 using the classification standards summarized in Table 2. From these maps, a final landcover composite ($LCOV_{dom}$ for all image pixels; Fig. 2) was then created based on a pixel-level, landcover-dominance evaluation, i.e.,

$$LCOV_{dom}\Big|_{for\ all\ pixels} = \underset{i=2000-2009}{Majority}(LCOV_i)\Big|_{for\ all\ pixels}, \quad (28)$$

where $LCOV_i$ and $LCOV_{dom}$ represent landcover at the pixel-level for individual years (2000-2009) and the dominant landcover over the same ten-year time period, respectively.

5.2. Study-area AET variation for different landcover types

Monthly AET for the study area was calculated using the complementary method of *Venturini et al.* [56] for the period of 2000-2009. Computed AET was compared with ET_o calculated from pan evaporation data (Fig. 3) corrected with season-specific coefficients reported in [91]. Pan evaporation coefficients were calculated by relating 50 years of pan evaporation data collected at 580 climate stations distributed across China to ET_o calculated for the same stations with the FAO Penman-Monteith equation. Pan evaporation coefficients by *Chen et al.* [91] varied from 0.45 to 0.54 from spring to winter. The scattergraphs in Fig. 3 show that modeled AET has very high positive correlation with ET_o for both watersheds ($R^2 > 0.78$; Fig. 3a for the Shiyang and Fig. 3b for the Hei River watershed). The line and error graphs in Fig. 3 are generated from AET-values extracted at the location of four meteorological stations within the Liangzhou, Minqin, and Zhangye oases (Fig. 1). The lines represent mean total monthly AET based on 2000-2009 data; error bars represent standard deviation of corresponding values. Modeled AET display seasonal patterns similar to those expressed in ET_o, but at substantially reduced levels (Fig. 3). This discrepancy can be rationalized by three important realities, i.e.,

- **i.** modeled AET represents the average conditions of individual image pixels (covering 62,500 m^2) and not points;
- **ii.** surface moisture and vegetation conditions and water-vapor-transfer resistances are openly different from those driving ET_o (i.e., open water vs. soil-plant environments); and
- **iii.** pan evaporation coefficients are based on seasonal averages independent of location.

Spatially-distributed average total monthly AET was calculated by averaging monthly AET-images generated for the 2000-2009 period (Fig. 4). Yearly total growing-season AET was calculated by summing AET from April-October of each year (Fig. 5). From Fig. 4, it is clear that AET in the watershed is very low (≤ 25 mm $month^{-1}$) during winter (January-March and October-December periods) and progressively higher in summer (> 75 mm $month^{-1}$). AET reaches its maximum during the June-August period and begins to decrease prior the start of winter. Both monthly and total growing-season AET reveals greatest AET in the oases and low-to-mid-slope positions of the Qilian Mountains and lowest in the deserts (not shown) and high-elevation portions of the Qilian Mountains. Within-year variation in AET also appears within the same elevation bands due to changes in vegetation type. Comparison of chronological-series of monthly AET for three landcover types is shown in Fig. 6. For most years, forest landcover is shown to contribute the most to AET in the Shiyang River watershed; crop cover, however, contributes the most in the Hei River watershed (Fig. 6). For the two watersheds, areas sparse of vegetation (as defined in Table 2) consistently contribute the least to AET during the growing seasons of 2000-2009.

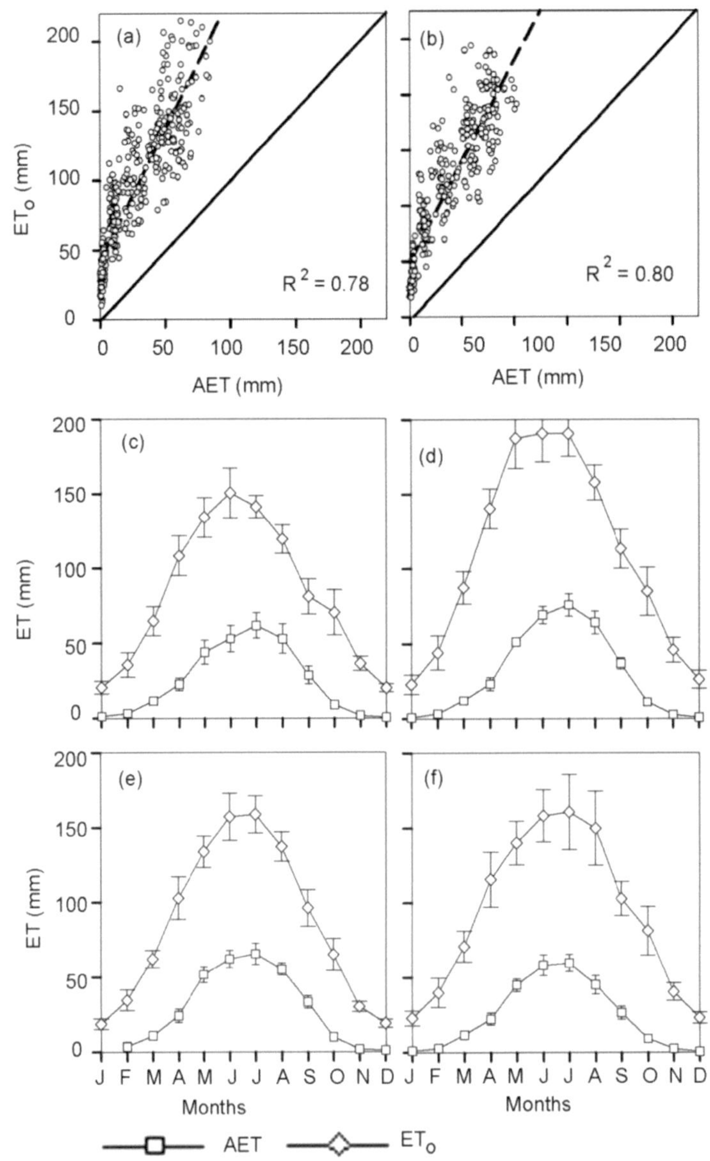

Figure 3. Comparison of monthly AET against ET_0 calculated from pan evaporation data over a ten-year (2000-2009) period. Dashed lines in (a) and (b) are lines of regression fitted to the ET_0-to-AET data pairs; R^2 is the coefficient of determination. Line and error graphs of monthly AET and ET_0 [i.e., (c)-(f)] are based on modeled AET-data extracted at the location of four climate stations, including Wuwei (c) and Minqin (d), representing the Shiyang River watershed, and Zhangye (e) and Shandan (f), representing the Hei River watershed and pan evaporation data.

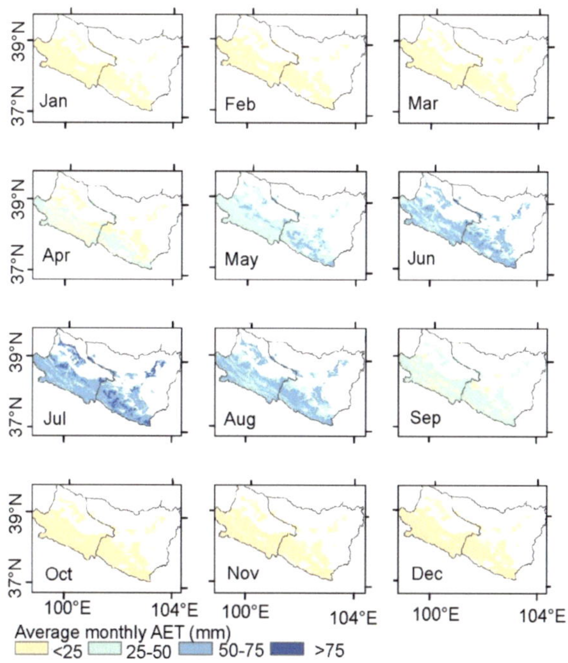

Figure 4. Study-area distribution of monthly AET averaged over the 2000-2009 period.

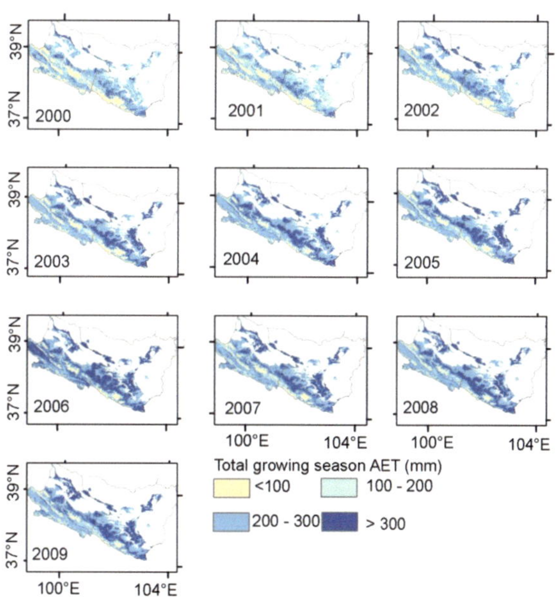

Figure 5. Study-area distribution of total growing-season (April-October period) AET for the 2000-2009 period.

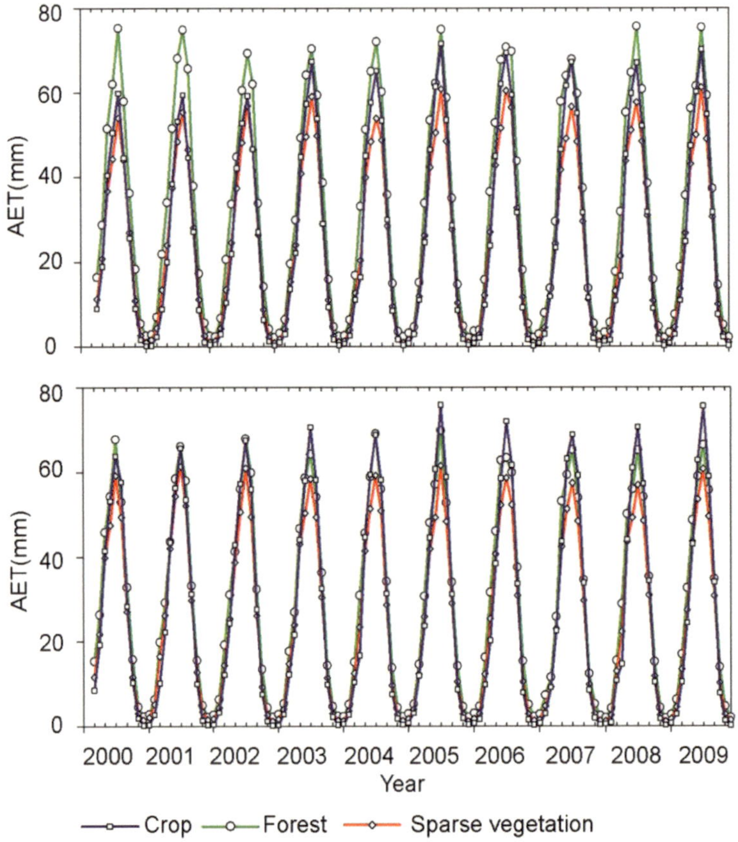

Figure 6. Monthly average AET (mm) for three landcover types; the upper graph gives monthly AET for the Shiyang River watershed, whereas the bottom graph gives it for the Hei River watershed.

5.3. Vegetation influence on AET

Using EVI as an indicator of vegetation density and vitality, monthly AET was compared with same-month EVI extracted at 220 randomly distributed points across the study area for different landcover types. Scattergraphs using all points and partitioned according to landcover type (Fig. 7) show that independent-evaluations of AET have an overall positive correlation with same-month EVI; the strength of correlation (R^2), however, varies with landcover type. Strongest correlation occurs for cropland areas and weakest, for areas sparse of vegetation.

http://dx.doi.org/10.5772/52812

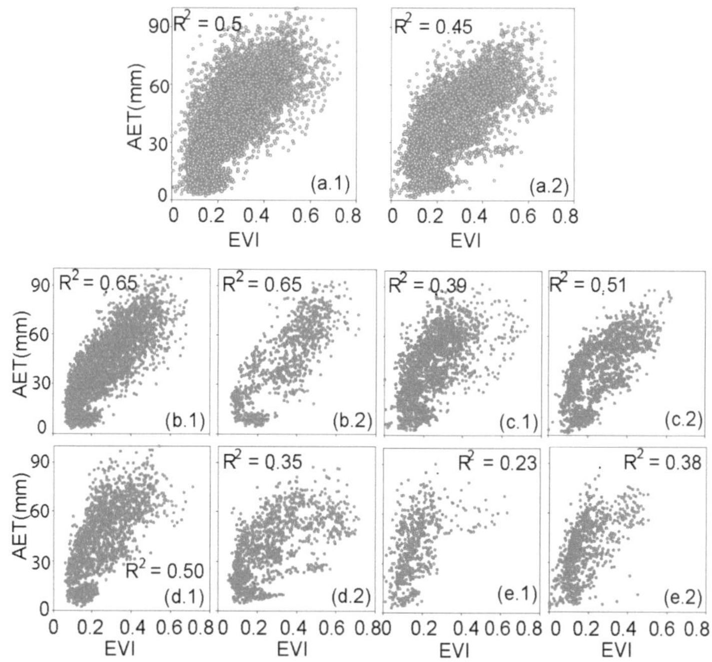

Figure 7. Scattergraphs showing AET as a function of same-month EVI for different vegetation covertypes for the two watersheds (labels with 1 refer to the Shiyang River watershed and 2, to the Hei River watershed): a.1 and a.2 represents all landcover types combined; b.1 and b.2 for crop cover; c.1 and c.2 for forest cover; d.1 and d.2 for dense grass cover; and e.1 and e.2 for sparse grass or shrub cover.

6. Conclusions

This chapter provides

i. a summary of methodologies available to estimate AET at point to regional scales, and

ii. case-study calculations of AET for two watersheds in NW China based on the complementary relationship of *Venturini et al.* [56].

The complementary relationship of *Venturini et al.* is considered appropriate for complex landscapes of NW China, because of the method's independence from wind velocity, its overall accurracy, and its ability to regionalize AET-calculations with assistance of RS-data as input. Because of the frequent acquisition of MODIS data (primary input datatype to the calculation of AET), AET-calculations can be updated frequently. For NW China, AET-calculations at 250-m resolution were carried out on a monthly interval over a ten-year time period (2000-2009). Based on a landcover map generated from decision-tree classification and landcover-dominance analysis, regional AET was partitioned along three vegetation-dominated landcover types. Forest and crop landcover types were shown to contribute the most to AET across the study area, particularly in lowland areas. Areas of sparse vegetation (among the three landcover types) contributed the least to regional AET. This supports the view that the state and abundance of vegetation (defined here by EVI), particularly in the lowlands and low-to-mid-slope positions of the Qilian Mountains, have an important influence on regional AET and on the water budget of the study area.

Author details

Mir A. Matin and Charles P.-A. Bourque*

*Address all correspondence to: cbourque@unb.ca

Faculty of Forestry and Environmental Management, University of New Brunswick, Canada

References

[1] Allen, R. G., Pereira, L. S., Raes, D., & Smith, M. (1998). Crop evapotranspiration-Guidelines for computing crop water requirements. *Rome: FAO- Food and Agriculture Organization of the United Nations.*

[2] Thornthwaite, C. W. (1948). An Approach toward a Rational Classification of Climate. *Geographical Review*, 38(1), 55-94.

[3] Curry, L. (1963). Thornthwaites Potential Evapotranspiration Term. *Annals of the Association of American Geographers*, 53(4), 585.

[4] Luo, X., Wang, K., Jiang, H., Sun, J., Xu, J., Zhu, Q., et al. (2010). Advances in research of land surface evapotranspiration at home and abroad. *Sciences in Cold and Arid Regions*, 2(2), 104-11.

[5] Mu, Q., Heinsch, F. A., Zhao, M., & Running, S. W. (2007). Development of a global evapotranspiration algorithm based on MODIS and global meteorology data. Remote sensing of environmentDec 28; , 111(4), 519-36.

[6] Verstraeten, W. W., Veroustraete, F., & Feyen, J. (2008). Assessment of evapotranspiration and soil moisture content across different scales of observation. Sensors. Jan; , 8(1), 70-117.

[7] Bourque, C. P.-A, & Hassan, Q. K. (2009). Vegetation control in the long-term self-stabilization of the Liangzhou Oasis of the Upper Shiyang river watershed of West-Central Gansu, Northwest China. *Earth Interactions*, 13, 1-22.

[8] Helldén, U. (2008). A coupled human-environment model for desertification simulation and impact studies. *Global and Planetary Change*, 64(3-4), 158-168.

[9] Pielke, R. A., & Avissar, R. (1990). Influence of landscape structure on local and regional climate. *Landscape Ecology*, 4(2), 133-55.

[10] Bonan, G. B. (2008). Ecological Climatology, Concepts and Applications. *2nd edition ed. New York: Cambridge University Press.*

[11] Greene, E. M., Liston, G. E., & Pielke, R. A. (1999). Relationships between landscape, snowcover depletion, and regional weather and climate. Hydrological processes. Oct; , 13(14-15), 2453-66.

[12] Pielke, R. A., Adegoke, J., Beltran-Przekurat, A., Hiemstra, C. A., Lin, J., Nair, U. S., et al. (2007). An overview of regional land-use and land-cover impacts on rainfall. Tellus Series B-Chemical and Physical Meteorology. Jul; , 59(3), 587-601.

[13] Pielke, R. A. (2001). Influence of the spatial distribution of vegetation and soils on the prediction of cumulus convective rainfall. *Review of Geophysics*, 39(2), 151-77.

[14] Raddatz, R. L. (2007). Evidence for the influence of agriculture on weather and climate through the transformation and management of vegetation: Illustrated by examples from the Canadian Prairies. *Agricultural and Forest Meteorology*, 142(2-4), 186-202.

[15] Rose, C. W., & Sharma, M. L. (1984). Summary and Recommendations of the Workshop on Evapotranspiration from Plant-Communities. *Agricultural Water Management*, 8(1-3), 325-42.

[16] Bowen, I. S. (1926). The ratio of heat losses by conduction and by evaporation from any water surface. Physical Review. Jun; , 27(6), 779-87.

[17] Baldocchi, D. D. (2003). Assessing the eddy covariance technique for evaluating carbon dioxide exchange rates of ecosystems: past, present and future. *Global Change Biology*, 9, 1-14.

[18] Rana, G., & Katerji, N. (2000). Measurement and estimation of actual evapotranspiration in the field under Mediterranean climate: a review. European Journal of Agronomy Aug; , 13(2-3), 125-153.

[19] Holmes, J. W. (1984). Measuring Evapotranspiration by Hydrological Methods. *Agricultural Water Management*, 8(1-3), 29-40.

[20] Wright, J. L. (1991). Using Weighing Lysimeters to Develop Evapotranspiration Crop Coefficients. *Lysimeters for Evapotranspiration and Environmental Measurements*, 191-9.

[21] Pieri, P., & Fuchs, M. (1990). Comparison of Bowen-Ratio and Aerodynamic Estimates of Evapotranspiration. Agricultural and Forest MeteorologyFeb; , 49(3), 243-56.

[22] Verma, S. B. (1987). Aerodynamic resistance to transfer of heat, mass and momentum. Estimation of aerial evapotranspiration. *Vancouver, B.C., Canada: International association of hydrological sciences.*

[23] Liu, S. M., Sun, Z. P., Wang, H. M., Li, X. W., & Han, L. J. (2003). An intercomparison study on models of estimating- The aerodynamic resistance. *Igarss: Ieee International Geoscience and Remote Sensing Symposium, Vols I- Vii, Proceedings*, 3341-3.

[24] Drexler, J. Z., Snyder, R. L., Spano, D., & Paw, K. T. U. (2004). A review of models and micrometeorological methods used to estimate wetland evapotranspiration. Hydrological processes. Aug 15; , 18(11), 2071-101.

[25] Denmead, O. T. (1984). Plant Physiological Methods for Studying Evapotranspiration- Problems of Telling the Forest from the Trees. *Agricultural Water Management*, 8(1-3), 167-89.

[26] Glenn, E. P., Huete, A. R., Nagler, P. L., Hirschboeck, K. K., & Brown, P. (2007). Integrating remote sensing and ground methods to estimate evapotranspiration. *Critical Reviews in Plant Sciences*, 26(3), 139-68.

[27] Bastiaanssen, W. G. M., Menenti, M., Feddes, R. A., & Holtslag, A. A. M. (1998). A remote sensing surface energy balance algorithm for land (SEBAL)- 1. Formulation. Journal of Hydrology Dec; , 213(1-4), 198-212.

[28] Kustas, W. P., & Norman, J. M. (1999). Evaluation of soil and vegetation heat flux predictions using a simple two-source model with radiometric temperatures for partial canopy cover. Agricultural and Forest Meteorology. Apr 1; , 94(1), 13-29.

[29] Su, Z. (2002). The surface energy balance system (SEBS) for estimation of turbulent heat fluxes. *Hydrology and Earth System Sciences*, Feb, 6(1), 85-99.

[30] Penman, H. L. (1948). Natural evaporation from open water, bare soil and grass. *Proceedings of the Royal Society of London Series A, Mathematical and Physical Sciences*, 193(1032), 120-45.

[31] Penman, H. L. (1956). Evaporation: an introductory survey. *Netherland Journal of Agricultural Science*, 4, 9-29.

[32] Priestley, C. H. B., & Taylor, R. J. (1972). On the assessment of surface heat flux and evaporation using large-scale parameters. *Monthly Weather Review*, 100(2).

[33] Bouchet, R. J. (1963). Evapotranspiration re ́elle et potentielle, signification climatique. *International Association of Hydrological Sciences, Proceedings of General Assmbly; Berkely, California: California Symposium Publication.*

[34] Granger, R. J. (1989). A complementary relationship approach for evaporation from nonsaturated surfaces. *Journal of Hydrology.*, Nov, 111(1-4), 31-8.

[35] Archer, M. D., & Barber, J. (2004). Molecular to Global Photosynthesis. *London: Imperial College Press.*

[36] Bourque-A, C. P., & Gullison, J. J. (1998). A technique to predict hourly potential solar radiation and temperature for a mostly unmonitored area in the Cape Breton Highlands. *Canadian Journal of Soil Science*, 78, 409-20.

[37] Bourque-A, C. P., Meng-R, F., Gullison, J. J., & Bridgland, J. (2000). Biophysical and potential vegetation growth surfaces for a small watershed in northern Cape Breton Island, Nova Scotia, Canada. *Canadian Journal of Forest Research*, 30(8), 1179-95.

[38] Dubayah, R., & Rich, P. M. (1995). Topographic solar-radiation models for GIS. International Journal of Geographical Information SystemsJul-Aug; , 9(4), 405-19.

[39] Fu, P., & Rich, P. M. (2000). The Solar Analyst 1.0 Manual.

[40] Li, Z. L., Tang, R. L., Wan, Z. M., Bi, Y. Y., Zhou, C. H., Tang, B. H., et al. (2009). A Review of Current Methodologies for Regional Evapotranspiration Estimation from Remotely Sensed Data. Sensors. May; , 9(5), 3801-53.

[41] Brutsaert, W. (1999). Aspects of bulk atmospheric boundary layer similarity under free-convective conditions. Reviews of GeophysicsNov; , 37(4), 439-51.

[42] Meneti, M. (1984). Physical Aspects and Determination of Evaporation in Deserts Applying Remote-Sensing Techniques- Menenti,M. *Ground Water*, 22(6), 801-2.

[43] Monteith, J. L. (1965). Evaporation and environment. *Symposia of the Society for Experimental Biology*, 19, 205-34.

[44] Cleugh, H. A., Leuning, R., Mu, Q. Z., & Running, S. W. (2007). Regional evaporation estimates from flux tower and MODIS satellite data. Remote Sensing of EnvironmentFeb 15; , 106(3), 285-304.

[45] Huete, A. R., Didan, K., Miura, T., rodriguez, E. P., Gao, X., & Fereira, L. G. (2002). Overview of the radiometric and biophysical performance of the MODIS vegetation indices. *Remote sensing of environment*, 83(1-2), 195-213.

[46] Mu, Q., Zhao, M. S., & Running, S. W. (2011). Improvements to a MODIS global terrestrial evapotranspiration algorithm. Remote sensing of environmentAug 15; , 115(8), 1781-800.

[47] Morton, F. I. (1969). Potential evaporation as a manifestation of regional evaporation. *Water Resources Research*, 5(6), 1244-55.

[48] Szilagyi, J., Hobbins, M. T., & Jozsa, J. (2009). Modified advection-aridity model of evapotranspiration. Journal of Hydrologic EngineeringJun; , 14(6), 569-74.

[49] Szilagyi, J. (2001). On Bouchet's complementary hypothesis. Journal of HydrologyJun 1 , 246(1-4), 155-8.

[50] Granger, R. J., & Gray, D. M. (1989). Evaporation from natural nonsaturated surfaces. Journal of Hydrology. Nov; , 111(1-4), 21-9.

[51] Brutsaert, W., & Stricker, H. (1979). An advection-aridity approach to estimate actual regional evapotranspiration. *Water Resources Research*, 15(2), 443-50.

[52] Morton, F. I. (1983). Operational estimates of aerial evaporation and their siginificance to science and practice of hydrology. *Journal of hydrology*, 66, 1-76.

[53] Hobbins, M. T., Ramirez, J. A., Brown, T. C., & Claessens, L. H. J. M. (2001). The complementary relationship in estimation of regional evapotranspiration: The Complementary Relationship Areal Evapotranspiration and Advection-Aridity models. Water Resources Research. May; , 37(5), 1367-87.

[54] Ramirez, J. A., Hobbins, M. T., & Brown, T. C. (2005). Observational evidence of the complementary relationship in regional evaporation lends strong support for Bouchet's hypothesis. *GEOPHYSICAL RESEARCH LETTERS.*, Aug 5;, 32(15).

[55] Xu, C. Y., & Singh, V. P. (2005). Evaluation of three complementary relationship evapotranspiration models by water balance approach to estimate actual regional evapotranspiration in different climatic regions. Journal of Hydrology. Jul 12; , 308(1-4), 105-21.

[56] Venturini, V., Islam, S., & Rodrigue, Z. L. (2008). Estimation of evaporative fraction and evapotranspiration from MODIS products using a complementary based model. Remote Sensing of EnvironmentJan 15; , 112(1), 132-41.

[57] Kalma, J. D., Mc Vicar, T. R., & Mc Cabe, M. F. (2008). Estimating Land Surface Evaporation: A Review of Methods Using Remotely Sensed Surface Temperature Data. Surveys in Geophysics. Oct; , 29(4-5), 421-69.

[58] Justice, C. O., Vermote, E., Townshend, J. R. G., Defries, R., Roy, D. P., Hall, D. K., et al. (1998). The Moderate Resolution Imaging Spectroradiometer (MODIS): Land remote sensing for global change research. IEEE TRANSACTIONS ON GEOSCIENCE AND REMOTE SENSING. Jul; , 36(4), 1228-49.

[59] Ramachandra, B., Justice, C. O., & Abrams, MJ. (2011). Land remote sensing and global environmental change: NASA's earth obeserving system and science of ASTER and MODIS. *New York: Springer*.

[60] Bisht, G., Venturini, V., Islam, S., & Jiang, L. (2005). Estimation of the net radiation using MODIS (Moderate Resolution Imaging Spectroradiometer) data for clear sky days. *Remote sensing of environment.*, Jul 15;, 97(1), 52-67.

[61] Wan, Z., Zhang, Y., Zhang, Q., & Li, Z. L. (2004). Quality assessment and validation of the MODIS global land surface temperature. International Journal of Remote SensingJan; , 25(1), 261-74.

[62] Wan, Z. M., & Dozier, J. (1996). A generalized split-window algorithm for retrieving land-surface temperature from space. IEEE TRANSACTIONS ON GEOSCIENCE AND REMOTE SENSING. Jul; , 34(4), 892-905.

[63] Petitcolin, F., & Vermote, E. (2002). Land surface reflectance, emissivity and temperature from MODIS middle and thermal infrared data. *Remote sensing of environment*, Nov;, 83(1-2), 112-34.

[64] Huete, A. R., Litu, H. Q., Batchily, K., & Leeuwen, Wv. (1997). A comparison of vegetation indices over a global set of TM images for EOS-MODIS. *Remote Sensing of Environment*, 59(3), 440-51.

[65] Wardlow, B. D., & Egbert, S. L. (2010). A comparison of MODIS 250-m EVI and NDVI data for crop mapping: a case study for southwest Kansas. *International Journal of Remote Sensing*, 31(3), 805-30.

[66] Davidson, A., & Wang, S. S. (2005). Spatiotemporal variations in land surface albedo across Canada from MODIS observations. Canadian Journal of Remote SensingOct; , 31(5), 377-90.

[67] Strahler, A. H., & Muller-P, J. (1999). NASA MODIS BRDF/Albedo product: algorithm and theoritical basis document, version 5.0.

[68] Schaaf, C. B., Gao, F., Strahler, A. H., Lucht, W., Li, X. W., Tsang, T., et al. (2002). First operational BRDF, albedo nadir reflectance products from MODIS. Remote sensing of environment. Nov; , 83(1-2), 135-48.

[69] Jimenez-Munoz, J. C., Sobrino, J. A., Mattar, C., & Franch, B. (2010). Atmospheric correction of optical imagery from MODIS and Reanalysis atmospheric products. Remote Sensing of Environment. Oct 15; , 114(10), 2195-210.

[70] Seeman, S. W., Borbas, E. E., Li, J., Menzel, W. P., & Gumley, L. E. (2012). Modis atmospheric profile retrieval, algorithm theoritical basis document version 6 reference number ATBD-MOD07. http://modis.gsfc.nasa.gov/data/atbd/atbd_mod07.pdf, accesses on June 20,)., *Madison, WI: Cooperative Institute for Meteorological Satellite Studies2006.*

[71] Li, X. Y., Xiao, D. N., He, X. Y., Chen, W., & Song, D. M. (2007). Factors associated with farmland area changes in arid regions: a case study of the Shiyang river basin, Northwestern China. *Frontiers in Ecology and the Environment*, Apr;, 5(3), 139-44.

[72] Gu, J., Li, X., & Huang, C. L. (2008). Land cover classification in Heihe river basin with time series MODIS NDVI data. Fifth International Conference on Fuzzy Systems and Knowledge Discovery Proceedings , 2, 477-81.

[73] Ma, M. G., & Frank, V. (2006). Interannual variability of vegetation cover in the Chinese Heihe river basin and its relation to meteorological parameters. *International Journal of Remote Sensing*, Aug 20;, 27(16), 3473-86.

[74] Ji, X. B., Kang, E. S., Chen, R. S., Zhao, W. Z., Zhang, Z. H., & Jin, B. W. (2006). The impact of the development of water resources on environment in arid inland river basins of Hexi region, Northwestern China. Environmental Geology. Aug; , 50(6), 793-801.

[75] Olson, D. M., Dinerstein, E., Wikramanayake, E. D., Burgess, N. D., Powell, G. V. N., Underwood, E. C., et al. (2001). Terrestrial ecoregions of the world: A new map of life on earth. *BioScience*, 51(11), 933-8.

[76] Zhao, C., Nan, Z., & Cheng, G. (2005). Methods for estimating irrigation needs of spring wheat in the middle Heihe basin, China. *Agricultural Water Management*, 75(1), 54-70.

[77] Carpenter, C. (2001a). Alashan Plateau semi-desert (PA1302). *World Wildlife Fund ©,* [accessed on 2011 13 December]; Available from:, https://secure.worldwildlife.org/wildworld/profiles/terrestrial/pa/pa1302_full.html.

[78] Wang, J. S., Feng, J. Y., Yang, L. F., Guo, J. Y., & Pu, Z. X. (2009). Runoff-denoted drought index and its relationship to the yields of spring wheat in the arid area of Hexi corridor, Northwest China. Agricultural Water Management. Apr; , 96(4), 666-76.

[79] Jin, X. M., Schaepman, M., Clevers, J., Su, Z. B., & Hu, G. C. (2010). Correlation between annual runoff in the Heihe river to the vegetation cover in the Ejina Oasis (China). *Arid Land Research and Management,* 24(1), 31-41.

[80] Roe, G. H. (2005). Orographic precipitation. *Annual Review of Earth and Planetary Sciences,* 33, 645-71.

[81] Zhu, Y. H., Wu, Y. Q., & Drake, S. (2004). A survey: obstacles and strategies for the development of ground-water resources in arid inland river basins of Western China. Journal of Arid EnvironmentsOct; , 59(2), 351-67.

[82] Kang, S., Su, X., Tong, L., Shi, P., Yang, X., Abe, Y., et al. (2009). The impacts of human activities on the water-land environment of the Shiyang River Basin, an arid region in Northwest China. *Hydrological Sciences Journal,* 49(3), 413-27.

[83] Liu, S., Zhang, C., Zhao, J., Wang, S., & Huang, Y. (2005). Study on the changes of water vapor over Hexi corridor and adjacent regions. *Acta Meteorologica Sinica,* 20, 108-21.

[84] Warner, T. T. (2004). Desert Meteorology. *London: Cambridge University Press.*

[85] Bourque, C. P. A., & Mir, M. A. (2012). Seasonal snow cover in the Qilian Mountains of Northwest China: Its dependence on oasis seasonal evolution and lowland production of water vapour. *Journal of Hydrology,* 454-455, 141-51.

[86] Friedl, M. A., Sulla-Menashe, D., Tan, B., Schneider, A., Ramankutty, N., Sibley, A., et al. (2010). MODIS Collection 5 global land cover: Algorithm refinements and characterization of new datasets. *Remote sensing of environment.,* Jan 15;, 114(1), 168-82.

[87] Liang, L., & Gong, P. (2010). An Assessment of MODIS Collection 5 Global Land Cover Product for Biological Conservation Studies. *18th International Conference on Geoinformatics.*

[88] Friedl, M. A., & Brodley, C. E. (1997). Decision tree classification of land cover from remotely sensed data. Remote sensing of environment. Sep; , 61(3), 399-409.

[89] Jin, X. M., Zhang-K, Y., Schaepman, M. E., Clevers, J. G. P. W., & Su, Z. (2008). Impact of elevation and aspect on the spatial distribution of vegetation in the qilian mountain area with remote sensing data. *The International Archives of the Photogrammetry, Remote Sensing and Spatial Information Sciences,* 37, Part B7).

[90] Carpenter, C. (2001b). Montane Grasslands and Shrublands. *World Wildlife Fund* ©, [accessed on 2011 13 Dec]; Available from:, http://www.worldwildlife.org/science/wildfinder/profiles/pa1015.html.

[91] Chen, D., Gao, G., Xu-Y, C., Guo, J., & Ren, G. (2005). Comparison of the Thornthwaite method and pan data with the standard Penman-Monteith estimates of reference evapotranspiration in China. *Climate Research*, 28, 123-32.

Chapter 7

Effect of Evapotranspiration on Hydrothermal Changes in Regional Scale

Tadanobu Nakayama

Additional information is available at the end of the chapter

http://dx.doi.org/10.5772/52808

1. Introduction

Evapotranspiration plays an important role not only on hydrologic cycle but also on thermal changes in various ways. In China, hydro-climate is diverse between north and south (Fig. 1). Semi-arid north is heavily irrigated and combination of increased food demand and declining water availability is creating substantial pressures in Yellow River (Brown and Halweil, 1998; Yang et al., 2004; Nakayama et al., 2006, 2010; Nakayama, 2011a, 2011b), whereas flood storage ability around lakes has decreased and impact of Three Gorges Dam (TGD) on flood occurrence in Changjiang downstream against original purpose is increasing problem in humid south (Shankman and Liang, 2003; Zhao et al., 2005; Nakayama and Watanabe, 2008b). Irrigation has a different impact on evapotranspiration changes at rotation between winter wheat and summer maize in the semi-arid region in the north (downstream of Yellow River), and double-cropping of rice in the humid south (middle of Changjiang River) in China. This mechanism changes greatly hydrologic cycle such as river discharge and groundwater, and in particular, affects extremes of flood and drought under climatic change (Nakayama, 2011a, 2011b, 2012c; Nakayama and Watanabe, 2006, 2008b; Nakayama et al., 2006, 2010).

On the other hand, urban heat island (Oke, 1987), where the urban temperature is higher than its rural surroundings, has become a serious environmental problem with the expansion of cities and industrial areas in the world (Fig. 1). Surfaces covered by concrete or asphalt can absorb a large amount of heat during the day and release it to the atmosphere at night. The evaporation of water provides an important counter to this effect, and so open parks and water surfaces are vital in urban areas for creating urban cool-island (Spronken-Smith and Oke, 1999; Chang et al., 2007). Recent researches showed that cooling effect of water-holding pavements made of new symbiotic material (consisting of porous asphalt and water-holding filler made of steel by-products based on silica compound) in addition to that of natural green area on hydrothermal cycle is effective to recover sound hydrologic cycle and to create thermally-pleasing environments in eco-conscious society (Nakayama and Fujita, 2010; Nakayama and Hashimoto, 2011; Nakayama et al., 2007, 2012).

In this way, the evapotranspiration plays an important role on hydrologic change in continental basins where water resources are vital for human activity, and effective management of water resource is powerful for decision-making and adaptation strategy for sustainable development. This chapter represents the improvement in process-based National Integrated Catchment-based Eco-hydrology (NICE) model series (Nakayama, 2008a, 2008b, 2009, 2010, 2011a, 2011b, 2012a, 2012b, 2012c; Nakayama and Fujita, 2010; Nakayama and Hashimoto, 2011; Nakayama and Watanabe, 2004, 2006, 2008a, 2008b, 2008c; Nakayama et al., 2006, 2007, 2010, 2012) with more complex sub-systems to develop coupled human and natural systems and to analyze impact of evapotranspiration on hydrothermal changes in regional scale.

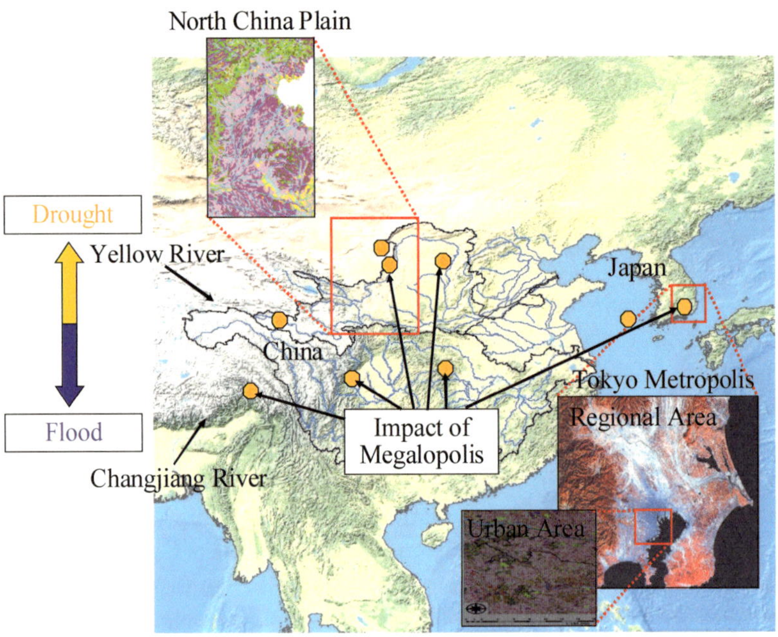

Figure 1. Study area in East Asia including the Changjiang and Yellow River basins in China, and the Tokyo Metropolis in Japan.

2. Material and methods

2.1. Coupling of process-based model with complex sub-systems

Previously, the author developed the process-based NICE model, which includes surface-unsaturated-saturated water processes and assimilates land-surface processes describing the variations of LAI (leaf area index) and FPAR (fraction of photosynthetically active radiation) from satellite data (Fig. 2) (Nakayama, 2008a, 2008b, 2009, 2010, 2011a, 2011b, 2012a, 2012b, 2012c; Nakayama and Fujita, 2010; Nakayama and Hashimoto, 2011; Nakayama and Watanabe, 2004, 2006, 2008a, 2008b, 2008c; Nakayama et al., 2006, 2007, 2010, 2012). The unsaturated layer divides canopy into two layers, and soil into three layers in the vertical dimension in the SiB2 (Simple Biosphere model 2) (Sellers et al., 1996). About the saturated layer, the NICE solves three-dimensional groundwater flow for both unconfined and confined aquifers. The hillslope hydrology can be expressed by the two-layer surface runoff model including freezing/thawing processes. The NICE connects each sub-model by considering water/heat fluxes: gradient of hydraulic potentials between the deepest unsaturated layer and the groundwater, effective precipitation, and seepage between river and groundwater.

In agricultural field, NICE is coupled with DSSAT (Decision Support Systems for Agrotechnology Transfer) (Ritchie et al., 1998), in which automatic irrigation mode supplies crop water requirement, assuming that average available water in the top layer falls below soil moisture at field capacity for cultivated fields (Nakayama et al., 2006). The model includes different functions of representative crops (wheat, maize, soybean, and rice) and simulates automatically dynamic growth processes. Potential evaporation is calculated on Priestley and Taylor equation (Priestley and Taylor, 1972), and plant growth is based on biomass formulation, which is limited by various reduction factors like light, temperature, water, and nutrient, et al. (Nakayama et al., 2006; Nakayama and Watanabe, 2008b; Nakayama, 2011a).

In urban area, NICE is coupled with UCM (Urban Canopy Model) to include the effect of hydrothermal cycle at various pavements, and with RAMS (Regional Atmospheric Modeling System) (Pielke et al., 1992) to include the hydrothermal interaction (Nakayama and Fujita, 2010; Nakayama and Hashimoto, 2011; Nakayama et al., 2012). In particular, the author expanded specific heat conductivity c_s and heat conductivity k_s in natural soil (Sellers et al., 1996) for engineered pavement in the following equations by including the effect of water amount on the heat characteristics in the material.

$$\begin{aligned} c_s &= \left[0.5\left(1-\theta_s\right)+\theta_s \cdot W_1\right] C_w \cdot \rho_w \quad \textit{for soil} \\ &= C_p \cdot \rho_p \left(1-\theta_s\right) + C_w \cdot \rho_w \cdot \theta_s \cdot W_1 \quad \textit{for engineered pavement} \end{aligned} \tag{1}$$

$$\begin{aligned} k_s &= 0.4186 \frac{1.5\left(1-\theta_s\right)+1.3\theta_s \cdot W_1}{0.75+0.65\theta_s-0.4\theta_s \cdot W_1} \quad \textit{for soil} \\ &= k_p\left(1-\theta_s\right)+k_w \cdot \theta_s \cdot W_1 \quad \textit{for engineered pavement} \end{aligned} \tag{2}$$

C_p (J/kg/K) is specific heat of pavement material; C_w (J/kg/K) is specific heat of water (=4.18×10^6); c_s (J/m^3/K) is specific heat conductivity including the effect of water; k_s (W/m/K) is heat conductivity including the effect of water; k_p (W/m/K) is heat conductivity of pavement material; k_w (W/m/K) is heat conductivity of water (=0.59); ρ_p (kg/m^3) is specific gravity of pavement material; ρ_w (kg/m^3) is density of water (=1,000), W_i is the soil moisture fraction of the i-th layer (=θ_i/θ_s); θ_s (m^3/m^3) is volumetric soil moisture in the i-th layer; θ_s (m^3/m^3) is the value of θ at saturation, respectively.

Downward short- and long-wave radiation, precipitation, atmospheric pressure, air temperature, air humidity, and wind speed simulated by the atmospheric model are input into the UCM, whereas momentum, sensible, latent, and long-wave flux simulated by the UCM are input into the atmospheric model at each time step. This procedure means that the feedback process about water and heat transfers between atmospheric region and land surface are implicitly included in the simulation process.

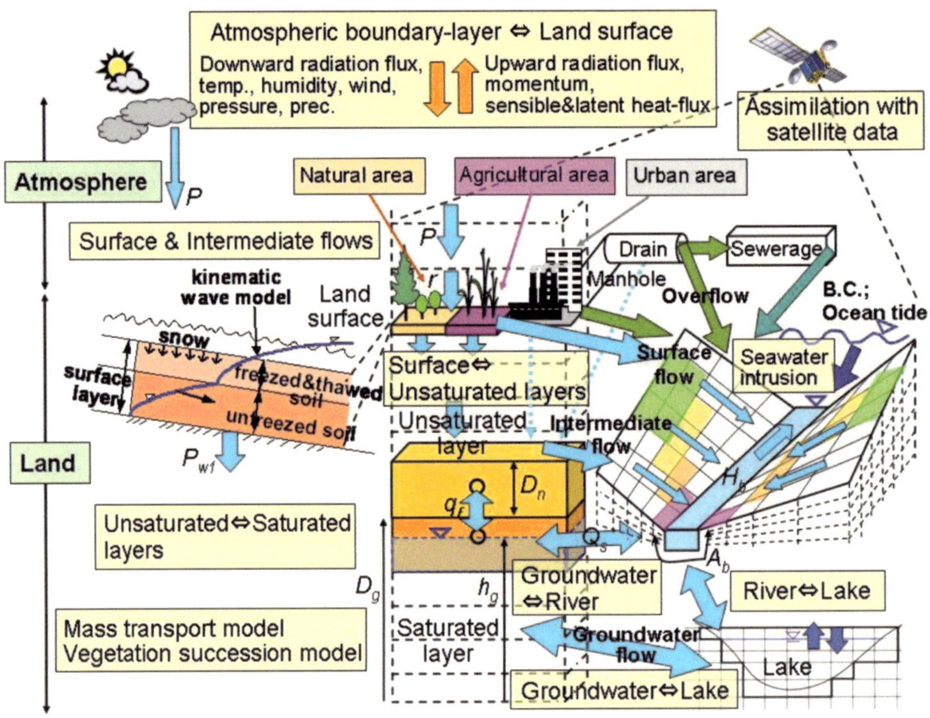

Figure 2. National Integrated Catchment-based Eco-hydrology (NICE) model.

2.2. Model input data and running the simulation

Six-hour re-analyzed data were input into the model after interpolation of ECMWF (European Centre for Medium-Range Weather Forecasts) in inverse proportion to the distance back-calculated in each grid. Because the ECMWF precipitation had the least reliability and underestimated observed peak values, rain gauge daily precipitation collected at meteorological stations were used to correct the ECMWF value (Nakayama, 2011a, 2011b, 2012c; Nakayama and Watanabe, 2006, 2008b; Nakayama et al., 2006, 2010). For a multi-scaled model in the Tokyo area, hourly observation data from AMeDAS (Automated Meteorological Data Acquisition System) data (Japan Meteorological Agency, 2005-2006) were assimilated with the model. At the lateral boundaries of regional area, some meteorological data are input to the model from the ECMWF (European Centre for Medium-Range Weather Forecasts) with a resolution of 1°x 1° and from the MSM (Meso Scale Model) with a resolution of 10km x 10km (Japan Meteorological Agency, 2005-2006). Mean elevation was calculated by using a global digital elevation model (DEM; GTOPO30) (U.S. Geological Survey, 1996). Digital land cover data were categorized such as forests, grasses, bushes, shrubs, paddy fields, and cultivated fields. About 50 vegetation and soil parameters were calculated on the basis of vegetation class and soil maps (Chinese Academy of Sciences, 1988; Digital National Land Information GIS data of Japan, 2002). The geological structures were divided into four types on the basis of hydraulic conductivity, the specific storage of porous material, and specific yield by scanning and digitizing the geological material (Geological Atlas of China, 2002; Nakayama et al., 2007) and core-sampling data. Artificial augmentation of waterworks and sewerage systems, and anthropogenic sensible, latent, and sewage heats generated by buildings and factories were input into the model (Nakayama and Fujita, 2010; Nakayama and Hashimoto, 2011; Nakayama et al., 2007, 2012).

At the upstream boundaries, a reflecting condition on the hydraulic head was used assuming that there is no inflow from the mountains in the opposite direction (Nakayama and Watanabe, 2004). Time-series of tidal level was input as a variable head at the sea boundary (Nakayama, 2011a; Nakayama et al., 2006, 2007, 2010). Vertical geological structures were divided into 10–20 layers by using sample database. The hydraulic head values parallel to the ground level were input as the initial conditions. In river grids decided by digital river network, inflows or outflows from the riverbeds were simulated at each time step depending on the difference in the hydraulic heads of groundwater and river. The hydraulic head values parallel to the observed ground level were input as initial conditions for the groundwater sub-model. In river grids decided by digital river network from topographic maps, inflows or outflows from the riverbeds were simulated at each time step depending on the difference in the hydraulic heads of groundwater and river.

The simulation area covered 3,000 km by 2,000 km with a grid spacing of 10 km, covering the entire Changjiang and Yellow River Basins (Fig. 1). The vertical layer was discretized in thickness with depth, with each layer increased in thickness by a factor of 1.1 (Nakayama, 2011b; Nakayama and Watanabe, 2008b; Nakayama et al., 2006). In the Tokyo area, the simulation was conducted with multi-scaled levels in horizontally regional area (260 km wide by 260 km long with a grid spacing of 2 km covering Kanto region) nesting with one way to urban area (36 km wide by 26 km long with 200 m grid covering the Kawasaki City) (Nakayama and Hashimoto, 2011; Nakayama et al., 2007, 2012) (Fig. 1). These areas were discretized with a grid spacing of 200 m – 5 km in the horizontal direction. The NICE simulation was conducted on a NEC SX-8 supercomputer. The first 6 months were used as a warm-up period until equilibrium water levels were reached, and parameters were estimated by a comparison of simulated steady-state value with that published in previous literatures. A time step of the simulation was changed from $\Delta t = 1.5$ sec to 1 h depending on spatial scale and the sub-model. Simulations were validated against various hydrothermal observed variables such as river discharge, soil moisture, groundwater level, air temperature, surface temperature, and heat-flux budget, et al.

3. Result and discussion

3.1. Effect of irrigation on hydrologic change

After the verification procedure (Nakayama, 2011b; Nakayama and Watanabe, 2008b), the model simulated effect of irrigation on evapotranspiration at rotation between winter wheat and summer maize in the downstream of Yellow River, and double-cropping of rice in the middle of Changjiang River (Fig. 3). Because more water is withdrawn during winter-wheat period due to small rainfall in the semi-arid north, the irrigation in this period affects greatly the increase in evapotranspiration (Fig. 3a), which is supplied by the limited water resources of river discharge and groundwater there. In particular, most of the irrigation is withdrawn from aquifer in the North China Plain (NCP) because surface water is seriously limited there (Nakayama, 2011a, 2011b; Nakayama et al., 2006). In the south, the irrigation is usually from the river to fill the paddy fiddle as ponding water depth (Nakayama, 2012c; Nakayama and Watanabe, 2008b), which increases evapotranspiration more in the drier season (Fig. 3b). This implies that energy supply is abundant relative to the water supply and the hydrological process is more sensitive to precipitation in the north, whereas the water supply is abundant relative to the energy supply and sun duration has a more significant impact in the south (Cong et al., 2010).

The model also simulated groundwater level in both Changjiang and Yellow River basins (Fig. 3c). The level decreases rapidly around the source area and the Qinghai Tibet Plateau, indicating that there are many sources of spring water in this region. The value is very low in the downstream because of the low elevation and overexploitation, in particular, in the NCP (Nakayama, 2011a, 2011b; Nakayama et al., 2006). This result indicates that hydrologic cycle including groundwater level is highly related not only to the topography but also to the irrigation water use. The NICE is effective to provide better evaluation of hydrological trends in longer period including 'evaporation paradox' (Roderick and Farquhar, 2002; Cong et al., 2010) together with observation networks because the model does not need the crop coefficient (depending on a growing stage and a kind of crop) for the calculation of actual evaporation and simulates it directly without detailed site-specific information or empirical relation to calculate effective precipitation (Nakayama, 2011a; Nakayama et al., 2006).

The mean TINDVI (Time-Integrated Normalized Difference Vegetation Index) gradients during 1982-1999 in various field crops (wheat, maize, and rice) at 4 stations were compared with trends of crop yields in the previous research (Tao et al., 2006) (Fig. 4a). The correlation of both values is relatively good ($r^2 = 0.986$) and the TINDVI gradient has a linear relation to the yield trend. The spatial pattern of the mean TINDVI gradient in agricultural fields shows a generally increasing tendency, particularly in the Yellow downstream and the NCP (Fig.

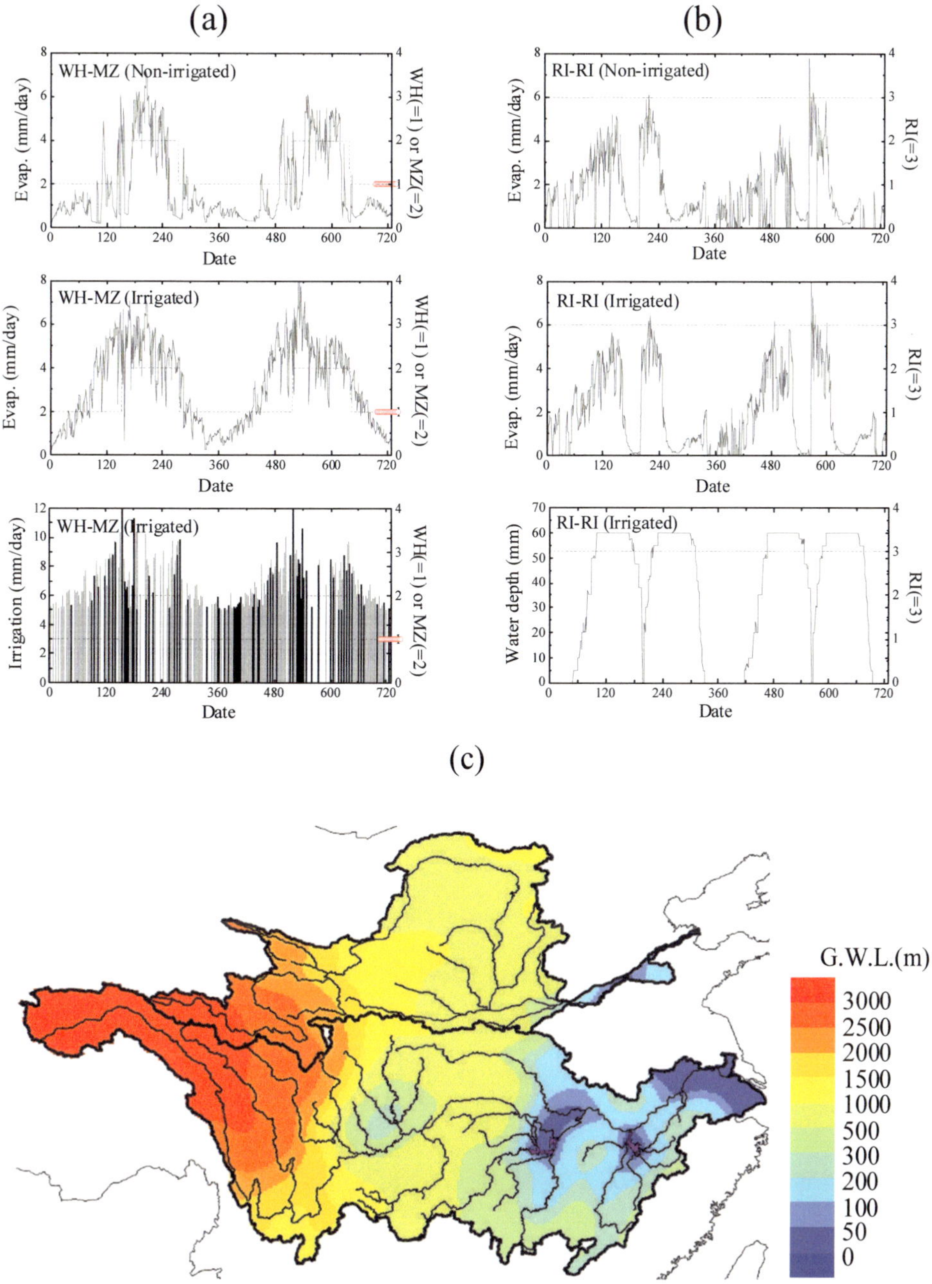

Figure 3. Impact of irrigation and ponding water depth on hydrological cycle; (a) evapotranspiration change at rotation between winter wheat and summer maize in the downstream of Yellow River, (b) evapotranspiration change at double-cropping of rice in the middle of Changjiang River, and (c) simulated results of annual-averaged groundwater level (a.s.l.) in the Changjiang and Yellow River basins. In Fig. 3a and 3b, right axis shows a period of each crop (WH; wheat, MZ; maize, and RI; rice, respectively).

4b), which is closely related to increasing tendency for winter wheat production in the downstream (U.S. Department of Agriculture, 1994), increase in irrigation water use (Yang et al., 2004) and chemical fertilizer, changes in crop varieties, improvements in technology such as agricultural machines, and other agronomic changes. On the other hand, it shows a generally decreasing tendency, particularly in the mid-lower reaches and around the lakes in the Changjiang River. This is caused mainly by an increase in lake reclamation, levee construction, and the resultant relative decrease in rice productivity in the lower reaches (Shankman and Liang, 2003; Zhao et al., 2005; Nakayama and Watanabe, 2008b). The decrease in the TINDVI gradient near the Bohai Sea, the East China Sea, and the Taihang Mountains was due to several effects including groundwater degradation, seawater intrusion, and rapid urbanization in the areas surrounding bigger cities (Brown and Halweil, 1998). Generally, these results suggest that the increase in irrigation water use is one of the reasons for the increase in crop production (Yang et al., 2004).

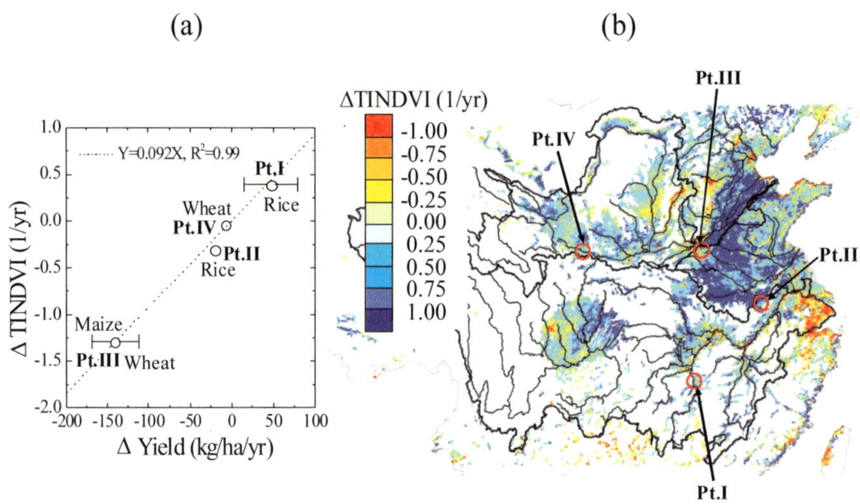

Figure 4. TINDVI over the agricultural fields by NOAA/AVHRR satellite images in the agricultural fields; (a) comparison of TINDVI with trend in observed yields at 4 stations (Pt.I; Changsha, Pt.II; Hefei, Pt.III; Zhengzhou, Pt.IV; Tianshui), and (b) spatial patterns in mean TINDVI gradient (per year) during 1982-1999.

3.2. Effective management of water resources to reduce urban heat island

The simulated heat-flux budget was compared at the infiltration and water-holding pavements after the verification with the analyzed value (Fig. 5). The water-holding pavement takes about 10.0 °C lower of the surface temperature than the infiltration pavement. The observed daily cycles of the road surface and air temperature at 1.5 m height were compared with those simulated by NICE after water irrigation in the same way as the previous research (Nakayama and Fujita, 2010). The model generally captured the observed water amount, and the associated diurnal cycle of the road surface and air temperature, and could estimate the temperature decrease trend with high accuracy. This cooling temperature is closely related to the promotion of vaporization by using the water-holding pavement, and this trend continues during 5 days after the fulfillment of water-holding pavement. The rapid increase of latent heat (438 W/m^2) relative to sensible heat (127 W/m^2) was reported in the field observation at sprinkling roads (Yamagata et al., 2008). The simulated maximum sensible and latent heat fluxes in the water-holding pavement just after water irrigation were 130 W/m^2 and 345 W/m^2, which had relatively good approximations of 161 W/m^2 and 337 W/m^2 in observed values and this previous research. The model could simulate reasonably the general trend that the latent heat in the infiltration pavement was smaller than that in the water-holding pavement because the infiltration was more predominant than the evaporation.

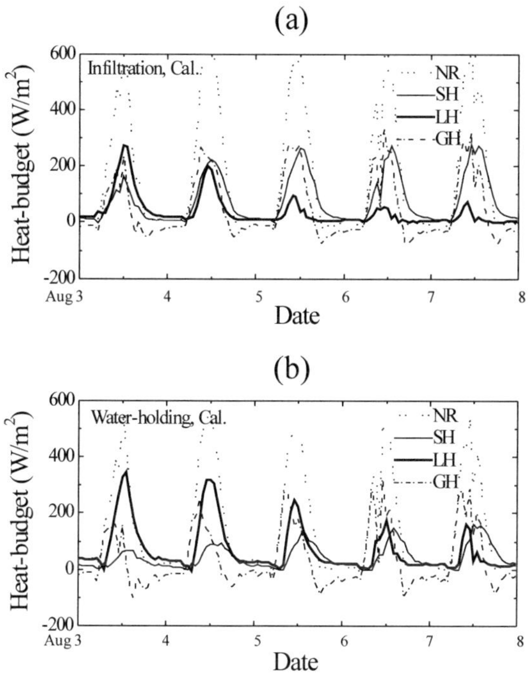

Figure 5. Simulated heat-flux budget in the symbiotic urban pavements after water irrigation at 3 August 2007; (a) infiltration pavement, and (b) water-holding pavement. Dotted line, net radiation (NR); solid line, sensible heat flux (SH); bold line, latent heat flux (LH); dash-dotted line, ground transfer heat flux (GH), respectively.

The author predicted the hydrothermal changes in the symbiotic urban scenarios (Fig. 6) (Nakayama and Hashimoto, 2011; Nakayama et al., 2012). NICE correctly predicted the much lower surface temperature of the water-holding block than those of the other pavements for several days after rainfall in comparison with simplified AUSSSM (Nakayama and Fujita, 2010), which was caused mainly by the rapid increase of evaporation after the rainfall (Fig. 6b). The predicted surface temperature on the scenario of water-holding pavement shows drastic decrease in the entire Kawasaki City (Fig. 6c). In particular, the business district beside the sea, where the urban heat island is predominant mainly due to the paved surface and the greater anthropogenic heat sources, has an effective cooling on this scenario.

The predicted groundwater level change in August is interesting in contrast to the simulated temperature (Fig. 6d). In the simulation on the scenario of water-holding pavement, all the necessary water to fill the water-holding pavement was automatically simulated by considering the difference between precipitation and evaporation, and withdrawn from the underneath groundwater in the NICE, which means that the pavement was always saturated. The groundwater level would decrease drastically at the commercial and industrial areas beside the sea and at the inland residential area, in exchange for a drastic cooling in the corresponding and the surrounding areas. The increase in recharge rate and the consequent increase in groundwater level on the scenario of a natural zone and green area are also effective to promote the groundwater resources in the urban area covered by impermeable pavement (Nakayama et al., 2007, 2012). The predicted temperature and groundwater level are greatly affected not only at the business district in the Kawasaki area but also at the Tokyo metropolitan area in the northern side of this study area. This result is very important from the political point of view, which indicates that we have to estimate more precisely the arrangement of symbiotic urban scenario in the study area together with the neighbouring administrations.

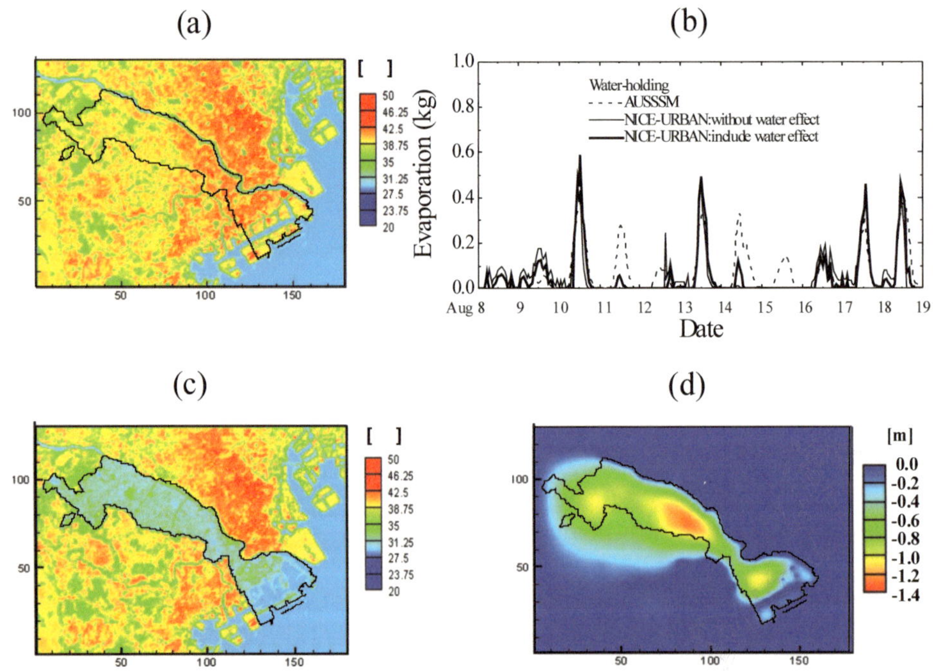

Figure 6. Prediction of hydrothermal changes at 5 August 2006; (a) present surface temperature, (b) evaporation simulated by NICE and simplified AUSSSM above water-holding pavement during 8-19 August 2006, and (c-d) predictions of surface temperature and groundwater level changes after water-holding pavement.

3.3. Discussion

Simulated results about the impact of irrigation on evapotranspiration change showed a clear difference between the Changjiang and Yellow Rivers. Because surface water and groundwater are administered separately by different authorities in China (Nakayama, 2011a), water management becomes further complicated if we consider it in both the Changjiang and Yellow Rivers. Any change in water accounting procedures may need to be negotiated through agreements brokered at relatively high levels of government, because surface water and groundwater are physically closely related to each other. The future development of irrigated and unirrigated fields and the associated crop production would affect greatly hydro-climate change and usable irrigation water from river and aquifer, and vice versa (Nakayama, 2011b). This research presented the lateral subsurface flow also has an important effect on the hydrologic cycle even in the continental scale, which extends traditional 'dynamic equilibrium' with atmospheric forcing (Maxwell and Kollet, 2008). From this point of view, nonstationarity models of relevant environmental variables have to be further developed to incorporate water infrastructure and water users including agricultural and energy sectors with a careful estimation of uncertainty (Milly et al., 2008).

This study also showed that the water-holding pavement scenario is effective for the urban area of floating subways, stations, and buildings in order to use moderately groundwater and to ameliorate the severe heat island through promotion of evaporation, whereas alternative land cover scenario of green area is also effective for the urban area of groundwater depression and seawater intrusion in order to promote the infiltration and to cool temperature. Therefore, the effective management of water resources including groundwater as a heat sink, particularly during the summer, would be attractive for both recovering a sound hydrologic cycle and tackling the urban heat island phenomenon (Ministry of Environment, 2004; Nakayama et al., 2007, 2012), so-called, 'Win-Win' approach. We are convinced that integrated management of both surface water and groundwater by using NICE in a political scenario for the effective selection and use of ecosystem service sites (Millennium Ecosystem Assessment, 2005) would play an important role in the creation of thermally-pleasing environments and the achievement in sustainable development in urban regions.

Recently, environmental pollution is becoming intertwined with various aspects (Fig. 7). Land subsidence is still a serious environmental problem around the foot of mountain near the Tokyo Metropolis due to delayed regulatory enforcement and continuation of extensive groundwater extraction. Water contaminant also shows some relation to this heterogeneity in addition to the thermal environment through complex chemical reaction, which implies further importance to evaluate these problems synthetically. Present result indicates effective management of water resource including evapotranspiration is also powerful for mitigation of heat island, which implies a possibility of achieving win-win solution about hydrothermal pollutions in eco-conscious society. The procedure to construct integrated assessment system would be also valuable for adaptation to climate change and urbanization in global scale, proposal of sustainability index, and providing eco-conscious society (Ministry of Environment, 2004; Nakayama and Hashimoto, 2011; Nakayama et al., 2007, 2012).

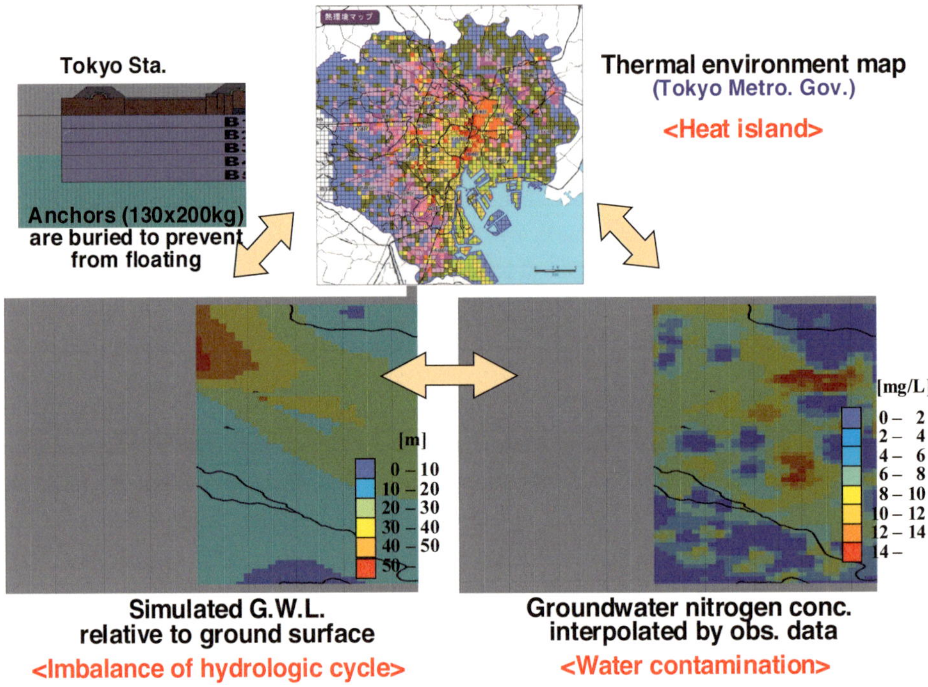

Figure 7. Intertwined environmental pollution in urban and surrounding areas.

4. Conclusion

This study coupled National Integrated Catchment-based Eco-hydrology (NICE) model series with complex sub-models to develop coupled human and natural systems and to analyze impact of evapotranspiration on hydrothermal changes in regional scale. The model includes different functions of representative crops (wheat, maize, soybean, and rice) and simulates automatically dynamic growth processes and biomass formulation. The simulated result showed impact of irrigation on eco-hydrological processes. The spatial pattern of Time-Integrated NDVI (TINDVI) gradient in agricultural fields indicated heterogeneous characteristics of crop yield, which implied the increase in irrigation water use is one of the reasons for the increase in crop production. NICE also reproduced reasonably the observed hydrothermal characteristics including the water and heat budgets in various pavements, evaluated the role of a new surface material (water-holding pavement) in promoting evaporation and cooling temperature to counter the urban heat island phenomenon, and predicted the hydrothermal changes under alternative land cover scenarios. These results suggest strongly the needs of trans-boundary and multi-disciplinary solutions of water management for sustainable development under sound socio-economic conditions contributory to national and global securities.

Acknowledgements

The author thanks Dr. M. Watanabe, Keio University, Japan, Dr. Y. Yang, Shijiazhuang Institute of Agricultural Modernization of the Chinese Academy of Sciences (CAS), China, and Dr. K. Xu, National Institute for Environmental Studies (NIES), Japan, for valuable comments about the study areas. Some of the simulations in this study were run on an NEC SX–8 supercomputer at the Center for Global Environmental Research (CGER), NIES. The support of the Environmental Technology Development Fund from the Japanese Ministry of Environment is also acknowledged.

Author details

Tadanobu Nakayama[1,2]

1 National Institute for Environmental Studies (NIES), Onogawa, Tsukuba, Ibaraki, Japan

2 Centre for Ecology & Hydrology (CEH), Crowmarsh Gifford, Wallingford, Oxfordshire, UK

References

[1] Brown, L. R, & Halweil, B. (1998). China's water shortage could shake world food security. *World Watch*, 0896-0615, 11, 10-18.

[2] Chang, C. R, Li, M. H, & Chang, S. D. (2007). A preliminary study on the local cool-island intensity of Taipei city parks. *Landscape Urban Plan.*, 0169-2046, 80, 386-395.

[3] Chinese Academy of Sciences(1988). *Administrative division coding system of the People's Republic of China*, Beijing

[4] Cong, Z, Zhao, J, Yang, D, & Ni, G. (2010). Understanding the hydrological trends of river basins in China. *J. Hydrol.*, doi:j.jhydrol.2010.05.013, 0022-1694, 388, 350-356.

[5] Digital National Land Information GIS Data of Japan(2002). *Database of groundwater in Japan.* Ministry of Land Infrastructure and Transport of Japan, http://www.nla.go.jp/ksj/

[6] Geological Atlas of China(2002). Geological Publisher, Beijing (in Chinese)

[7] Japan Meteorological Agency (JMA)(2005). *AMeDAS (Automated Meteorological Data Acquisition System) Annual Reports 20052006*Japan Meteorological Business Support Center (CD-ROM)

[8] Japan Meteorological Agency (JMA)(2005). *MSM (Meso Scale Model) Objective Analysis Data 20052006*Japan Meteorological Business Support Center (CD-ROM)

[9] Maxwell, R. M, & Kollet, S. J. (2008). Interdependence of groundwater dynamics and land-energy feedbacks under climate change. *Nat. Geosci.*, doi:ngeo315, 1752-0894, 1, 665-669.

[10] Millennium Ecosystem Assessment(2005). *Strengthening capacity to manage ecosystems sustainability for human well-being*, http://www.millenniumassessment.org/en/index.aspx

[11] Milly, P. C. D, Betancourt, J, Falkenmark, M, Hirsch, R. M, Kundzewicz, Z. W, Lettenmaier, D. P, & Stouffer, R. J. (2008). Stationarity is dead: Whither water management ?. *Science*, doi:science.1151915, 0036-8075, 319, 573-574.

[12] Ministry of Environment(2004). *Report on heat-island measures by controlling anthropogenic exhaustion heat in the urban area*, http://www.env.go.jp/air/report/hin Japanese)

[13] Nakayama, T, & Watanabe, M. (2004). Simulation of drying phenomena associated with vegetation change caused by invasion of alder (Alnus japonica) in Kushiro Mire. *Water Resour. Res.*, W08402, doi:WR003174, 0043-1397, 40

[14] Nakayama, T, & Watanabe, M. (2006). Simulation of spring snowmelt runoff by considering micro-topography and phase changes in soil layer. *Hydrol. Earth Syst. Sci. Discuss.*, 1027-5606, 3, 2101-2144.

[15] Nakayama, T, Yang, Y, Watanabe, M, & Zhang, X. (2006). Simulation of groundwater dynamics in North China Plain by coupled hydrology and agricultural models. *Hydrol. Process.*, doi:hyp.6142, 0885-6087, 20, 3441-3466.

[16] Nakayama, T, Watanabe, M, Tanji, K, & Morioka, T. (2007). Effect of underground urban structures on eutrophic coastal environment. *Sci. Total Environ.*, doi:j.scitotenv.2006.11.033, 0048-9697, 373, 270-288.

[17] Nakayama, T. (2008a). Factors controlling vegetation succession in Kushiro Mire. *Ecol. Model.*, doi:j.ecolmodel.2008.02.017, 0304-3800, 215, 225-236.

[18] Nakayama, T. (2008b). Shrinkage of shrub forest and recovery of mire ecosystem by river restoration in northern Japan. *Forest Ecol. Manag.*, doi:j.foreco.2008.07.017, 0378-1127, 256, 1927-1938.

[19] Nakayama, T, & Watanabe, M. (2008a). Missing role of groundwater in water and nutrient cycles in the shallow eutrophic Lake Kasumigaura, Japan. *Hydrol. Process.*, doi:hyp.6684, 0885-6087, 22, 1150-1172.

[20] Nakayama, T, & Watanabe, M. (2008b). Role of flood storage ability of lakes in the Changjiang River catchment. *Global Planet. Change*, doi:j.gloplacha.2008.04.002, 0921-8181, 63, 9-22.

[21] Nakayama, T, & Watanabe, M. (2008c). Modelling the hydrologic cycle in a shallow eutrophic lake. *Verh. Internat. Verein. Limnol.*, 0368-0770, 30, 345-348.

[22] Nakayama, T. (2009). Simulation of Ecosystem Degradation and its Application for Effective Policy-Making in Regional Scale, In: *River Pollution Research Progress*, Mattia

N. Gallo & Marco H. Ferrari (Eds.), Nova Science Publishers, Inc., 978-1-60456-643-7New York, 1-89.

[23] Nakayama, T. (2010). Simulation of hydrologic and geomorphic changes affecting a shrinking mire. *River Res. Appl.*, doi:rra.1253, 1535-1459, 26, 305-321.

[24] Nakayama, T, & Fujita, T. (2010). Cooling effect of water-holding pavements made of new materials on water and heat budgets in urban areas. *Landscape Urban Plan.*, doi:j.landurbplan.2010.02.003, 0169-2046, 96, 57-67.

[25] Nakayama, T, Sun, Y, & Geng, Y. (2010). Simulation of water resource and its relation to urban activity in Dalian City, Northern China. *Global Planet. Change*, doi:j.gloplacha. 2010.06.001, 0921-8181, 73, 172-185.

[26] Nakayama, T. (2011a). Simulation of complicated and diverse water system accompanied by human intervention in the North China Plain. *Hydrol. Process.*, doi:hyp.8009, 0885-6087, 0885-6087.

[27] Nakayama, T. (2011b). Simulation of the effect of irrigation on the hydrologic cycle in the highly cultivated Yellow River Basin. *Agr. Forest Meteorol.*, doi:j.agrformet. 2010.11.006, 0168-1923, 151, 314-327.

[28] Nakayama, T, & Hashimoto, S. (2011). Analysis of the ability of water resources to reduce the urban heat island in the Tokyo megalopolis. *Environ. Pollut.*, doi:j.envpol. 2010.11.016, 0269-7491, 159, 2164-2173.

[29] Nakayama, T. (2012a). Visualization of missing role of hydrothermal interactions in Japanese megalopolis for win-win solution. *Water Sci. Technol.*, doi:wst.2012.205, 0273-1223, 66, 409-414.

[30] Nakayama, T. and regime shift of mire ecosystem in northern Japan. *Hydrol. Process.*, doi:hyp.9347, 0885-6087, 26, 2455-2469.

[31] Nakayama, T. (2012c). Impact of anthropogenic activity on eco-hydrological process in continental scales. *Proc. Environ. Sci.*, doi:j.proenv.2012.01.008, 1878-0296, 13, 87-94.

[32] Nakayama, T, Hashimoto, S, & Hamano, H. (2012). Multi-scaled analysis of hydrothermal dynamics in Japanese megalopolis by using integrated approach. *Hydrol. Process.*, doi:hyp.9290, 0885-6087, 26, 2431-2444.

[33] Oke, T. R. (1987). *Boundary layer climates*, Methuen Press, 978-0-41504-319-9London.

[34] Pielke, R. A, Cotton, W. R, Walko, R. L, Tremback, C. J, Lyons, W. A, Grasso, L. D, Nicholls, M. E, Moran, M. D, Wesley, D. A, Lee, T. J, & Copeland, J. H. (1992). A omprehensive meteorological modeling system-RAMS. *Meteorol. Atmos. Phys.*, 0177-7971, 49, 69-91.

[35] Priestley, C. H. B, & Taylor, R. J. (1972). On the assessment of surface heat flux and evaporation using large-scale parameters. *Mon. Weather Rev.*, 0027-0644, 100, 81-92.

[36] Ritchie, J. T, Singh, U, Godwin, D. C, & Bowen, W. T. (1998). Cereal growth, development and yield, In: *Understanding Options for Agricultural Production*, Tsuji, G.Y., Hoogenboom, G. & Thornton, P.K. (Eds.), Kluwer, 0-79234-833-8Britain, 79-98.

[37] Roderick, M. L, & Farquhar, G. D. (2002). The cause of decreased pan evaporation over the past 50 years. *Science*, 0036-8075, 298, 1410-1411.

[38] Sellers, P. J, Randall, D. A, Collatz, G. J, Berry, J. A, Field, C. B, Dazlich, D. A, Zhang, C, Collelo, G. D, & Bounoua, L. (1996). A revised land surface prameterization (SiB2) for atomospheric GCMs. Part I : Model formulation. *J. Climate*, 0894-8755, 9, 676-705.

[39] Shankman, D, & Liang, Q. (2003). Landscape changes and increasing flood frequency in China's Poyang Lake region. *Prof. Geogr.*, 0033-0124, 55, 434-445.

[40] Spronken-smith, R. A, & Oke, T. R. (1999). Scale modeling of nocturnal cooling in urban parks. Boundary-Layer Meteorology 0006-8314, 93, 287-312.

[41] Tao, F, Yokozawa, M, Xu, Y, Hayashi, Y, & Zhang, Z. (2006). Climate changes and trends in phenology and yields of field crops in China, 1981-2000. *Agr. Forest Meteorol.*, 0168-1923, 138, 82-92.

[42] U.S. Department of Agriculture (USDA). (1994). *Major World Crop Areas and Climatic Profiles*. World Agricultural Outlook Board, USDA, Agricultural Handbook No.664, http://www.usda.gov/oce/weather/pubs/Other/MWCACP/MajorWorldCropAreas.pdf

[43] U.S. Geological Survey. (1996). *GTOPO30 Global 30 Arc Second Elevation Data Set*, USGS, http://www1.gsi.go.jp/geowww/globalmap-gsi/gtopo30/gtopo30.html

[44] Yamagata, H, Nasu, M, Yoshizawa, M, Miyamoto, A, & Minamiyama, M. (2008). Heat island mitigation using water retentive pavement sprinkled with reclaimed wastewater. *Water Sci. Technol.*, doi:wst.2008.187, 0273-1223, 57, 763-771.

[45] Yang, D, Li, C, Hu, H, Lei, Z, Yang, S, Kusuda, T, Koike, T, & Musiake, K. (2004). Analysis of water resources variability in the Yellow River of China during the last half century using historical data. *Water Resour. Res.*, W06502, doi:WR002763, 0043-1397, 40

[46] Zhao, S, Fang, J, Miao, S, Gu, B, Tao, S, Peng, C, & Tang, Z. (2005). The 7-decade degradation of a large freshwater lake in central Yangtze River, China. *Environ. Sci. Technol.*, 0001-3936X, 39, 431-436.

Chapter 8

A Parametric Model for Potential Evapotranspiration Estimation Based on a Simplified Formulation of the Penman-Monteith Equation

Aristoteles Tegos, Andreas Efstratiadis and
Demetris Koutsoyiannis

Additional information is available at the end of the chapter

http://dx.doi.org/10.5772/52927

1. Introduction

The estimation of evapotranspiration is a typical task in several disciplines of geosciences and engineering, including hydrology, meteorology, climatology, ecology and agricultural engineering. According to the specific area of interest, we distinguish between three different aspects of evapotranspiration, i.e. actual, potential and reference crop. In particular, the actual evapotranspiration refers to the quantity of water that is actually removed from a surface due to the combined processes of evaporation and transpiration. At the global scale, it is estimated that about 60% of precipitation reaching the Earth's terrain returns to the atmosphere by means of water losses due to evapotranspiration [1]. Obviously this percentage is not constant but exhibits significant variability both in space and time. In particular, in semi-arid basins the annual percentage of evapotranspiration losses may be 70-80% of precipitation, while in cold hydroclimates this percentage is lower that 30%.

The actual evapotranspiration is one of the most difficult processes to measure in the field, except for experimental, small-scale areas, for which lysimeter observations can be assumed representative. For the spatial scale of practical applications (e.g., river basin), there is a growing interest on remote sensing technologies, which attempt to estimate actual evapotranspiration at regional scales, by combining ground measurements with satellite-derived data [2]. Otherwise, the only reliable method is based on the calculation of the water balance of the basin. Apparently, this approach is valid only when actual evapotranspiration is the single unknown component of the water balance. In addition, it can be implemented only for large time scales (annual and over-annual), for which storage regulation effects can be neglected. Alternatively, the spatiotemporal representation of the water balance can be evaluated through hydrological models. The latter simulate the main hydrological mechanisms of the river basin, using areal rainfall and potential evapotranspiration as input data. In order to obtain the actual evapotranspiration, which is output of the model, it is therefore essential to estimate the potential evapotranspiration across the basin.

Potential, as well as and crop reference evapotranspiration, are two key concepts, for which different definitions have been proposed, often leading to confusing interpretations [3]. The term "potential evapotranspiration", introduced by Thornthwaite [4], generally refers to the maximum amount of water that could evaporate and transpire from a large and uniformly vegetated landscape, without restrictions other than the atmospheric demand [5, 6]. On the other hand, the reference crop evapotranspiration (or reference evapotranspiration) is strictly defined as the evapotranspiration rate from a reference surface, not short of water, of a hypothetical grass crop with a height of 0.12 m, a fixed surface resistance of 70 s/m and an albedo of 0.23. This description closely resembles a surface of green, well-watered grass of uniform height, actively growing and completely shading the ground [7].

Potential and reference evapotranspiration can be theoretically retrieved on the basis of mass transfer and energy balance approaches. In this respect, a wide number of methods and models have been developed, of different levels of complexity. Lu *et al.* [6] report that there exist over 50 of such methods, which can be distinguished according to their mathematical framework and data requirements. The most integrated approaches are the analytical (also called combination) ones, which combine energy drivers, i.e. solar radiation and temperature, with atmospheric drivers, i.e. vapour pressure deficit and surface wind speed, towards a physically-based representation of the phenomenon [8]. In particular, Penman [9] developed the classical method named after him, which combines the energy balance and aerodynamic processes to predict the evaporation through open water, bare soil and grass. Later, Monteith [10] expanded the Penman equation to also account for the role of vegetation in controlling transpiration, particularly through the opening and closing of stomata.

The subsequently referred to as the Penman-Monteith method is by far the most recognized among all evapotranspiration models, as reported in numerous investigations worldwide. For instance, this approach provided the optimal estimates on the daily and monthly scales and was the most consistent across all locations [11]. It was also indicated that the Penman-Monteith method exhibited excellent performance in both arid and humid climates [12] and was the most accurate in estimating the water needs of turfgrass [13]. Its suitability was confirmed even when relative humidity and wind velocity data are missing [14].

The Penman-Monteith method requires complex calculations and it is data demanding. In this respect, a number of radically simplified methods have been proposed that require fewer meteorological variables. The latter can be divided into two categories: radiation-based approaches, which account for both the variability of solar radiation and temperature [15, 16, 17, 18], and elementary empirical methods that use as single meteorological input temperature [4, 19, 20]. A comprehensive review of them is provided in [21] and [22]. Yet, the predictive capacity of most of these models remains questionable, since they are based on empirical assumptions that do not ensure consistency with the physics of the natural phenomenon. Moreover, their applicability is usually restricted to specific geographical locations and climatic regimes, because their parameters are derived from experiments employed at the local scale.

Except for temperature, long records of input time series for the Penman-Monteith method are rarely available, especially in older meteorological stations. Thus, a typical problem is the estimation of evapotranspiration, in case of missing meteorological data. The Food and Agriculture Organization (FAO) emphatically suggests employing the Penman-Monteith method even under limited data availability. In this context, it provides a number of procedures for dealing with missing information, while it discourages the use of empirical approaches [7]. Yet, most of the proposed procedures still require meteorological data that are barely accessible, or require arbitrary assumptions with regard to the climatic regime of the study area.

Here we propose an alternative parametric approach, which is founded on a simplified parsimonious expression of the Penman-Monteith formula. The method can be assigned to radiation-based approaches, yet its key difference is the use of free variables (i.e. parameters) instead of constants, which are fitted to local climatic conditions. The method is parsimonious both in terms of meteorological data requirements (temperature) and with regard to the number of parameters (one to three).

For the identification of parameters, it is essential to have reference time series of satisfactory accuracy, preferably estimated through the Penman-Monteith method. Under this premise, the model parameters can be obtained through calibration, thus by minimizing the departures between the modelled evaporation data and the reference data. This task was applied in 37 meteorological stations that are maintained by the National Meteorological Service of Greece. As will be shown, the proposed method is clearly superior to commonly-used empirical approaches. Moreover, by mapping the parameter values over the entire Greek territory, using typical interpolation tools, we provide a flexible framework for the direct and reliable estimation of evapotranspiration anywhere in Greece.

The article is organized as follows: In section 2, we review the Penman-Monteith method and its simplifications, which estimate evapotranspiration on the basis of temperature and radiation data. In section 3 we present the new parametric model, which compromises the requirements for parsimony and consistency. In section 4, we calibrate the model at the point scale, using historical meteorological data, and evaluate it against other empirical approaches. In addition, we investigate the geographical distribution of its parameters over Greece. Finally, in section 5 we summarize the outcomes of our research and discuss next research steps.

2. The Penman-Monteith method and its simplifications

2.1. The Penman method and its physical background

Evaporation can be viewed both as energy (heat) exchange and an aerodynamic process. According to the energy balance approach, the net radiation at the Earth's surface ($R_n = S_n - L_n$, where S_n and L_n are the shortwave—solar—and longwave—earth—radiation, respectively) is mainly transformed to latent heat flux, Λ, and sensible heat flux to the air, H. The evaporation rate, expressed in terms of mass per unit area and time (e.g. kg/m²/d), is given by the ratio $E' := \Lambda / \lambda$, where λ is the latent heat of vaporization, with typical value 2460 kJ/kg. By ignoring fluxes of lower importance, such as soil heat flux, the heat balance equation is solved for evaporation, yielding:

$$E' = (R_n - H) / \lambda = R_n / \lambda(1 + b) \tag{1}$$

where $b := H / \Lambda$ is the co-called Bowen ratio. The estimation of b requires the measurement of temperature at two levels (surface and atmosphere), as well as the measurement of humidity at the atmosphere. On the other hand, the estimation of the net radiation R_n is based on a radiation balance approach to determine the components S_n and L_n. Typical input data required (in addition to latitude and time of the year), are solar radiation (direct and diffuse, or, in absence of them, sunshine duration data or cloud cover observations), temperature and relative humidity. The net radiation also depends of surface properties (i.e. albedo) and topographical characteristics, in terms of slope, aspect and shadowing. Recent studies proved that the impacts of topography are important at all spatial scales, although they are usually neglected in calculations [23].

From the aerodynamic viewpoint, evaporation is a mass diffusion process. In this context, the rate of evaporation is related to the difference in the water vapor content of the air at two levels above the evaporating surface and a function of the wind speed $F(u)$ in the diffusion equation. Theoretically, $F(u)$ can be computed on the basis of elevation, wind velocity, aerodynamic resistance and temperature. Yet, for simplicity it is usually given by empirical formulas, derived through linear regression, for a standard measurement level of 2 meters.

Penman [9] combined the energy balance with the mass transfer approaches, thus allowing the use of temperature, humidity and wind speed measurements at a single level. His classical formula for computing evaporation from an open water surface is written as:

$$E' = \frac{\Delta}{\Delta + \gamma} \frac{R_n}{\lambda} + \frac{\gamma}{\Delta + \gamma} F(u) D \tag{2}$$

where Δ is the slope of vapor pressure/temperature curve at equilibrium temperature (hPa/K), γ is a psychrometrcic coefficient, with typical value 0.67 hPa/K, and D is the vapor pressure deficit of the air (hPa), defined as the difference between the saturation vapor pressure e_a and the actual vapor pressure e_s, which are functions of temperature and relative humidity. We remind that (2) estimates the evaporation rate (mass per unit area per day), which is expressed in terms of equivalent water depth by dividing by the water density ρ (1000 kg/m³). Next we will use symbols E' for evaporation rates, and $E := E' / \rho$ for equivalent depths per unit time.

Summarizing, the Penman equation requires records of solar radiation (alternatively, sunshine duration or cloud cover), temperature, relative humidity and wind speed, as well as a number of parameters accounting for surface characteristics (latitude, Julian day index, albedo, elevation, etc.).

2.2. The Penman-Monteith method

Penman's method was extended to cropped surfaces, by accounting for various resistance factors, aerodynamic and surface. As mentioned in the introduction, Monteith [10] introduced the concept of the so-called "bulk" surface resistance that describes the resistance of vapor flow through the transpiring crop and evaporating soil surface [7]. This depends on multiple factors, such as the solar radiation, the vapor pressure deficit, the temperature of leafs, the water status of the canopy, the height of vegetation, etc. Combining the two resistance components, the following generalized expression of the Penman equation is obtained, referred to as Penman–Monteith formula:

$$E' = \frac{\Delta}{\Delta+\gamma'}\frac{R_n}{\lambda} + \frac{\gamma}{\Delta+\gamma'}F(u)D \tag{3}$$

where:

$$\gamma' = \gamma(1 + r_s / r_a) \tag{4}$$

In the above relationship, r_s and r_a are the surface and aerodynamic resistance factors, respectively. For water, the surface resistance is zero, thus eq. (3) is identical to the Penman formula. For any other type of surface, the evapotranspiration can be analytically computed, provided that the two resistance parameters are known. In practice, it is extremely difficult to obtain parameter r_s on the basis of typical field data, even more so because this is time-varying. Thus, its evaluation is only possible for some specific cases.

In this context, FAO proposed the application of the Penman–Monteith method for the hypothetical reference crop, thus introducing the concept of reference evapotranspiration. With standardized height for wind speed, temperature and humidity measurements at 2.0 m and the crop height of 0.12 m, the aerodynamic and surface resistances become $r_a = 208 / u_2$ (where u_2 is the wind velocity, in m/s) and $r_s = 70$ s/m. The experts of FAO suggested using the Penman–Monteith method as the standard for reference evapotranspiration and advised on procedures for calculation of the various meteorological inputs and parameters [7].

2.3. Radiation-based approaches

The complexity of the Penman–Monteith method stimulated many researchers to seek for alternative, simplified expressions, based on limited and easily accessible data. Given that the main sources of variability in evapotranspiration are temperature and solar radiation, the two variables are introduced in a number of such models, typically referred as radiation-based. We note that the classification to "radiation-based" instead of the more complete "radiation-temperature-based" is made for highlighting the difference with even simpler approaches that use temperature as single input variable.

A well-known simplification is the Priestley–Taylor formula [17], which is expressed in terms of equivalent depth (mm/d):

$$E = \alpha_e \frac{\Delta}{\Delta + \gamma} \frac{R_n}{\lambda \rho} \tag{5}$$

where a_e is a numerical coefficient, with values from 1.26 to 1.28. Thus, in the above expression, the energy term of the Penman–Monteith equation is increased by about 30%, in order to skip over the aerodynamic term. This assumption allows for omitting the use of wind velocity and surface resistance in evapotranspiration calculations.

Despite its simplifications, the Priestley–Taylor method is still physically-based, as opposed to most of radiation-based approaches that are rather empirical. In addition, many of them use either total solar radiation or even extraterrestrial radiation, instead of net radiation data. For instance, Jensen & Haise [15], based on evapotranspiration measurements from irrigated areas across the United States, proposed the following expression:

$$E = \frac{R_a T_a}{40 \lambda \rho} \tag{6}$$

This equation only uses average daily temperature, T_a, and extraterrestrial radiation, R_a. Later, McGuiness & Bordne [16] suggested a slight modification of the Jensen–Haise expression, known as McGuinness model:

$$E = \frac{R_a (T_a + 5)}{68 \lambda \rho} \tag{7}$$

Another widely used approach is the Hargreaves model [18] that estimates the reference evapotranspiration at the monthly scale by:

$$E = 0.0023 \, (R_a / \lambda) \, (T_a + 17.8)(T_{max} - T_{min})^{0.5} \tag{8}$$

where T_a is the average monthly temperature and $T_{max} - T_{min}$ is the difference between the maximum and minimum monthly temperature (°C). This method provides satisfactory results, with typical error of 10-15% or 1.0 mm/d, and it is suggested when only temperature data is available [24].

Many studies proved the superiority of radiation-based approaches against temperature-based ones. For instance, Lu *et al.* [6] compared three models from each category across SE

United States and showed that radiation-based methods produced clearly better results, with reference to pan evaporation measurements. Oudin *et al.* [25] investigated the suitability of such approaches, as input to rainfall-runoff models. In this context, they evaluated a number of evapotranspiration methods, on the basis of precipitation and streamflow data from a large number of catchments in U.S., France and Australia. Within their research, they provided a generalized expression of the Jensen–Haise and McGuinness models, i.e.

$$E = \frac{R_a (T_a + K_2)}{K_1 \lambda \rho} \tag{9}$$

where K_1 (°C) is a scale parameter and K_2 (°C) is a parameter related to the threshold for air temperature, i.e. the minimum value of air temperature for which evapotranspiration is not zero. The researchers tested several values of K_1 and K_2 for four well-known hydrological models, and kept the combination that gave the best streamflow simulations over the entire catchment sample. After extended analysis, they concluded that the parameters are model-specific. As general recommendation, they proposed the values $K_1 = 100$°C and $K_2 = 5$°C.

2.4. Temperature-based approaches

The most elementary approaches only use temperature data as input, while the radiation term is indirectly accounted for, by means of empirical or semi-empirical expressions. The most recognized method of this category is the Thornthwaite formula [4]:

$$E = 16 \left(\frac{10 T_a}{I}\right)^a \left(\frac{d}{12}\right)\left(\frac{N}{30}\right) \tag{10}$$

where T_a is the average monthly temperature, d is the average number of daylight hours per day for each month, N is the number of days in the month, I is the so-called annual heat index, which is estimated on the basis on monthly temperature values, and a is another parameter, which is function of I. Initially, the method was applied in USA, but later it was also tested in a number of regions worldwide [26]. Today, this method is not recommended, since it results in unacceptably large errors. In dry climates, evapotranspiration is underestimated, while in wet climates it is significantly overestimated [27].

Another widely known temperature-based method is the Blaney & Criddle formula [19]:

$$E = 0.254 p (32 + 1.8 T_a) \tag{11}$$

where p is the mean daily percentage of annual daytime hours. The method estimates the reference crop evapotranspiration on a monthly basis, and has been widely applied in Greece for the assessment of irrigation needs. Yet, the Blaney–Criddle method is only suitable for rough estimations. Especially under "extreme" climatic conditions, the method is inaccurate: in windy, dry, sunny areas, the evapotranspiration is underestimated by up to 60%, while in calm, humid, clouded areas it is overestimated up to 40% [26]. For this reason, several improvements of the method have been proposed. A well-known one is the modified Blaney-Criddle formula by Doorenbos & Pruitt [28], which however require additional meteorological inputs, thus loosing the advantage of data parsimony.

Linacre [20] has also developed simplified formulas to estimate open water evaporation and reference evapotranspiration on the basis of temperature. For the calculation of evaporation, the Linacre formula is:

$$E = \frac{700\,(\mathrm{T} + 0.006z)\,/\,(100\ - \varphi) + 15\,\left(T - T_d\right)}{\left(80\ - T\right)} \tag{12}$$

where z is the elevation, φ is the latitude and T_d is the dew point, which is also approximately derived though air temperature. For the reference evapotranspiration, the constant 700 in numerator is replaced by 500. In Greece, the Linacre method generally seriously overestimates the evapotranspiration, and it is not recommended [33]. Its major drawback is the assumption of a one-to-one relationship between evaporation and radiation, which is not realistic.

We notify that all the above approaches are only valid for monthly or daily scales, which are sufficient for most of engineering applications. For shorter time scales (e.g. hourly), the estimation of reference evapotranspiration requires additional astronomical calculations as well as finely-resolved meteorological data. Empirical formulas for such time scales are also available in the literature, e.g. in [29].

3. Development of parsimonious parametric formulas for evapotranspiration modelling

3.1. The need for a parametric model

A typical shortcoming of most of simplified approaches (i.e. empirical models that use solar radiation and temperature data) for evapotranspiration modeling is their poor predictive capacity against the entire range of climatic regimes and locations. For, these models are built and validated in specific conditions. Thus, while they work well in the areas and over the periods for which they have been developed, they provide inaccurate results when they are applied to other climatic areas, without reevaluating the constants involved in the empirical formulas [30]. Hence, the key idea behind the development of a parametric model is the replacement of some of the constants that are used in such formulas by free parameters, which are regionally-varying. The optimal values of parameters can be obtained through calibration, i.e. by fitting model outputs to "reference" evapotranspiration time series. Nowadays, calibration is a rather straightforward task, thanks to the development of powerful global optimization techniques, such as evolutionary algorithms. The reference data can be provided either directly (e.g.

through remote sensing) or estimated by well-established methods (preferably the Penman-Monteith model). Optimization is carried out against a goodness-of-fit criterion, which aggregates the departures between the simulated and reference data (e.g. coefficient of efficiency).

In this context, a parametric expression can be used as a generalized regression formula, for infilling and extrapolating evapotranspiration records. This is of high importance, since in many conventional meteorological stations, the temperature records usually overlap the records of the rest of variables that are required by the Penman-Monteith method. The use of a parametric model, with parameters calibrated on the basis of reliable evapotranspiration data, is obviously preferable to empirical models, which cannot account for the whole available meteorological information, i.e. the past observations of humidity, solar radiation and wind velocity.

3.2. Ensuring parsimony and consistency

The need for parsimonious models is essential in all aspects of water engineering and management [31]. Generally, the concept of parsimony refers to both the model structure, which should be as simple as possible, and the input data, which should be as fewer as possible and easily available. Yet, as indicated in the Section 2, many of the empirical models for evapotranspiration estimations, although they are data parsimonious, fail to provide realistic results (see also Section 4 below), since they do not account for important aspects of the natural phenomenon. Therefore, before building the model it is crucial to ensure that simplicity will not be in contrast with consistency. This requires: (a) the choice of the most important explanatory variables in the equations, and (b) the determination of the number of parameters, which should be identified through calibration.

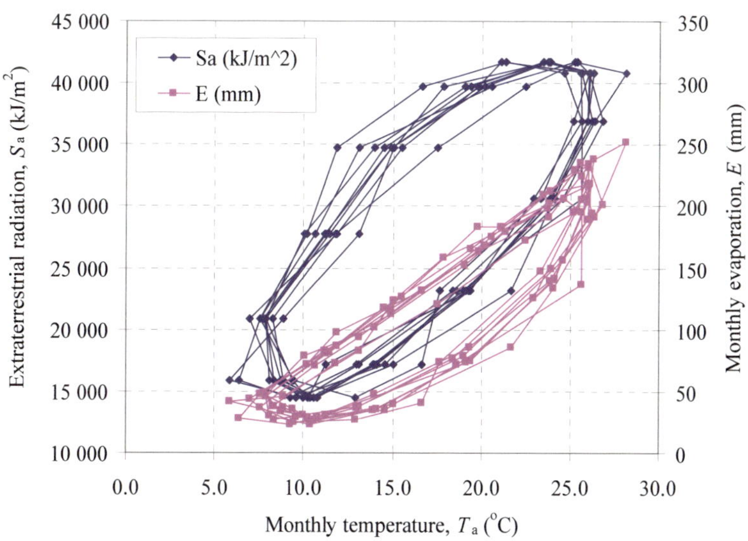

Figure 1. Scatter plots of mean monthly temperature data vs. extraterrestrial solar radiation and Penman evaporation (data from Agrinio station)

It is well-known that the variability of evapotranspiration is mainly explained by the variability of solar radiation and temperature. Yet, instead of measured solar radiation, which is rarely available, and following the practice of most of the empirical methods, we decided to use extraterrestrial radiation, together with temperature, as explanatory variables in the new formula. In order to justify this option, we examined several combinations of variables, using data from a number of meteorological stations in Greece. In the example of Figure 1 are illustrated the scatter plots of mean monthly temperature T_a, against extraterrestrial solar radiation S_a, and against monthly evaporation E, which is estimated through the Penman method. The data are from the Agrinio station, Western Greece, and cover a ten-year period (1979-1988). The graphical representation of E vs. T_a and S_a vs. T_a exhibit characteristic loop shapes, which clearly indicate that there is no one-to-one relationship between evapotranspiration and temperature, due to the influence of thermal inertia. Thus, to ensure consistent estimations, the knowledge of the temperature is not sufficient, but requires extraterrestrial radiation as additional explanatory variable.

Next, in order to determine the essential number of parameters, we took advantage of the associated experience with conceptual hydrological models. For, in lumped rainfall-runoff models, it is argued that no more than five to six parameters are adequate to describe the variability of streamflow [32]. Since the variability of monthly evapotranspiration is clearly less than of runoff, in the proposed model we restricted the maximum number of parameters to three. This number is reasonable, also because three are the missing meteorological variables. Within the case study, we also examine even more parsimonious formulations, using two or even one parameter.

3.3. Model formulation and justification

By dividing both the numerator and the denominator by Δ, the Penman-Monteith equation (now expressed in mm/d) is written in the form:

$$E = \frac{1}{\lambda\rho}\frac{R_n + \gamma\lambda F(u)D}{1 + \gamma' / \Delta} \tag{13}$$

In the above expression, the numerator is the sum of a term related to solar radiation and a term related to the rest of meteorological variables, while the denominator is function of temperature. Koutsoyiannis & Xanthopoulos [33] proposed a parametric simplification of the Penman-Monteith formula, where the numerator is approximated by a linear function of extraterrestrial solar radiation, R_a (kJ/m^2), while the denominator is approximated by a linear descending function of temperature T_a (°C) i.e.:

$$E = \frac{aR_a + b}{1 - cT_a} \tag{14}$$

The formula contains three parameters, i.e. *a* (kg/kJ), *b* (kg/m^2) and *c* (°C^{-1}), to which a physical interpretation can be assigned. Since extraterrestrial solar radiation is the upper bound of net shortwave radiation, the dimensionless term $a^* = a / \lambda\rho$ represents the average percentage of the energy provided by the sun (in terms of R_a) and, after reaching the Earth's terrain, is transformed to latent heat, thus driving the evapotranspiration process. Parameter *b* lumps the missing information associated with aerodynamic processes, driven by the wind and the vapour deficit in the atmosphere. Finally, the expression $1 - c\, T_a$ approximates the more complex $1 + \gamma' / \Delta$. We remind that γ' is function of the surface and aerodynamic resistance (eq. 4) and Δ is the slope vapour pressure curve, which is function of T_a. As shown in Fig. 2, T_a and $-(1 / \Delta)$ are highly correlated, thus justifying the adoption of the linear approximation.

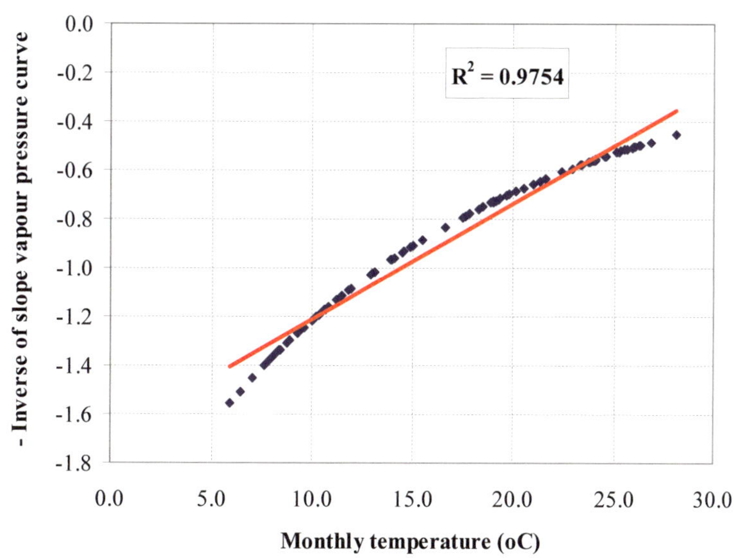

Figure 2. Scatter plot of T_a and $- 1 / \Delta$ (data from Agrinio station)

4. Application of the parametric formula over Greece

4.1. Meteorological data and computational tools

We used monthly meteorological data from 37 stations well-distributed over Greece, run by the National Meteorological Service of Greece. The locations of the stations are shown in Figure 3 and their characteristics are summarized in Table 1. Most of stations are close to the sea. Their latitudes range from 35.0° to 41.5°, while their elevations range from 2 to 663 m. The original data include mean temperature, relative humidity, sunshine percent (i.e. ratio of actual sunshine duration to maximum potential daylight hours per month) and wind velocity records. These records cover the period 1968-1989 with very few missing values, which have not been considered in calculations. At all study locations, the monthly potential evapotranspiration was calculated with the Penman-Monteith formula, assumed as reference model for the evaluation of the examined methodologies. The mean annual potential evapotranspiration values, shown in Table 2, range from 912 mm (Florina station, North Greece, altitude 662 m) to 1628 mm (Ierapetra station, Crete Island, South Greece).

The handling of the time series and the related calculations were carried out using the Hydrognomon software, a powerful tool for the processing and management of hydrometeorological data [34]. The software is open-access and freely available at http://www.hydrognomon.org/.

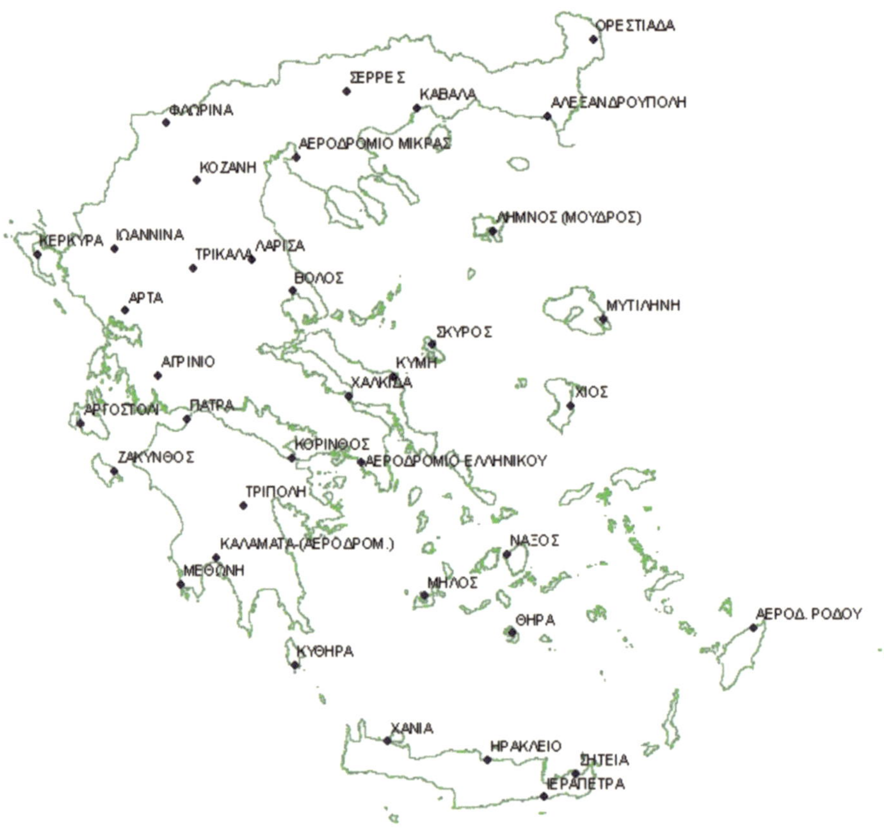

Figure 3. Locations of meteorological stations

No	Station	φ (°)	z (m)	No	Station	φ (°)	z (m)
1	Agrinio	38.37	46	20	Larisa	39.39	74
2	Alexandroupoli	40.51	4	21	Limnos	39.54	17
3	Argostoli	38.11	2	22	Methoni	36.50	34
4	Arta	39.10	38	23	Milos	36.41	183
5	Chalkida	38.28	6	24	Mytiline	39.30	5
6	Chania	35.30	63	25	Naxos	37.60	9
7	Chios	38.20	4	26	Orestiada	41.50	48
8	Florina	40.47	662	27	Patra	38.15	3
9	Hellinico	37.54	15	28	Rhodes	36.24	37
10	Heracleio	35.20	39	29	Serres	41.40	35
11	Ierapetra	35.00	14	30	Siteia	35.12	28
12	Ioannina	39.42	484	31	Skyros	38.54	5
13	Kalamata	37.40	8	32	Thera	36.25	208
14	Kavala	40.54	63	33	Thessaloniki	40.31	5
15	Kerkyra	39.37	2	34	Trikala	39.33	116
16	Korinthos	38.20	15	35	Tripoli	37.32	663
17	Kozani	40.18	626	36	Volos	39.23	7
18	Kymi	38.38	221	37	Zakynthos	37.47	8
19	Kythira	36.10	167				

Table 1. List of meteorological stations (latitude, φ, and elevation, z)

4.2. Model calibration and validation

The entire sample was split into two control periods, 1968-1983 (calibration) and 1984-1989 (validation). At each station, the three parameters of eq. (14) were calibrated against the reference potential evapotranspiration time series. This task was automatically employed via a least square optimization technique, embedded in the evapotranspiration module of Hydrognomon.

The optimized values of a, b and c were next introduced into the parametric model, which ran for the six-year validation period, in order to evaluate its predictive capacity against the Penman-Monteith method (detailed results are provided in [36]).

No	MAPET (mm)>	CE (cal.)	CE (val.)	No	MAPET (mm)	CE (cal.)	CE (val.)
1	1108.4	0.989	0.975	20	1039.0	0.987	0.980
2	1164.0	0.971	0.970	21	1345.5	0.964	0.971
3	1277.9	0.982	0.980	22	1286.4	0.962	0.970
4	1195.0	0.981	0.871	23	1461.9	0.972	0.980
5	1274.5	0.950	0.953	24	1458.6	0.988	0.970
6	1296.0	0.973	0.963	25	1459.0	0.975	0.980
7	1412.9	0.919	0.953	26	1028.4	0.981	0.970
8	911.7	0.967	0.960	27	1205.4	0.987	0.960
9	1476.6	0.983	0.980	28	1551.9	0.972	0.970
10	1521.8	0.980	0.980	29	997.8	0.982	0.970
11	1628.4	0.962	0.940	30	1477.9	0.985	0.990
12	939.5	0.987	0.980	31	1297.9	0.926	0.910
13	1221.0	0.983	0.980	32	1437.1	0.971	0.940
14	962.4	0.983	0.980	33	1157.8	0.983	0.980
15	1089.5	0.989	0.990	34	1118.7	0.973	0.970
16	1483.1	0.957	0.820	35	1133.5	0.942	0.950
17	979.7	0.982	0.980	36	1240.6	0.981	0.910
18	1307.3	0.975	0.910	37	1370.4	0.974	0.980
19	1508.2	0.957	0.970				

Table 2. Mean annual potential evapotranspiration (MAPET) and coefficients of efficiency (CE) for calibration and validation periods

For both calibration and validation, we used as performance measure the coefficient of efficiency, given by:

$$CE = 1 - \frac{\sum_{t=1}^{N}(E_{PM}(t) - E(t))^2}{\sum_{t=1}^{N}(E_{PM}(t) - E_{mean})^2} \tag{15}$$

where $E_{PM}(t)$ and $E(t)$ are the potential evapotranspiration values of month t, computed by the Penman-Monteith method and the parametric relationship, respectively, E_{mean} is the monthly average over the common data period, estimated by the Penman-Monteith formula, and N is the sample size.

The coefficients of efficiency for all stations, during the two control periods are given in Table 2. For all stations the model fitting is very satisfactory, since the CE values are very high. Specifically, the average CE values of the 37 stations are 97.2% during calibration and 95.9% during validation, while their minimum values are 91.9 and 82.0%, respectively.

4.3. Comparison with other empirical methods

At all stations, we compared the performance of the parametric method (in terms of coefficients of efficiency) with two radiation-based approaches, i.e. the McGuiness model (eq. 7), and the generalized eq. (9), proposed by Oudin *et al.* [25]. In the latter, we applied the recommended values K_1 = 100°C and K_2 = 5°C. The distribution of the CE values for the two control periods are summarized in Table 3. The suitability of the parametric model is obvious, while the other two methods exhibit much less satisfactory performance. Moreover, the improvement of the empirical expression by Oudin *et al.* against the classical McGuiness model is rather marginal.

	Parametric		**McGuiness**		**Oudin *et al.***	
CE	Cal.	Val.	Cal.	Val.	Cal.	Val.
>0.95	37	30	0	2	5	2
0.90-0.95	0	6	8	9	5	9
0.70-0.90	0	1	12	19	12	15
0.50-0.70	0	0	15	6	12	7
<0.50	0	0	2	1	3	4

Table 3. Distribution of CE values of radiation-based approaches

We also compared the performance of the new parametric approach against the most widely used methods in Greece, i.e. Thornthwaite, Blaney-Criddle, and Hargreaves. We remind that the first two are temperature-based, while the latter also uses extraterrestrial radiation and monthly average minimum and maximum temperature as inputs. The calculations of monthly potential evapotranspiration, using the Hydrognomon software, were made for ten representative stations, with different climatic characteristics. For each station Table 4 gives the coefficient of efficiency and the bias, i.e. the relative difference of the monthly average against the average of the Penman-Monteith method (both refer to the entire control period 1968-1988). The proposed formula is substantially more accurate, while all other commonly used approaches do not provide satisfactory results across all locations. The suitability of each method depends on local conditions. Thus, at each location, one method may be satisfactorily accurate, while another method may result to unacceptably large errors. In general, the Thornthwaite formula underestimates potential evapotranspiration by 30%, while the Blaney-Criddle method overestimates it by 30%. In stations that are far from the sea (Florina, Larisa), the deviation exceeds 50%. The Hargreaves method, although it uses additional inputs, totally fails to predict the potential evapotranspiration at some stations, resulting even in negative values of CE. On the other hand, the parametric method is impressively accurate, as indicated by both the CE values and the practically zero bias.

Station	Thornthwaite		Blaney-Criddle		Hargreaves		Parametric	
	CE	Bias	CE	Bias	CE	Bias	CE	Bias
Alexandroupoli	0.747	-0.287	0.645	0.321	0.935	-0.055	0.973	-0.006
Chios	0.440	-0.384	0.725	0.153	0.568	-0.274	0.898	0.006
Florina	0.883	-0.200	0.339	0.557	0.842	0.200	0.966	-0.002
Heracleio	0.068	-0.402	0.807	0.113	0.170	-0.341	0.981	-0.001
Kythira	0.094	-0.421	0.815	0.087	-0.114	-0.446	0.961	0.011
Larisa	0.920	-0.167	0.413	0.504	0.843	0.227	0.986	-0.004
Milos	0.180	-0.406	0.806	0.116	0.105	-0.398	0.974	-0.009
Patra	0.686	-0.258	0.381	0.379	0.942	0.046	0.987	0.009
Thessaloniki	0.816	-0.250	0.614	0.363	0.937	0.003	0.983	0.002
Tripoli	0.711	-0.303	0.651	0.325	0.957	0.088	0.948	-0.008
Average	0.555	-0.308	0.620	0.292	0.619	-0.095	0.966	0.000

Table 4. Performance characteristics (efficiency, bias) of various evapotranspiration models at representative meteorological stations

4.4. Investigation of alternative parameterizations

In order to provide a further parsimonious expression, we investigated two alternative parameterizations by simplifying eq. (14). In the first approach, we omitted parameter *b*, thus using a two-parameter relationship of the form:

$$E = \frac{aR_a}{1 - cT_a} \tag{16}$$

The two parameters of eq. (16) were calibrated using as reference data the Penman-Monteith estimations. Next, we provided a single-parameter expression, in which we applied the spatially average value of *c* (= 0.00234), i.e.:

$$E = \frac{aR_a}{1 - 0.00234T_a} \tag{17}$$

The two models (16) and (17) were evaluated on the basis of CE, for the entire control period (results not shown). Compared to the full expression (14), the reduction of CE is negligible when parameter *b* is omitted, while the use of the simplest form (17) occasionally leads to a considerably reduced predictive capacity (in particular, in two out of 37 stations). This is because the spatial variability of *c* is very low (<10%), thus allowing for substituting the parameter by a constant, in terms of the average value of all stations. The statistical characteristics of the model parameters (average, standard deviation) for the three alternative expressions are given in Table 5.

Parameter	Equation (14)		Equation (16)		Equation (17)	
	Average	St. dev.	Average	St. dev.	Average	St. dev.
a (10^{-5} kg/kJ)	5.486	1.487	6.362	0.913	6.257	0.735
b (kg/m^2)	0.1973	0.4330	–	–	–	–
c (°C^{-1})	0.0241	0.0024	0.0234	0.0020	–	–

Table 5. Statistical characteristics of model parameters, for the three alternative expressions (for the 37 study locations)

4.5. Spatial variability of model parameters

For the simplified model (17), we investigated the spatial variability of its single parameter *a* (kg/kJ) over the Greek territory [35, 36]. Initially, we examined whether this parameter is correlated with two characteristic properties of the meteorological stations, i.e. latitude and elevation (Fig. 4). From the two scatter plots it is detected that parameter *a* is not correlated with elevation, while there is a notable reduction of the parameter value with the increase of latitude.

Next, we visualized the variation of the parameter over the Greek territory, using spatial integration procedures that are embedded in the ESRI ArcGIS toolbox. After preliminary investigations, we selected the Inverse Distance Weighting (IDW) method to interpolate between the known point values at the 37 meteorological stations, using a cell dimension of 0.5 km. The IDW method estimates the parameter values at the grid scale as the weighted sum of the point values, where the weights are decreasing functions of the distance between the centroid of each cell from the corresponding station. Since the parameter variability is not correlated with elevation, it was not necessary to employ more complex integration methods, such as co-kriging, which also account for the influence of altitude.

The mapping of parameter *a* over Greece through the IDW approach is illustrated in Fig. 5. We detect that the parameter values increase from SE to NW Greece. The physical interpretation of this systematic pattern is the increase of both the sunshine duration and the wind velocity as we move from the continental to insular Greece. A similar pattern appears for parameter *a* when accounting for the two-parameter expression (eq. 16). On the other hand, as shown in Fig. 6, the spatial distribution of *c* is irregular, which makes impossible to assign a physical interpretation. Obviously, this parameter is site-specific.

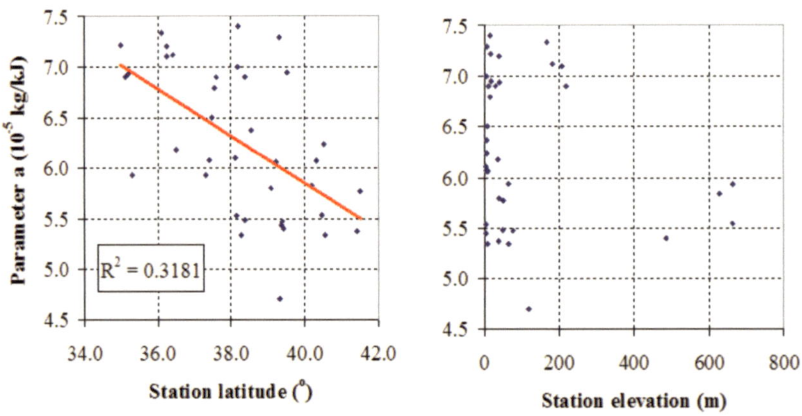

Figure 4. Scatter plots of parameter *a* (eq. 17) vs. latitude and elevation

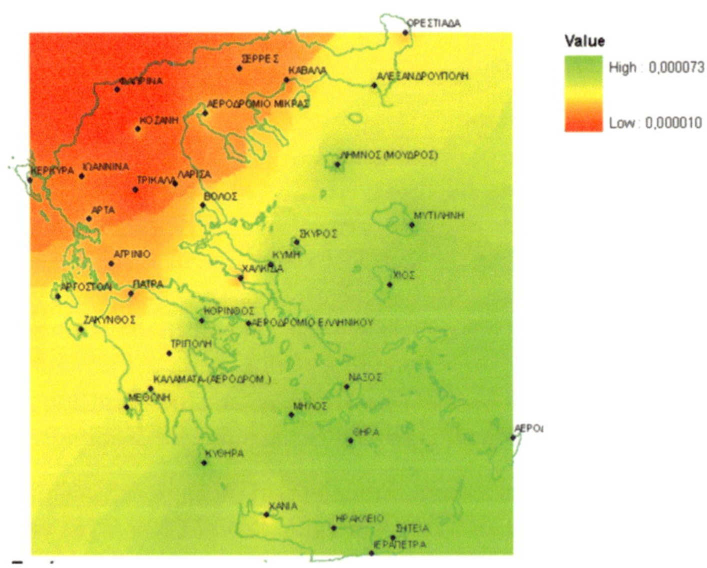

Figure 5. Geographical distribution of parameter *a* (eq. 17) over Greece

http://dx.doi.org/10.5772/52927

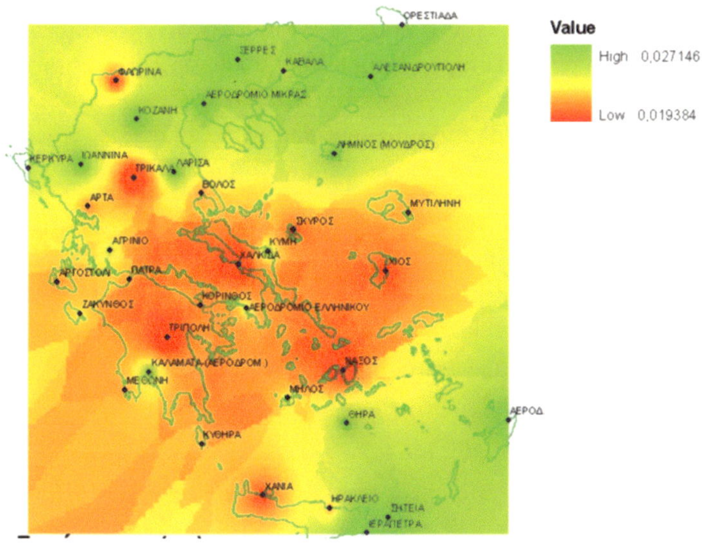

Figure 6. Geographical distribution of parameter *c* (eq. 16) over Greece

5. Synopsis and discussion

A parametric, radiation-based model for the estimation of potential evapotranspiration has been developed, which is parsimonious both in terms of data requirements and number of parameters. Its mathematical expression originates from consecutive simplifications of the Penman-Monteith formula, thus being physically-consistent. In its full form, the model requires three parameters to be calibrated. Yet, simpler formulations, with either one or two parameters, were also examined, without significant degradation of the model performance.

The model parameters were optimized on the basis of monthly evapotranspiration data, estimated with the Penman-Monteith method at 37 meteorological stations in Greece, using historical meteorological data for a 20-year period (1968-1988). The model exhibits excellent fitting to all locations. Its appropriateness is further revealed through extensive comparisons with other empirical approaches, i.e. two radiation-based methods and three temperature-based ones. In particular, common temperature-based methods that have been widely used in Greece provide rather disappointing results, while the new parametric model retains a systematically high predictive capacity across all study locations. This is the great advantage of parametric approaches against empirical ones, since calibration allows the coefficients that are involved in the mathematical formulas to be fitted to local climatic conditions.

The key model parameter was spatially interpolated throughout Greek territory, using a typical method in a GIS environment. The geographical distribution of the parameter exhibits a systematic increase from SE to NW, which is explained by the increase of sunshine duration and wind velocity as we move from the continental to insular Greece. Similar patterns were not found for the rest of model parameters. The practical value of this analysis is important, since it allows for establishing a specific formula for the estimation of potential evapotranspiration at any point. Thus, a minimalistic model of high accuracy is now available everywhere in Greece, which requires monthly temperature data as the only meteorological input. The other inputs are the monthly extraterrestrial radiation, which is function of latitude and time of the year, and a map of the parameter distribution.

To improve the model, it is essential to implement additional investigations, using a denser network of meteorological stations (especially stations in mountainous areas). In particular, it is necessary to examine the sensitivity of the various hydrometeorological variables that are involved in the evapotranspiration process, and provide alternative formulations for various combinations of missing meteorological data.

Author details

Aristoteles Tegos, Andreas Efstratiadis and Demetris Koutsoyiannis

Department of Water Recourses & Environmental Engineering, School of Civil Engineering, National Technical University of Athens, Zographou, Greece

References

[1] Seckler D, Amarasinghe U, Molden D, de Silva R, Barker R. World Water Demand and Supply, 1990-2025: Scenarios and Issues. Research Report 19. International Water Management Institute, Colombo, Sri Lanka; 1998.

[2] Tsouni A, Contoes C, Koutsoyiannis D, Elias P, Mamassis N. Estimation of actual evapotranspiration by remote sensing: Application in Thessaly Plain, Greece. Sensors 2008; 8(6): 3586–3600.

[3] Lhomme JP. Towards a rational definition of potential evaporation. Hydrology and Earth System Sciences 1997; 1: 257-264.

[4] Thornthwaite CW. An approach towards a rational classification of climate. Geographical Review 1948; 38: 55-94.

[5] Brutsaert W. Evaporation into the Atmosphere: Theory, History, and Applications, Springer; 1982.

[6] Lu J, Sun G, McNulty SG, Amatya DM. A comparison of six potential comparison evapotranspiration methods for regional use in the South-eastern United States. Journal of the American Water Resources Association 2005; 41: 621–633.

[7] Allen RG, Pereira LS, Raes D, Smith M. Crop evapotranspiration: Guidelines for computing crop water requirements. FAO Irrigation and Drainage Paper 56, Rome; 1998.

[8] Fisher JB, Whittaker RJ, Malhi Y. ET come home: Potential evapotranspiration in geographical ecology. Global Ecology and Biogeography 2011; 20(1): 1–18.

[9] Penman HL. Natural evaporation from open water, bare soil and grass. In: Proceedings of the Royal Society of London 1948; 193: 120–145.

[10] Monteith JL. Evaporation and the environment: The state and movement of water in living organism. XIXth Symposium. Cambridge University Press, Swansea; 1965.

[11] Allen RG, Jensen ME, Wright JL, Burman RD. Operational estimates of reference evapotranspiration. Agronomy Journal 1989; 81: 650–662.

[12] Jensen ME, Burman RD, Allen RG. Evapotranspiration and irrigation water requirement. ASCE Manuals and Reports on Engineering Practice, 70; 1990.

[13] Mecham BQ. Scheduling turfgrass irrigation by various ET equations. Evapotranspiration and irrigation scheduling. In: Proceedings of the Irrigation International Conference. 3–6 November, San Antonio; 1996.

[14] Sentelhas PC, Gillespie TJ, Santos EA. Evaluation of FAO Penman-Monteith and alternative methods for estimating reference evapotranspiration with missing data in Southern Ontario, Canada. Agricultural Water Management 2010; 97(5): 635–644.

[15] Jensen M, Haise H. Estimating evapotranspiration from solar radiation. Journal of the Irrigation and Drainage Division. Proceedings of the ASCE 1963; 89(1R4): 15–41.

[16] McGuinness JL, Bordne EF. A comparison of lysimeter-derived potential evapotranspiration with computed values. Technical Bulletin 1452. Agricultural Research Service, US Department of Agriculture, Washington, DC; 1972.

[17] Priestley CHB, Taylor RJ. On the assessment of surface heat fluxes and evaporation using large-scale parameters. Monthly Weather Review 1972; 100: 81–92.

[18] Hargreaves GH, Samani ZA. Reference crop evaporation from temperature. Applied Engineering in Agriculture 1985; 1(2) 96–99.

[19] Blaney HF, Criddle WD. Determining water requirements in irrigated areas from climatological irrigation data. Technical Paper No. 96. US Department of Agriculture, Soil Conservation Service, Washington, DC; 1950.

[20] Linacre ET. A simple formula for estimating evaporation rates in various climates, using temperature data alone. Agricultural and Forest Meteorology 1977; 18: 409–424.

[21] Xu CY, Singh VP. Evaluation and generalization of radiation-based methods for calculating evaporation. Hydrological Processes 2000; 14: 339–349.

[22] Xu CY, Singh VP. Evaluation and generalization of temperature-based methods for calculating evaporation. Hydrological Processes 2001; 15: 305–319.

[23] Mamassis N, Efstratiadis A, Apostolidou E. Topography-adjusted solar radiation indices and their importance in hydrology. Hydrological Sciences Journal 2012; 57(4): 756–775.

[24] Shuttleworth WJ. Evaporation. In: Maidment DR. (ed.) Handbook of Hydrology, McGraw-Hill, New York; 1993: Ch. 4.

[25] Oudin L, Hervieu F, Michel C, Perrin C, Andréassian V, Anctil F, Loumagne C. Which potential evapotranspiration input for a lumped rainfall–runoff model? Part 2 —Towards a simple and efficient potential evapotranspiration model for rainfall–runoff modelling. Journal of Hydrology 2004; 303(1–4): 290–306.

[26] Ward RC, Robinson M. Principles of Hydrology. 3rd edition, McGraw Hill, London; 1989.

[27] Shaw EM. Hydrology in Practice. 3rd edition, Chapman and Hall, London; 1994.

[28] Doorenbos J, Pruit WO. Crop water requirements. Irrigation and Drainage Paper No 24, FAO, Rome; 1977.

[29] Alexandris S, Kerkides P. New empirical formula for hourly estimations of reference evapotranspiration. Agricultural Water Management 2003; 60(3): 157–180.

[30] Hounam CE. Problems of Evaporation Assessment in the Water Balance. Report on WMO/IHP Projects No. 13, World Health Organization, Geneva; 1971.

[31] Koutsoyiannis D. Seeking parsimony in hydrology and water resources technology. EGU General Assembly 2009, Geophysical Research Abstracts, Vol. 11, Vienna, 11469, European Geosciences Union; 2009 (http://itia.ntua.gr/en/docinfo/906/).

[32] Wagener T, Boyle DP, Lees MJ, Wheater HS, Gupta HV, Sorooshian S. A framework for development and application of hydrological models. Hydrology and Earth System Sciences 2001; 5(1): 13–26.

[33] Koutsoyiannis D, Xanthopoulos T. Engineering Hydrology, 3rd ed., National Technical University of Athens, Athens; 1999.

[34] Kozanis S, Christofides A, Mamassis N, Efstratiadis A, Koutsoyiannis D. Hydrognomon – open source software for the analysis of hydrological data, EGU General Assembly 2010, Geophysical Research Abstracts, Vol. 12, Vienna, 12419, European Geosciences Union; 2010 (http://itia.ntua.gr/en/docinfo/962/).

[35] Tegos A. Simplification of evapotranspiration estimation in Greece. Postgraduate Thesis. Department of Water Resources and Environmental Engineering – National Technical University of Athens, Athens; 2007 (in Greek; http://itia.ntua.gr/en/docinfo/820/).

[36] Tegos A, Mamassis N, Koutsoyiannis D. Estimation of potential evapotranspiration with minimal data dependence. EGU General Assembly 2009, Geophysical Research Abstracts, Vol. 11, Vienna, 1937, European Geosciences Union; 2009.

Chapter 9

Data Driven Techniques and Wavelet Analysis for the Modeling and Analysis of Actual Evapotranspiration

Zohreh Izadifar and Amin Elshorbagy

Additional information is available at the end of the chapter

http://dx.doi.org/10.5772/52809

1. Introduction

Evapotranspiration (ET) is one of the important components of the hydrological cycle, which its modeling and analysis is vital for better understanding of watersheds hydrology and efficient water resource designs and managements. Evapotranspiration (ET) is a combined term including the transport of water to the atmosphere in the form of evaporation from the soil surfaces and from the plant tissues as a result of transpiration. Evapotranspiration is considered as a major cause for water loss around the world (Dingman, 2002).

ET is basically a complex and not fully understood mechanism, which varies over temporal and spatial scales. ET can be conceptually expressed either in the form of potential or actual evapotranspiration. Potential evapotranspiration (PET) describes the maximum loss of water under specific climatic conditions when unlimited water is available. The actual evapotranspiration (AET) is the rate at which water is actually removed to the atmosphere from a surface due to the evapotranspiration process. The influence of soil moisture on the AET has made its physical modeling more complicated than the PET. Complexity of AET has also imposed some limitations on the previously developed estimation models. Although the AET is the preferred form of ET in the hydrological analysis, vast majority of the previous studies have investigated the modeling of PET. As a result, there is a vital need for modeling and analysis of AET mechanism. Complexity of the AET physics, limitations of the currently available AET estimation approaches, such as requirement of extensive information and reasonable estimation of models parameters has led to the investigation of some techniques/tools that can model/analyze such complicated mechanism without having a complete understanding of it.

Data driven techniques can provide a model to predict and investigate the process without having a complete understanding of it. Inductive modeling approach is also interesting because of its knowledge discovery property. Using data driven models, one can extract useful implicit information from a large collection of data and improve the understanding of the investigated process. Machine learning (ML) techniques are modern data driven modeling methods that originated from the advances in computer technologies and mathematical algorithms. These techniques are usually employed for characterizing complicated systems, which cannot be easily understood, analyzed, and modeled. Artificial neural networks (ANNs) and genetic programming (GP) are two robust ML techniques, which apply artificial intelligence for the modeling of complex systems. ANNs are computational models that can be used for the modeling of complex relationships by simulating the functional aspects of biological neural networks. GP is an evolutionary-based technique inspired by the biological evolution to generate computer programs (e.g. models) for solving a user-defined problem.

Both ANNs and GP technique has been examined for the modeling of ET. Kumar et al. (2002) developed an ANN model for the prediction of reference evapotranspiration (ET_0) and compared its performance with that of a conventional method (Penman-Monteith equation) to examine the capabilities of ANNs in ET_0 prediction compared to the PM method. The results of the study showed that the ANN model can predict ET_0 better than the conventional method for the considered local case study. The utility of ANNs for the estimation of reference and crop evapotranspiration (ET_c) of wheat crop was examined by Bhakar et al. (2006) and it was revealed that the ANN model was suitable for the prediction of ET_0 and ET_c. Zanetti et al. (2007) found that by using ANNs, it was possible to estimate ET_0 just as a function of maximum and minimum air temperature. The results of a study conducted by Jain et al. (2008) indicated that ANNs can efficiently estimate ET_0 from the limited meteorological variables of temperature and radiation only. Landeras et al. (2008) developed seven ANNs with different input combinations and then compared ANNs to locally calibrated empirical and semi-empirical equations of ET_0. Their proposed ANNs performed better than the locally calibrated equations particularly in situations where appropriate meteorological inputs were lacking. Dai et al. (2009) investigated the predictive ability of ANNs for the prediction of ET_0 in arid, semi-arid, and sub-humid areas of Mongolia, China, and conducted a comparison between the estimated ET_0 values from ANNs and MLR. The results showed that regional ET_0 can be satisfactorily estimated using ANN models and conventional meteorological variables.

In the majority of the conducted studies, researchers have focused on the modeling of potential and reference crop evapotranspiration but not actual evapotranspiration (AET). To the knowledge of the authors, the only publications reporting the application of ANNs for the modeling of AET include the studies conducted by Sudheer et al. (2003) and Parasuraman et al. (2006; 2007). Sudheer et al. (2003) estimated the lysimeter-measured AET of rice crop using RBF-ANNs. The results demonstrated that ANNs can successfully estimate the AET. Parasuraman et al. (2006) developed spiking modular neural networks (SMNNs) for modeling the dynamics of EC-measured hourly latent heat flux. The results demonstrated that although the SMNNs are computationally intensive, they can perform better than regular feed forward neural networks (FFNNs) in modeling evaporation flux. Parasuraman et al. (2007) developed a regular three-layered FFNN model for the estimation of EC-measured hourly AET as a function of net radiation, ground temperature, air temperature, wind speed, and relative humidity. Their results indicated that the ANN model performed better than the currently used PM method in northern Alberta, Canada.

Among the various published studies on the application of GP in hydrological modeling, only a few studies examined the applicability and robustness of GP for modeling of the evapotranspiration process. To the best knowledge of the authors, the only publications that investigated the application of GP for modeling the evapotranspiration mechanism are the studies conducted by Parasuraman et al. (2007), Parasuraman and Elshorbagy (2008), and El-Baroudy et al. (2009). Parasuraman et al. (2007) employed equation-based GP for modeling the hourly actual evapotranspiration process as a function of net radiation, ground temperature, air temperature, wind speed, and relative humidity. The performance of the evolved GP model was compared with that of ANN model and the traditional Penman Monteith (PM) method. It was noted that GP and ANN models had comparable performances and both predicted AET values with better closeness to the measured AET than the PM method. Their analysis also indicated that the effect of net radiation and ground temperature on the AET dominated over other variables. Parasuraman and Elshorbagy (2008) investigated a GP-based modeling framework for quantifying and analyzing the model structure uncertainty on an AET case study. The results of the study demonstrated the capability of the ensemble-based GP in quantifying the uncertainty associated with the hourly AET model structure. El-Baroudy et al. (2009) did not develop a new GP model for AET, but rather developed models using a technique called evolutionary polynomial regression (EPR), and then compared its performance to the ANN and GP models developed by Parasuraman et al. (2007). With the exception of Parasuraman et al. (2007), Parasuraman and Elshorbagy (2008), and El-Baroudy et al. (2009), no other publication was observed that reports an explicit equation for the prediction of AET.

Understanding of the not fully understood mechanism of AET as well as its correlation with the interacting meteorological variables can be improved by exploiting the available time series data and some data mining tools. New digital signal processing tool, namely wavelet analysis (WA), has a robust property for providing multiresolution representation of hydrological time series. Representation of the time series data into time and scale domains makes it possible to extract useful information about temporal cyclic events existing in the underlying signal. In addition, the correlation structure of time series data, in terms of temporal cyclic variations, can be investigated using extensions of wavelet analysis such as cross wavelet analysis. In the field of hydrology, wavelet has been increasingly used for the analysis of spatial-temporal variability of hydrological processes and systems as well as their interactions with climatic variations. WA has been frequently applied for feature extraction of discharge time series data (Saco and Kumar, 2000; Kirkup et al., 2001; Cahill, 2002; Lafreniere and Sharp, 2003; Coulibaly and Burn, 2004; Labat, 2006; Schaefli et al., 2007; Labat, 2008), and characterization of temporal variability of rainfall (Gupta and Waymire, 1990; Kumar and Foufoula-Georgiou, 1993a, b; Kirkup et al., 2001; Coulibaly, 2006; Westra and Sharma, 2006, Miao et al., 2007; Chen and Liu, 2008). In the above-mentioned studies, the utility of WA was mainly employed for detecting and analyzing different periodic events existing in the time series and correlated meteorological signals. Coulibaly and Burn (2004) investigated the temporal and spatial variability of

Canadian streamflows. The results exhibited different period bands of significant activities in the streamflow time series, which were found to be correlated to the considered climatic patterns at some spatial locations. Coulibaly (2006) employed wavelet and cross wavelet analysis to investigate both spatial and temporal variability in seasonal precipitation and its relationship with climatic modes in the Northern Hemisphere. The results revealed striking climatic-related cyclic features in the precipitation time series and, in the temporal-spatial variability of the relationship between precipitation and climate throughout Canada. There are very limited case studies in the literature that investigated the application and capability of WA in analyzing the variability of the evapotranspiration process. Kaheil et al. (2008) used discrete wavelet transform (DWT) for decomposing and reconstructing processes involving the AET phenomenon at various spatial scales, and to find the relationship between the inputs and outputs using support vector machines technique. To the best knowledge of the authors, no effort has been made, in the literature, which benefited from the capability of WA in the temporal scaling of AET variations. Time-scale analysis of the AET signal seems to be an effective approach in improving the understanding of the AET process as well as the efficiency and predictive ability of AET prediction models. Temporal variations of AET and meteorological variables, as well as their correlations, can be examined using wavelet analysis. Wavelet-provided information can improve the understanding of AET temporal variations, its relationship with influential meteorological variables, and hopefully improve the modeling of the AET mechanism.

This chapter presents the ANNs and GP modeling of AET, and the WA of the AET and meteorological signals. The rest of this chapter is organized as follow. In sections 2 and 3, the three techniques of ANNs, GP, and WA are described. Section 4 presents the application of the ANNs, GP, and WA techniques for the modeling and analysis of AET in a specific case study. The hourly eddy covariance (EC)-measured AET is modeled as a function of five meteorological variables; net radiation (R_n), ground temperature (T_g), air temperature (T_a), relative humidity (RH), and wind speed (W_s), using the ANNs and GP techniques, and their performances are compared. The advantage of the investigated data driven models for revealing some information about the AET function and its most influential variables are also examined. Temporal variability of the AET and associated contribution of the meteorological variables is also examined using the wavelet analysis as an approach to modeling input determination. Conclusions of the results and analysis and possible future research are provided in section 5.

2. Data driven modelling

2.1. Artificial neural network

Artificial Neural Networks (ANNs) (Swingler, 1996) are massive networks of parallel information processing systems resembling (simulating) the human brain's analytical function, and they have an inherent ability to learn and recognize highly nonlinear and complex relationships by experience. ANNs learn from empirical examples, which make them a non-rule-based technique, like statistical methods (Maier and Dandy, 2000). Each neuron (information-processing unit) in ANNs consists of input connection links, a central processing unit, and output connection links (Fig.1a). Input signals are received through the connection links from the outside environment or other neurons. Each connection link is assigned a synaptic weight (w) representing the strength of the connection between two nodes in characterizing input-output relationship (ASCE, 2000). Received information is processed in the central processing unit (neuron body), by adding up the weighted inputs and bias (Eq. 1), and passed through the activation function (Eq. 2). Bias (b) is the threshold value, which must be exceeded before the node (neuron) can be activated (ASCE, 2000). Activation function forms the output of the node and enables the nonlinear transformation of inputs to outputs. The log-sigmoid activation function is one of the two most commonly used activation functions in the literature because it is continuous, relatively easy to compute, its derivatives are simple (during the training process), maps the outputs away from extremes, and provide nonlinear response (ASCE, 2000).

$$t = \sum_{i=1}^{n} w_i x_i + b \tag{1}$$

$$f(t) = \frac{1}{1 + e^{-t}} \tag{2}$$

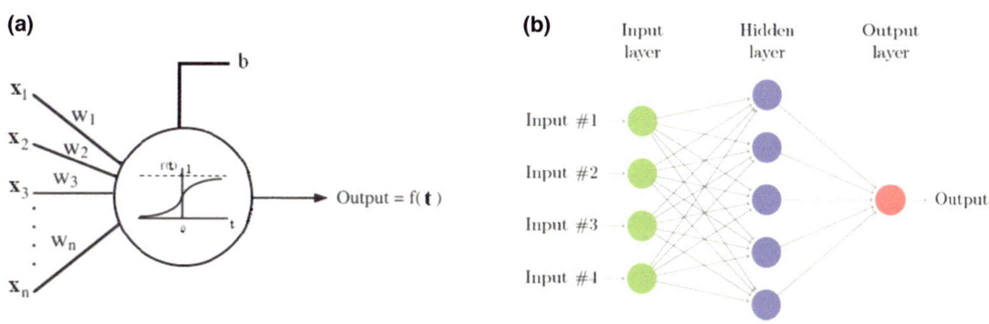

Figure 1. a) Schematic diagram of an artificial neuron, (b) Simple configuration of three-layer feed forward ANN (from Fauske, 2006).

One of the popular types of ANNs, in water resource problems, is the feed forward neural networks (FFNNs) in which the neurons are arranged in layers; input layer, one or more hidden layers, and output layer. The information in FFNNs flows and is processed in one direction from input layer, through hidden layer(s), to the output layer (Fig.1b). Each of the neurons in the hidden layer receives the input signals from the input layer. Received information is processed individually in each of the hidden layer neurons and the outputs are passed to the output layer neuron(s) to release the final response of the network. A simple configuration of three-layer feed forward ANNs is shown in Fig.1b.

It was observed in the literature that a single hidden layer has been usually sufficient for the approximation of conventional hydrological processes (Maier and Dandy, 2000), and it was noted also, in particular, for the process of evapotranspiration (Kumar et al., 2002; Parasuraman et al., 2007). The number of hidden layers and hidden neurons is specified, based on the complexity of the problem, using different methods (usually trial-and-error procedure (ASCE, 2000)). ANNs with single hidden sigmoid layer and linear output layer are the most popular network architectures in the field of water resources (Cybenko, 1989; Hornik et al., 1989).ANNs learn the pattern of the investigated process by adjusting the connection weights and bias values using the provided examples of input-output relationship (namely, training samples). A training algorithm is employed to optimize the weight matrices and bias vectors, which minimize the value of a predetermined error function. Minimum error function results in an ANN model that can generate the most similar output vector to the target vector.

Back propagation algorithm is the most common type of training algorithm in the FFNNs in water related problems (Maier and Dandy, 2000). The network starts with random weight and bias values and generates the output of the network using the given input data; this step is called the forward step (ASCE, 2000). The network output is compared with the desired target output, and the associated error value is computed. The error is propagated backward through the network and the connection weights are adjusted accordingly. The forward and backward steps, together called an epoch, are implemented repeatedly for several times until the error function reaches its minimum value and the optimum weight and bias values are achieved.

One of the problems that threaten the learning process is over-fitting. It usually occurs when the network has memorized the training examples, but it has not learned to generalize to new situations. Various techniques can be employed to avoid over training and improve network generalization ability such as; regularization and early stopping (Neural Network Toolbox User's Guide, 2009). Regularization attempts to smooth the network response by keeping the size of the network weights adequately small (MacKay, 1992) using the modified form of the error function, which considers network weights and biases (Neural Network Toolbox User's Guide, 2009). Through early stopping approach, an independent test set, namely cross-validation, can be used to monitor the performance of the model on a set of not-yet-encountered examples at some stages of the training process. Training is stopped when error on the cross-validation dataset begins to rise to prevent the model from being over-trained (Neural Network Toolbox User's Guide, 2009). Levenberg-Marquardt (Levenberg, 1944; Marquardt, 1963) and Bayesian-regularization (MacKay, 1992) are two of the common training algorithms in ANNs. Levenberg-Marquardt is one of the high-performance algorithms that appear to be the fastest method for training moderate-sized FFNNs (Neural Network Toolbox User's Guide, 2009). Bayesian-regularization algorithm is an automated regularization algorithm. This algorithm also keeps the network size as small as possible (Neural Network Toolbox User's Guide, 2009).

2.2. Genetic programming (GP)

The origins of evolutionary computation traced back to the late 1950's (Box, 1957; Friedberg, 1958; Friedberg et al., 1959; Bremermann, 1962) when it was proposed for the first time. Genetic programming (GP) was first recognized as a different and new development in the world of evolutionary algorithms in the seminal monograph of Genetic Programming by Koza (1992).

http://dx.doi.org/10.5772/52809

Genetic algorithms (GA) belong to the family of evolutionary algorithms, and are generally considered as an optimization method for searching global optimum of a function using natural genetic operators. Genetic programming (GP), which was introduced by Koza (1992), is an extension of GA for inducing computer programs, as solutions for problems at hand, using an intelligent and adaptive search. This type of search uses the information gained from the performance (fitness) of individual computer programs, in the search space, for modifying and improving the current programs. Depending on the particular problem, computer programs of the GP search space may be different, e.g. Boolean-valued models and symbolic mathematical models (Koza, 1992). Symbolic regression GP evolves computer programs in the form of mathematical expressions in which both functional form and numerical coefficients of the regression symbolic model are optimized through the evolutionary process of GP. This application of GP can be adopted for obtaining explicit mathematical AET models.

In the first step of GP implementation, a population of computer programs is randomly generated. This initial population is called first generation. Symbolic regression models are represented by structured parse trees, which are composed of functional and terminal sets appropriate to the problem. A functional set can be a set of mathematical arithmetic operators such as {+, -, *, /}, mathematical functions, Boolean and conditional operators, and any other user-defined functions where the number of arguments of each function is specified. The terminal set, which is associated with the nodes that terminate a branch of a tree in tree-based GP (Banzhaf et al., 1998), is defined as independent variables; i.e. the terminal set $z=\{x,y\}$ where x and y are independent variables (Sette and Boullart, 2001).

GP begins to search in the search space of randomly generated models of initial generation. The fitness measure is used to evaluate how well each individual in the population performs. Fitness is usually measured by the errors produced by individual models. Each model in the population is run using a number of provided data instances (training dataset) to measure the performance of each individual over a variety of representative different situations (Koza, 1992). A scalar fitness value is assigned to each individual using the defined fitness evaluation function. Base on the assigned fitness values, some individuals in the population perform better than others with smaller error values, which means that they have higher chance to be selected for the next step of GP.

In the second step, genetic operators are used to create the next generation. Individuals with better performance are allowed to survive and be reproduced in the next population, called mating pool. In the mating pool, two other operations are performed on the reproduced individuals, namely crossover and mutation. Crossover acts on specific percentage of the mating pool population, *crossover probability* (P_c), and results in the creation of new individuals in the population. Crossover exploits two individuals (parents), selected based on their fitness, and splits each parent at the crossover point into two fragments (sub trees), which are swapped between the parents to create two new offspring (Fig. 2). The offspring (new models) are improved individuals, compared to their parents, which carry some genetic properties from each of them.

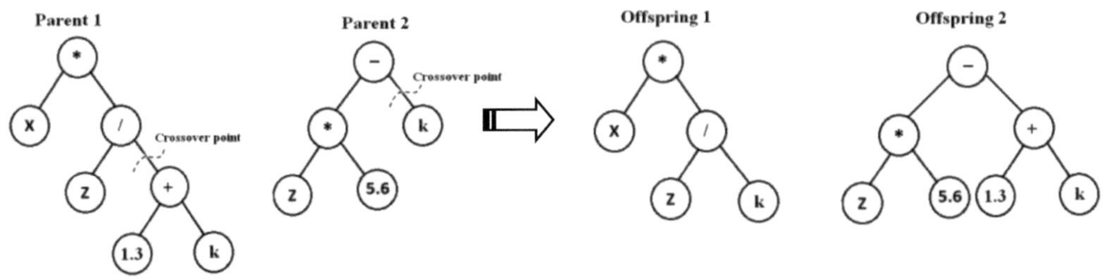

Figure 2. Crossover operation on two selected individuals.

Mutation operates on the population individuals in proportion to the *mutation probability* (P_m). A string is randomly selected from the mating pool and it undergoes some changes at the randomly selected mutation point (Fig. 3). The mutation operation also results in new individuals, which increases the genetic diversity of the population (Koza, 1992). Simply reproduced individuals from the mating pool and newly created individuals resulted from genetic operations of reproduction, crossover, and mutation form the next generation of GP search space. The described evolutionary process is performed iteratively over several generations until some *termination criterion* is satisfied. The termination criterion might be a maximum number of generations or some measure of the goodness of the generated solution and stop the algorithm once the solution is found (Koza, 1992). The result of the GP algorithm, which is a GP-evolved model for the investigated problem, based on the termination criterion, is either the best found model or the best individual of the last GP generation.

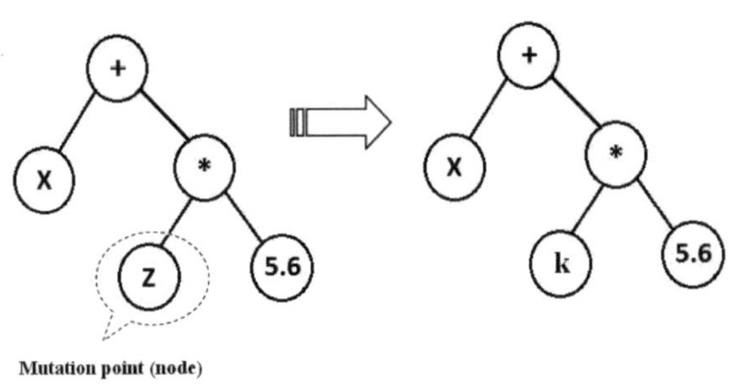

Figure 3. Mutation operation on a selected individual.

3. Wavelet analysis

Natural functions, e.g. meteorological and hydrological processes, operate over a wide range of spatial and temporal scales leading to spatial/temporal variability of interacting mechanisms. AET is a hydrometeorological signal interacting with several temporally/spatially variable meteorological signals. Evaluation of dominant cyclic variations in the AET and correlated meteorological signals improves the understanding of the mechanism as well as its modeling. Temporal cyclic variations of natural processes are not usually stationary and contain several localized and transient frequency events. Therefore, conventional frequency domain analysis such as Fourier transform cannot reveal the localized natural cyclic events. Wavelet analysis (WA) provides a tool for decomposing the variations of a time series signal into time and scale (frequency) domains; allowing the identification and analysis of dominant temporal cyclic events. The basic component of WA is the wavelet transformation in which the studied function is represented by wave-like oscillating functions. The choice of the wavelet function is of high importance within the wavelet transformation. Wavelet functions are defined in different forms, namely mother wavelets, to have specific properties for information extraction of different types of signals. Figure 4 shows some examples of mother wavelets.

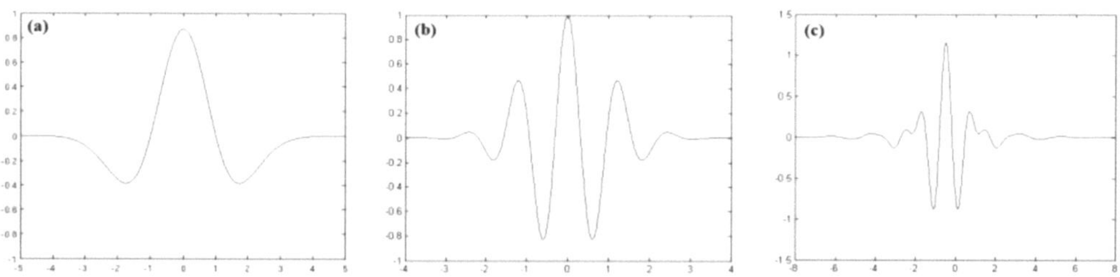

Figure 4. Examples of mother wavelet functions; (a) Mexican Hat, (b) Morlet, and (c) Meyer.

The term wavelet function generally refers to two types of wavelet functions, namely orthogonal and non-orthogonal (Torrence and Compo, 1998). Orthogonal wavelets are mainly used for decomposition of a signal into specific (preferably minimum) frequency bands (Polikar, 1996). This type of wavelet analysis is usually referred to as discrete wavelet transformation, which may not provide a physically meaningful analysis all the time (Si, 2008). Non-orthogonal wavelets are usually used for continuous wavelet transformation (CWT) of time series signals in which a continuous set of frequencies are examined. CWT results in a highly redundant time-scale resolution of the signal, which in one hand induces some uncertainties in the reconstruction of the signal and, on the other hand, provides better scale analysis of the time series (Si, 2003; He et al., 2007). Because of the wide range of possible dominant frequencies that can be obtained using CWT, Coulibaly and Burn (2004) indicated that the CWT is more appropriate for analysis of geophysical and hydrological time series.

3.1. Continuous wavelet analysis

As it was mentioned earlier, the choice of wavelet function is an important component in the wavelet transformation. Wavelet function can be a real or complex function. Complex wavelet functions make it possible to extract the information of both amplitude and phase, which is more suitable for analyzing the signal's oscillatory behavior (Torrence and Compo, 1998). Morlet, Mexican Hat, and Haar are some of the mother wavelets usually employed in the CWT. Morlet is a complex and non-orthogonal wavelet that provides sufficient resolution in time and scale domains (Grinsted et al., 2004; Si, 2008). Morlet function, with non-dimensional frequency parameter (ω_0) equal to 6, has been shown to successfully work for the analysis of observed time series in different hydrological applications (Lafreniere and Sharp, 2003; Anctil and Tape, 2004; Coulibaly and Burn, 2004; Labat et al., 2005; Si and Zelek, 2005; Coulibaly, 2006). This Morlet wavelet is an exponential oscillatory function defined as (Torrence and Compo, 1998):$\tau_0(\eta)=\pi^{-1/4}e^{iw_o\eta}\ e^{-\eta^2/2}$,where η and ω_0 are non-dimensional time and frequency parameters. The CWT of a discrete time series data of x_i (i=1,2,...,N) is defined as the inner product of time series signal with the scaled and translated version of mother wavelet function, $\psi_o(\eta)$, according to a specific scale (s) and time location (τ), which is given as:

$$CWT(\tau,\ s)=\sum_{i=1}^{N} x_i(t).\psi^*_{\tau,s}(t) \tag{3}$$

where $\psi_{\tau,s}(t)$ is the normalized wavelet function and (*) represents the complex conjugate. Normalized wavelet function ensures that the wavelet transform at each scale is not weighted by the magnitude of the scale, which makes a direct comparison of wavelet co-efficients at different scales possible (Torrence and Compo, 1998). Normalized wavelet function is defined as:

$$\psi_{\tau,s}(t)=\frac{1}{\sqrt{s}}\psi_{o\tau,s}(t) \tag{4}$$

where τ and s are associated with the time location and scale resolution at which the wavelet transformation is performed. Localization of the time series signal into time and scale domains is implemented, first, by modulating the mother wavelet, corresponding to the current scale, and shifting the scaled wavelet through the signal to the end and performing the convolution at each discrete time location. This results in the time localization of the signal. The procedure is repeated, in the second step, for each scaled wavelet to localize the signal in the scale domain. Wavelet coefficients are computed for all time and scale steps (τ,s) to give the multiresolution representation (or CWT) of the signal. Scaled and translated wavelet at scale s and time location τ is computed by:

$$\psi_{\tau,s}(t)=\psi\left(\frac{t-\tau}{s}\right) \tag{5}$$

According to the mathematical definition of CWT, WA investigates the resemblance of the wavelet function with the in hand signal in the sense of frequency content (Polikar, 1996). In other words, "if the signal has a major component of the frequency corresponding to the current scale, then the wavelet at the current scale will be similar or close to the signal at the particular location where this frequency component occurs. Therefore, the CWT coefficient at this point in the time-scale plane will be a relatively large number" (Polikar, 1996) and will spike in the contour plot of CWT spectrum.

http://dx.doi.org/10.5772/52809

For implementing CWT, it is required to identify the set of analyzed scales a priori. In continuous wavelet analysis, the investigated scales must be incremented continuously to create a complete picture of the wavelet transform. Theses set of scales (s) can be generated using fractional powers of two (Torrence and Compo, 1998); $s_j = s_0\, 2^{j\delta j}$, $j=0,1,2,...,J$, where s_0 is the smallest scale and J determines the maximum number of scales to be investigated. δj is the scale step size whose value depends on the selected wavelet function (Torrence and Comp, 1998). Complex wavelet function, e.g. Morlet, results in complex wavelet coefficients constitute of real and imaginary parts or amplitude, $|CWT(\tau, s)|$, and phase, $tan^{-1}[Im\{CWT(\tau, s)\} / Re\{CWT(\tau, s)\}]$, respectively. For convenient description of time series cyclic variations, it is common to use wavelet power spectrum, defined as, $|CWT(\tau, s)|^2$, instead of continuous wavelet spectrum. The obtained wavelet power spectrum is also normalized; divided by the variance of the time series (σ^2), $|CWT(\tau, s)|^2 / \sigma^2$, for easier comparison with different wavelet spectra (Torrence and Compo, 1998). Cone of Influence (COI) has been defined in the wavelet spectrum to clarify the areas that are considerably affected by the zero paddings at the ends of the time series signal. Time series data are padded by zeros at both edges to overcome the problem caused by their finite lengths. These zero values decrease the magnitude of wavelet power at the areas close to the edge from which the COI distinguishes regions that are not or negligibly influenced. Length of COI is estimated for each examined scale using a mathematical expression, which is defined as a function of scale. For Morlet wavelet, the length of COI at each scale (s) was defined as $\sqrt{2}s$.

3.2. Statistical significance test

Most of the natural processes (e.g. geophysical and hydrological) are affected with background color noise (white or red noise). The effect of noise is reflected on the signal's wavelet power spectrum. It is essential to identify the powers caused by the background noise and distinguish them from the actual wavelet power peaks. Torrence and Compo (1998) developed a statistical significance test for wavelet power spectra to establish significant levels. Following Torrence and Compo (1998), a statistical significance test is implemented by modeling the appropriate background noise (either white or red) and then testing the significance of the power spectrum peaks against the modeled background noise at certain statistical significance level. Significance test investigates if the peaks of the wavelet spectrum represented some true cyclic features or they are just caused by noise. Most of the geophysical time series are contaminated with red noise background signals (Grinsted et al., 2004). Red noise refers to the temporal fluctuations that have higher amplitude at lower frequencies and lose the magnitude as the frequency increases

According to Hasselmann (1976), lag-1 auto regressive process (AR [1]) is a suitable background noise for many climatological applications. A simple theoretical AR [1] red noise model for modeling the background time series red noise (x_n) is given by (Torrence and Compo, 1998):

$$x_n = \alpha x_{n-1} + z_n \tag{6}$$

where $x_0 = 0$, z_n is the Gaussian white noise, and α is the lag-1 autocorrelation coefficient that can be estimated from observed time series (Allen and Smith, 1996).

It was shown by Torrence and Compo (1998) that the local wavelet power spectrum of the theoretical red noise, at every randomly selected time location, is on average identical to the Fourier transform of the noise time series. In the described statistical significance test, it is assumed that the time series variables have random normal distribution. Fourier power spectrum of the theoretical noise, which is the square of the normally distributed spectrum, has chi-square (χ^2) distribution with two degrees of freedom, X_2^2, corresponding to the real and imaginary parts. Statistical significance test can be performed at 95% confidence level. To perform the test, the 95% line is developed by multiplying the red noise spectrum by the 95th percentile value of X_2^2. Wavelet peaks are compared with this 95% line and the peaks that are above this confidence line are identified as cyclic features that are significantly different from background red noise at 95% confidence level.

3.3. Cross wavelet analysis

Cross wavelet analysis is an extension to WA, which examines the linear correlation between two time series. Cross wavelet spectrum between two processes, X and Y, is estimated by (Torrence and Comp, 1998):

$$W^{XY}(\tau, s) = CWT^{X}(\tau, s) CWT^{Y*}(\tau, s) \tag{7}$$

where $CWT^{X}(\tau, s)$ and $CWT^{Y}(\tau, s)$ are the continuous wavelet transforms of the investigated time series, X and Y, and (*) indicates the complex conjugate. Cross wavelet spectrum is complex and can be decomposed into amplitude and phase. Local relative phase between X and Y is estimated by the complex argument (phase), $tan^{-1}[Im\{W^{XY}(\tau, s)\} / Re\{W^{XY}(\tau, s)\}]$ and the cross wavelet power is also defined as $|W^{XY}(\tau, s)|$. The phase information in the cross wavelet spectrum gives the phase angel difference between the components of the two-time series. Using cross wavelet spectrum, cyclic features at which the underlying time series are co-varying can be detected. The co-variations of two signals demonstrate the existence of a link, in some way, between the underlying processes and also the fact that the information of one process is capable of predicting the other process. This information is very useful when it is of interest to find out the processes that have correlation (or strong correlation) with a target time series, e.g. AET here. The signals, which are showing to have high common power with the target signal in the cross wavelet spectrum, can be used as predictors in the estimation of temporal variations of the target time series. This is important information in the modeling of complex processes, e.g. hydro-meteorological processes, in which determination of important predictors is essential and a challenging task. Statistical significance test of the cross wavelet power spectrum can be conducted using the theoretical Fourier spectra of the two underlying time series. More description on the development of cross wavelet significance test can be found in Torrence and Comp (1998).

4. Application of data driven modelling and wavelet analysis in characterizing AET in a case study

This section describes the application of the previously explained data driven modeling techniques; ANNs and GP, and wavelet analysis for the modeling and analysis of AET in a case study.

4.1. Research scope and experimental data

The experimental data, which were used in this study, were collected from the South West Sand Storage (SWSS) site, located at Mildred lake mine north of Fort McMurray, Alberta, Canada. The SWSS facility is an active tailing disposal facility (dam), which covers an area of about 23 km^2, holding approximately 435×10^6 m^3 of materials, with 40 m higher than the surrounding landscape and an overall side slop of 5%. The soil cover system within the SWSS consists of a 45 cm thick peat/secondary mineral soil with a clay loam texture overlying the tailing sand. Vegetation cover system varies across the SWSS site including the dominant groundcover of horsetail (*Equisetum arvense*), fireweed (*Epilobium angustifolia*), sow thistle (*Sonchus arvense*), and white and yellow sweet clover (*Melilotus alba, Melilotus officinalis*), and tree and shrub species including Siberian larch (*Larix siberica*), hybrid poplar (*Populus* sp. hybrid), trembling aspen (*Populus tremuloides*), white spruce (*Picea glauca*) and willow (*Salix* sp.) (Parasuraman et al., 2007). The latent heat flux data were originally measured on a continuous basis (Baldocchi et al., 1988) using the eddy covariance technique, and the mean fluxes were recorded every 30 minutes on a data logger. In this study, the hourly Eddy Covariance latent heat (LE) flux (Wm^{-2}) data from May 3 to September 21, 2005 and from May 27 to September 8, 2006 were used. The day-time data, which were used for modeling purpose, were only associated with the period of 8:00 AM to 8:00 PM. The data of net radiation (R_n; Wm^{-2}) were also recorded using net radiometer. Air temperature (T_a; °C), ground temperature (T_g; °C), relative humidity (*RH*), and wind speed (W_s; m s^{-1}) constituted the rest of the meteorological data, which were measured by the weather station located at the site. The LE and R_n fluxes were originally recorded in the unit of Wm^{-2} on half hourly basis. For convenient interpretation, the latent heat flux (Wm^{-2}) was converted to the equivalent depth of water (mm m^{-2}). Since the hourly data were desired to be used in the modeling procedure, conversion of the recorded half-hourly data to hourly data was also implemented in the pre-processing step.

In the first step, the data of the year 2006 were used for modeling purposes with the two proposed techniques (ANNs and GP). Disregarding the missing data, the total number of available instances for modeling in year 2006 is 1207, which were randomly divided into three datasets consisting of 604 instances (50%), 201 instances (17%), and 402 instances (33%) of the data, for training, cross-validation, and testing purposes, respectively. To obtain three statistically consistent subsets, a population of 100 groups of three sub-datasets was randomly generated by sampling from the dataset. The statistical characteristics of the data, i.e. mean and standard deviation, were determined for every subset of each group. Then, the group possessing three subsets with relatively similar statistical characteristics was selected for this study. Aside from the described modeling procedure, a rigorous test was also implemented in the second step, using the data of 2005, to assess the generalization ability of the developed models in a more realistic way. Disregarding the missing data in 2005, 1600 instances are available. The 2005 dataset has different statistical properties from the data of 2006, which are discussed later in this study.

Multiresolution analysis of the AET and meteorological signals (wavelet analysis) was conducted using the data of the year 2006. The total number of instance that was available for wavelet analysis of the 2006 data is 2520, which constitute the hourly time series data from May 27 to September 9. All of the observed time series data were pre-treated before performing the WA to have zero mean and unite standard deviation.

4.2. ANN modeling

Three-layer FFNNs were adopted in this study for the modeling of AET process. The input layer contained five nodes providing the information of predictor variables; R_n, T_g, T_a, RH, and W_s, to the network and the output layer consisted of a single neuron representing the model output (predicted AET values). Activation functions adopted here include the log-sigmoid and linear functions for the hidden layer and output layer neurons, respectively. The commonly used trial-and-error procedure was employed and different number of hidden neurons ranging from 1 to 14 was investigated for finding the optimum number of hidden neurons. Regularization and early stopping approaches were employed with the examined training algorithms; Levenberg-Marquardt (Levenberg, 1944; Marquardt, 1963) and Bayesian-regularization (MacKay, 1992). Neural Network Toolbox in MATLAB (MALAB® Software, 2003) was used to develop the ANN models to predict AET based on five inputs of meteorological variables, R_n, T_g, T_a, W_s, and RH. The data pool of 2006 was randomly divided into three subsets of training, cross validation, and testing using the approach explained earlier. The training subset was used for optimizing the connection weights and bias of the network. The cross-validation subset was used for early stopping. Once the network was trained, the generalization and predictive ability of the network was evaluated using a completely unseen subset of 2006 called testing subset. The data subsets were normalized so that data fell between 0 and 1. Such scaling of data smoothness the solution space and averages out some of the noise effects (ASCE, 2000). Based on the training subset, different ANN models were trained using Levenberg-Marquardt and Bayesian-regularization training algorithms, using different number of hidden neurons ranging from 1 to 14. For each examined network architecture the training process was repeated several times, each time started with different random initial weight matrices, until satisfactory optimal network (with minimum errors) was obtained. The ANN model with the best performance measures associated with the cross-validation subset was selected as the optimal predictive network. The performance and generalization ability of the trained model was evaluated on the testing subset, which determines how well the ANN model performs on the dataset that have not been seen during the training process (Cheng and Titterington, 1994).

ANNs, as a data driven technique, have the ability to determine the critical model inputs (Maier and Dandy, 2000). In this study, the ANN modeling technique was used to identify the important meteorological variables affecting the AET process. In this approach, no prior knowledge was assumed about the physics of AET mechanism and the relationships among variables. All possible combinations of input variables, 26 combinations, were considered to be examined as ANN model input sets. Separate optimal ANN models were developed and trained for each input combination set using the model development approach explained earlier. The developed ANN models were compared based on their prediction accuracy in order to identify the most appropriate and efficient combinations of inputs for the estimation of AET. This approach is commonly referred to as trial-and-error procedure, which is under the category of heuristic approaches. The possible combination sets of five available input variables include; five-input combination, "R_n, T_g, T_a, RH, W_s", four-input combinations, "R_n, T_g, RH, W_s"; "R_n, T_g, T_a, RH"; "R_n, T_g, T_a, W_s"; "R_n, T_a, RH, W_s"; "T_g, T_a, RH, W_s"; three-input combinations, "R_n, T_g, RH"; "R_n, T_g, W_s"; "R_n, T_g, T_a"; "R_n, RH, W_s"; "R_n, T_a, RH"; "R_n, T_a, W_s"; "T_g, RH, W_s"; "T_g, T_a, W_s"; "T_g, T_a, RH"; "T_a, RH, W_s"; and two-input combinations, "R_n, T_g" ; "R_n, RH" ; "R_n, W_s" "R_n, T_a" ; "T_g, RH" ; "T_g, W_s" ; "T_g, T_a" ; "T_a, RH" ; "T_a, W_s" ; "RH, W_s".

4.3. GP modeling

Major steps in the implementation of GP to solve a problem, e.g. evolution of AET models in the current study, include determination of functional and terminal sets, fitness measure, initializing method, selection method, levels of GP parameters over the run (crossover and mutation probabilities, population size), and the termination criterion. The functional set, which was introduced to GP, included {+, -,*, /}. The terminal set was defined as $\{R_n, T_g, T_a, W_s, RH\}$. Root mean squared error (RMSE) was selected as the fitness function for evaluating individual performance and further fitness-based selection. Ramped-half-and-half method was adopted for initializing the first generation tree structures. Descriptions of initializing methods can be found in Koza (1992) and Banzhaf et al. (1998). The next important issue in the implementation of GP is the fitness-based selection method. Selection method determines the manner by which the individuals are selected based on the assigned fitness values for further GP operations (e.g. crossover, mutation). Roulette wheel selection method was employed here for implementing selection operation in the GP runs. Roulette wheel method is the simplest selection scheme that follows a stochastic algorithm. Several different levels of GP parameters; crossover and mutation probabilities, number of evaluated generations, and the size of population, were executed for obtaining symbolic regression AET models using the training subset. The termination criterion for each GP run was the identified maximum number of generations. The performances of the generated symbolic equations were assessed using the cross-validation subset to select the best equation (model). The selected symbolic equation was then tested using the unseen testing subset to evaluate the predictive accuracy and generalization ability of the proposed model. Data subsets that were used with the GP technique were exactly the same as those used with the ANNs. The data were normalized by dividing the values of variables by their corresponding maximum values. In this way, all variables could have dimensional consistency during the GP implementation (Parasuraman et al., 2007). In this study, GPLAB (Silva, 2005), GP toolbox for MATLAB, was used for the implementing of the GP technique and generating mathematical models based on the datasets where AET is a dependent variable as a function of the five independent variables: R_n, T_g, T_a, W_s, and RH.

The performances of the ANNs and GP models were evaluated to compare their predictive accuracies based on three statistical criteria: Pearson's correlation coefficient (R), root mean squared error (RMSE), and mean absolute relative error (MARE), which were calculated as follows:

$$R = \frac{\sum_{i=1}^{N}\left(O_i - \bar{O}\right)\left(P_i - \bar{P}\right)}{\left[\sum_{i=1}^{N}\left(O_i - \bar{O}\right)^2\right]^{0.5}\left[\sum_{i=1}^{N}\left(P_i - \bar{P}\right)^2\right]^{0.5}} \quad (8)$$

$$MARE = \frac{1}{N}\sum_{i=1}^{N}\frac{|O_i - P_i|}{|O_i|} \quad (9)$$

$$RMSE = \sqrt{\frac{\sum_{i=1}^{N}(O_i - P_i)^2}{N}} \quad (10)$$

where O_i, P_i, $\bar{P}$, and $\bar{O}$ are observed values, simulated values, mean of simulated, and mean of observed values, respectively. N is the number of instances in the dataset.

4.4. Wavelet analysis

In this study, only temporal scaling of the variables time series was investigated whereas the spatial variability of the AET and meteorological signals was not considered. Since scale analysis of the time series data were of interest, the CWT was employed for the analysis. The Morlet wavelet with non-dimensional frequency parameter (ω_0) equal to 6 was adopted as the mother wavelet for the current wavelet transformation. For the current analysis, the scale step size of $\delta_j = 0.083$ and the maximum examined scales of $S_J = 16$ and 48 hours were selected for performing the transformation. The smallest scale (S_0) was selected as approximately equal to $2\delta t$, where δt is the time step of the measured time series data. The time step of the AET and meteorological variables is an hour ($\delta t = 1$) and subsequently $S_0 = 2$ hours. A simple theoretical AR [1] red noise model was adopted for describing the background noise. The meteorological variables, whose covariations with the AET time series were investigated in this study, include R_n, T_g, T_a, RH, and W_s. The statistical significance test was performed at 95% confidence level.

Both continuous and cross wavelet analysis were implemented using the software package developed for MATLAB and provided on-line by Grinsted et al. (2004) (http://www.pol.ac.uk/home/research/waveletcoherence/). Wavelet and cross wavelet analysis were basically of interest to examine the temporal cyclic variations occurring during day-time (8:00 AM to 8:00 PM) of the AET and meteorological time series. However, wavelet transformation can only be performed on complete (continuous) time series but not non-continuous time series such as day-time data. To obtain accurate wavelet analysis, which were also associated only with the day-time variations, wavelet and cross wavelet analysis were performed using the complete time series data (day-time and night-time data). Then, the wavelet coefficients (spectrum segments), which were associated with night-time data were cut out to give the spectrum of the day-time only time series data. Wavelet spectra provided in the next section are all associated with the day-time only time series.

4.5. Results and discussions

4.5.1. ANN model and performance analysis

Figure 5 illustrates the influence of number of hidden neurons on the performance measures for two training algorithms; Levenberg-Marquardt and Bayesian-regularization. It appears that Levenberg-Marquardt training algorithm is more sensitive to the number of hidden neurons, represented by larger fluctuations in the error measures with respect to the number of hidden neurons than the Bayesian-regularization algorithm. Figure 5a indicates that the Levenberg-Marquardt algorithm leads to lower values of correlation coefficient (R) for all numbers of hidden neurons compared to the Bayesian-regularization algorithm. Figs. 5b and 5c show that Levenberg-Marquardt results in higher values of RMSE and MARE than Bayesian-regularization for all number of hidden neurons. It indicates that the Bayesian-regularization training algorithm performs more efficiently than the Levenberg-Marquardt algorithm on the dataset under consideration. This might be attributed to some hindrance caused by the use of redundant network parameters (weights and biases) in the output estimation of the network trained by Levenberg-Marquardt algorithm, while the network trained by Bayesian-regularization training algorithm only use the effective network parameters for computing the output. Among the 28 assessed ANN models, the ANN model with eight hidden neurons trained by Bayesian-regularization algorithm resulted in relatively better statistical measures; R of 0.89, RMSE of 0.06, and MARE of 0.28 when evaluated using the cross-validation dataset. Therefore, eight hidden neurons were adopted for the ANN model for the rest of the modeling process.

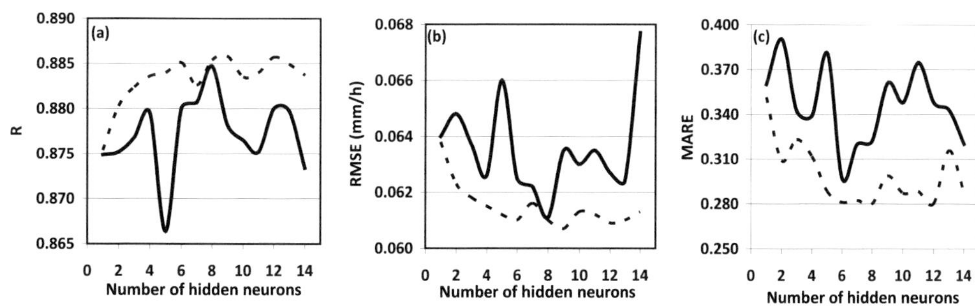

Figure 5. The influence of number of hidden neurons on the network performance for two training algorithms using the cross-validation subset: –, Levenberg-Marquardt; ---, Bayesian-regularization.

Applying the selected ANN model with eight hidden neurons to the testing dataset results in R, RMSE, and MARE values of 0.86, 0.07, and, 0.31, respectively, which indicates reasonably low values of RMSE and MARE, and high value of R associated with the testing dataset, which imply that the ANN model has good generalization ability for predicting AET based on the unseen testing dataset. For the five available meteorological variables, R_n, T_g, T_a, RH, and W_s, 26 different input combinations could be assessed, which were described earlier in this chapter. In order to examine the importance of each input combination, the associated optimum ANN model was developed. The primary results indicated that net radiation (R_n) is a crucial factor in the estimation of AET; its exclusion from the input set causes serious deterioration of the performance of the ANN models. For instance, ANN model with the predictors set of T_g, T_a, RH, and W_s (excluding R_n) resulted in the performance measures of 0.11 mm/h, 0.69, and 0.54 for the RMSE, MARE, and R, respectively, when applied to the testing subset. The significant role of net radiation, as the main source of energy, in the AET mechanism is expected based on the physics of the AET process. As a result, the rest of the analysis was performed only for the input subsets, which include R_n as one of the predictors. Consequently, the total number of investigated input combinations decreased from 26 to 16. Table 1 shows the performance statistics of ANN models trained using 16 different combinations of inputs.

The best performance of ANNs was obtained when all five meteorological variables were used for the modeling of AET; however, ANN models, which employed the predictor combinations of "R_n, T_g, RH, W_s"; "R_n, T_g, T_a, RH"; "R_n T_g, T_a, W_s"; "R_n, T_a, RH, W_s"; "R_n, T_g, RH"; and "R_n, T_g, W_s", also resulted in comparable performances. Among the input combinations of two factors only, the ANN model with predictor set of R_n and T_g performed fairly well, which shows the possibility of using fewer number of predictors for estimating AET in an efficient and parsimonious way. Obtaining acceptable prediction accuracies from different combinations of inputs demonstrates the difficulty of determining the significant input variables for modeling the AET process. Thus, the trial-and-error procedure using the ANN technique might not be the best approach for identifying the important AET predictors. This difficulty can also be associated with the complexity of the AET process itself. The interaction among multiple processes and variables involving the AET makes it possible, for ANN model, to sufficiently capture the variations of AET by using different combinations of variables. It is understood from the results that determination of a unique set of meteorological variables might not be necessary for the estimation of AET. Instead, the effort can be concentrated on the determination of the most efficient and parsimonious set of predictor variables.

4.5.2. GP modeling and performance analysis

Using GPLAB (Silva, 2005) several equation-based GP models were generated at 42 different levels of GP parameters including crossover probability, mutation probability, number of generations, and population size. Table 2 presents the values of RMSE, MARE, and R along with the associated GP parameters obtained with the best eight models generated by GP. The optimum GP models that resulted in the best statistics associated with the cross-validation subset are given below (Equations 11-18):

$$AET = -0.013 + 0.148R_n + 0.01T_g - 0.104RH \tag{11}$$

Input combination	Training			Cross-validation			Testing		
	RMSE*	MARE	R	RMSE	MARE	R	RMSE	MARE	R
R_n,T_g,T_a,RH,W_s	0.06	0.40	0.89	0.06	0.28	0.89	0.07	0.31	0.86
R_n,T_g,RH,W_s	0.06	0.43	0.88	0.06	0.28	0.88	0.07	0.33	0.86
R_n,T_g,T_a,RH	0.06	0.43	0.88	0.06	0.31	0.88	0.07	0.32	0.86
R_n T_g,T_a,W_s	0.07	0.44	0.87	0.07	0.29	0.87	0.07	0.33	0.85
R_n,T_a,RH,W_s	0.07	0.49	0.85	0.07	0.30	0.87	0.07	0.36	0.83
T_g,T_a,RH,W_s	0.12	0.87	0.61	0.10	0.71	0.62	0.11	0.69	0.54
R_n, T_g,RH	0.06	0.44	0.88	0.06	0.30	0.88	0.07	0.34	0.86
R_n,T_g,W_s	0.07	0.42	0.87	0.07	0.29	0.86	0.07	0.32	0.85
R_n,T_g,T_a	0.07	0.48	0.85	0.07	0.31	0.87	0.07	0.35	0.84
R_n,RH,W_s	0.07	0.47	0.85	0.07	0.29	0.86	0.07	0.37	0.83
R_n,T_a,RH	0.07	0.54	0.84	0.07	0.35	0.86	0.07	0.40	0.83
R_n,T_a,W_s	0.07	0.54	0.83	0.07	0.32	0.86	0.07	0.37	0.83
R_n,T_g	0.07	0.57	0.85	0.06	0.34	0.87	0.07	0.42	0.84
R_n,RH	0.07	0.53	0.82	0.07	0.36	0.87	0.07	0.49	0.82
R_n,W_s	0.08	0.53	0.79	0.08	0.35	0.82	0.09	0.45	0.77
R_n,T_a	0.08	0.51	0.83	0.07	0.34	0.85	0.08	0.43	0.82

*RMSE in mm/h

Table 1. Performance statistics of ANN models with different combinations of inputs.

$$AET = -0.018 + 5.54\times10^{-3}T_g + 9.49\times10^{-3}R_nT_g \tag{12}$$

$$AET = 0.0784 + 9.2\times10^{-3}R_nT_g - 3.5\times10^{-3}R_n^{\,2}T_g + 2.7\times10^{-4}R_nT_gT_a - 8.64\times10^{-7}T_a^{\,2} \tag{13}$$

$$AET = 0.039 + 0.063R_n + 1.88\times10^{-3}T_g + 7.37\times10^{-3}R_nT_g \tag{14}$$

$$AET = 0.0696 + 7.836\times10^{-3}R_nT_g + 2.569\times10^{-3}T_g \tag{15}$$

$$AET = 0.0633 + 3.1 \times 10^{-3} T_g + 0.011 \times R_n + 6.85 \times 10^{-3} T_g R_n \quad (16)$$

$$AET = 0.0775 + 2.23 \times 10^{-3} T_a + 6.35 \times 10^{-3} R_n T_a \quad (17)$$

$$AET = 0.129 R_n + 0.005 T_a \quad (18)$$

where, AET, R_n, T_g, T_a, and RH are the rate of actual evapotranspiration [mm h^{-1}], net radiation [MJ], ground temperature [°C], air temperature [°C], and relative humidity [fraction], respectively.

Model	Crossover prob.	Mutation prob.	No.of generation	Population size	RMSE (mm/h)	MARE	R
Eq. 11	0.6	0.2	50	60	0.06	0.37	0.88
Eq. 12	0.5	0.2	60	70	0.07	0.34	0.86
Eq. 13	0.6	0.3	60	60	0.07	0.37	0.86
Eq. 14	0.7	0.5	50	300	0.07	0.35	0.85
Eq. 15	0.5	0.2	200	100	0.07	0.43	0.86
Eq. 16	0.7	0.4	200	50	0.07	0.44	0.86
Eq. 17	0.8	0.3	100	40	0.07	0.46	0.85
Eq. 18	0.6	0.3	50	80	0.08	0.40	0.85

Table 2. The best generated GP-based models using various GP parameters for the cross-validation subset.

The optimum GP-evolved models are structurally simple, characterizing the variation of AET as semi-linear functions of meteorological variables, since the models are linear in parameters. Most (six out of eight) of the presented GP models contain R_n and T_g as AET predictors. The appearance of RH (one out of eight times) and T_a (three out of 8 times) was limited in the developed models. Interestingly, W_s never came up as an important predictor in the presented optimum AET models, which means that GP did not find wind speed to be an effective component in the estimation of hourly AET. The simplicity of the models seems to be interesting, especially when the error measures also indicate relatively good generalization ability of the models based on the testing subset (Table 3). It is perceived from the GP models that the AET mechanism can be characterized by structurally simple models, which are also not physically complex. This can be considered as a strong advantage of the GP technique that searches for any possible combination of predictors that can properly model the AET process.

Model	RMSE (mm/h)	MARE	R
Eq. 11	0.07	0.35	0.85
Eq. 12	0.08	0.32	0.83
Eq. 13	0.07	0.32	0.82
Eq. 14	0.08	0.32	0.82
Eq. 15	0.07	0.39	0.83
Eq. 16	0.07	0.40	0.83
Eq. 17	0.07	0.41	0.81
Eq. 18	0.09	0.36	0.79

Table 3. Performance statistics of the GP-based models using testing subset.

Based on the equation-based GP models, the contribution of each meteorological variable in the estimation of AET can also be discussed. This is only possible by using the normalized form of the equations in which all input variables receive their values from a consistent range (e.g. less than 1). Then the contribution of each input variable or factor to the AET can be assessed based on the associated coefficient's magnitude. A selective set of models, which includes only GP models with different physical structures, was identified and the input variables were normalized for further analysis. The selected models, Eq. (11), (12), (13), (14), and (17), are rewritten, in order, as follow:

$$AET = -0.013 + 0.385R'_n + 0.285T'_g - 0.095RH' \tag{19}$$

$$AET = -0.018 + 0.15T'_g + 0.53R'_nT'_g \tag{20}$$

$$AET = 0.0784 + 0.514R'_nT'_g - 0.49R'^2_nT'_g + 0.424R'_nT'_gT'_a - 0.976\times10^{-3}T'^2_a \tag{21}$$

$$AET = 0.039 + 0.164R'_n + 0.051T'_g + 0.412R'_nT'_g \tag{22}$$

$$AET = 0.0775 + 0.075T'_a + 0.396R'_nT'_a \tag{23}$$

In these normalized equations, input variables, which are associated with the models' linear coefficients (e.g. T'_g, $T'_gR'_nT'_a$, and $R'^2_nT'_g$), are normalized inputs by dividing each of them by its corresponding maximum values and AET is the rate of actual evapotranspiration [mm h^{-1}]. These normalized models were only developed and used for interpreting the contribution of different inputs to the estimation of AET.

Equation (19) indicates that AET can be estimated as a simple linear function of R'_n, T'_g, and RH', which is highly dominated by the net radiation and ground temperature variables. Equation (20) also has a simple structure describing the AET process as a nonlinear function, of only net radiation and ground temperature, which is dominated by the two-factor interaction of R'_n and T'_g. The interaction factor of $R'_nT'_g$ has larger contribution to the estimation of AET than the T'_g, individually, with the average contribution magnitude of 0.15 and 0.10 mm/h for the terms $0.53R'_nT'_g$ and $0.15T'_g$, respectively. This indicates that when some factors are interacting, their interactions influence the individual contribution of each variable to the AET mechanism. Consequently, the interaction term is more responsible for AET variations than the individual variables. In Eq. (21), the air temperature (T'_a) variable has been included in addition to the R'_n and T'_g. The air temperature has appeared both as an individual variable and as an interacting factor in the three-factor interaction term of $R'_nT'_gT'_a$. According to the coefficients associated with these variables in Eq. (21), T'_a can affect the rate of AET only through the influence it might have on the R'_n and T'_g (interacting coefficient of 0.424 compared to that of air temperature, 0.000976). The structure of Eq. (22) also confirms the importance of interaction effects of multiple variables rather than the individual processes. The combined component of $R'_nT'_g$ is more responsible for the variation of AET than the R'_n and T'_g individually. The average contribution magnitude of each of the terms $0.412R'_nT'_g$, $0.164R'_n$, and $0.051T'_g$ in the estimation of AET values are 0.12, 0.05, and 0.03 mm/h, respectively. Equation (23) demonstrates that the AET mechanism can even be characterized as a simple semi-linear function of R'_n and T'_a only, which are commonly available meteorological measurements. Again, the AET model is dominated by the interaction factor of the two variables. The interaction and individual terms of $0.396R'_nT'_a$ and $0.075T'_a$ are contributing to the estimation of AET by the average magnitude of 0.11 and 0.05 mm/h, respectively. Although the generated models, based on error measures, are performing well and relatively similar, they are using different combinations of inputs with different mathematical structures. This demonstrates that precise identification of the meteorological variables driving the AET process is not an easy and straightforward task, where different combinations of inputs may result in relatively good AET estimation. The results obtained from the GP-evolved models indicate that the hourly AET process can be estimated by both linear and nonlinear relationships. Using the above-listed GP models, one may choose one of them for estimating the rate of AET based on the meteorological data that are available. Thus, the proposed GP models can suit different conditions of data availability.

4.5.3. Comparison among ANN and GP models

The performances of the models from the two proposed techniques; ANNs and GP, were compared based on the testing subset. It can be seen (Table 4) that the equation evolved by GP resulted in slightly larger MARE value compared to that of ANN model. Despite discrepancies among the statistics, the differences are small, which implies that the models have comparable performances for estimating AET based on the meteorological variables. Errors produced by the ANN and GP models have similar statistical characteristics and follow the same probability distribution of LogLogistic (Figure 6).

Model	RMSE (mm/h)	MARE	R
ANN	0.07	0.31	0.86
GP (Eq. 20)	0.07	0.35	0.85

Table 4. Performance statistics of different models using testing subset..

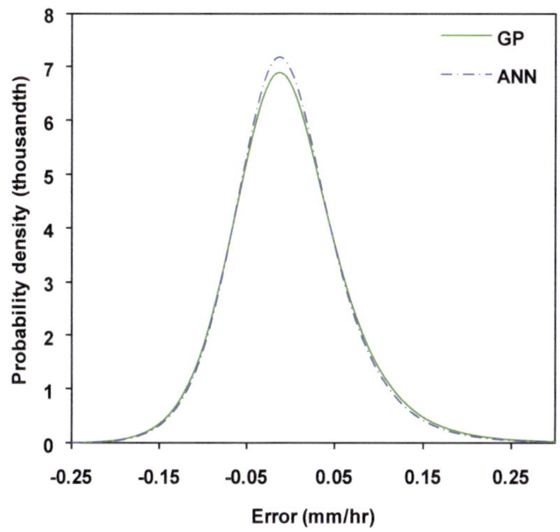

Figure 6. Probability distribution of the data driven models errors.

Figure 7 illustrates the scatter plots of the predicted AET values by ANNs and GP, respectively, versus observed data, using the testing subset. Based on the visual comparison, no substantial difference can be observed among the predictive abilities of the proposed models.

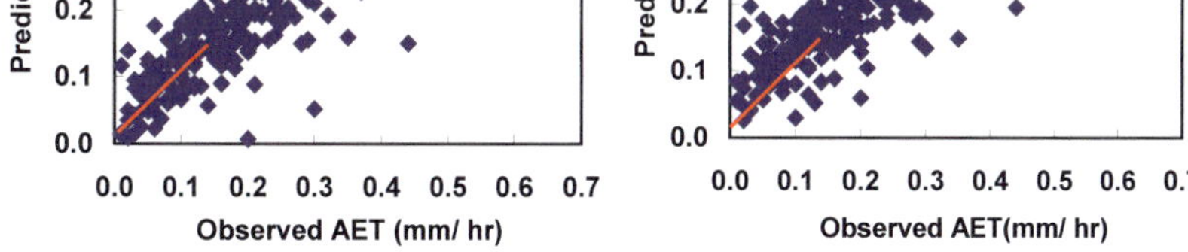

Figure 7. Scatter plots of predicted actual evapotranspiration (AET) versus observed AET by (a) ANN, (b) GP (Eq. 11) using testing subset.

In order to evaluate the generalization ability of the developed models in a more realistic and rigorous way, the optimum models obtained from the ANNs and GP techniques, using the 2006 data, were employed for the prediction of actual evapotranspiration of a different year (2005). This assessment was also conducted to identify the possible superiority of any of the proposed models for future prediction applications. The 2005 dataset, which was used for implementing the rigorous generalization test, has different statistical properties from the year 2006 dataset, which was employed for training and testing during model development (Table 5). Applying the developed models to a statistically different dataset helps to evaluate and compare the models' predictive abilities on instances from a statistically different population. The hourly meteorological variables of R_n, T_g, T_a, W_s, and RH of the year 2005 were used as the inputs of the optimum models, including ANN and GP (Eq. 11 and Eq. 12) to estimate the AET.

Year	Minimum (mm/h)	Maximum (mm/h)	Average (mm/h)	Median (mm/h)	SD * (mm/h)	Coefficient of variation
2005	-0.05	0.68	0.18	0.16	0.12	0.65
2006	-0.06	0.67	0.24	0.23	0.13	0.55

*Standard deviation

Table 5. Statistical characteristics of AET data of the years 2005 and 2006.

A comparison among the performance statistics of the data driven models, used for the prediction of AET in 2005, is given in Table 6. It is apparent that GP model is performing better than the ANN model with lower error measure values and higher correlation coefficients. GP technique was able to capture the semi-linearity of the AET process and to characterize it by simple equations. However, ANN model, because of its structure, tried to fit a complex non-linear model to the AET process, which was unnecessary and resulted in its poor performance in generalization.

Model	RMSE (mm/h)	MARE	R
ANN	0.10	0.91	0.78
GP			
- Equation (11)	0.07	0.55	0.82
- Equation (12)	0.06	0.47	0.85

Table 6. Performance statistics of different models using 2005 data.

For better interpretation of models' performances, further analysis was conducted. Scatter plots of the predicted AET values by ANN and GP models versus observed values, using the 2005 dataset, are illustrated in Fig. 8. It can be seen in Fig.8a that the ANN model is overestimating most of the AET values in 2005, which was expected from the large value of MARE.

Figure 8b indicates that the GP model is performing well on the estimation of AET in 2005. This demonstrates the superiority of GP model over the ANN with regard to the generalization ability.

The two different types of comparison discussed above highlighted the importance and also the reliability of the approach one may use for comparison purposes in the modeling process. In the first approach, the unseen testing subset, which was coming from the same year (statistical population) that was used for developing (training) the models, was employed for testing the generalization ability of the models. Based on this comparison, no considerable difference was observed among the two proposed data driven models in terms of models' generalization ability. However, using the second approach; testing the developed models using data from a different year, led to a more realistic assessment of the predictive accuracy, which revealed the discrepancies among the data driven models much better. Consequently, the choice of the testing dataset on which the generalization ability of the models is evaluated is important, and in the case of inappropriate and/or insufficient testing data, incorrect conclusions might be made.

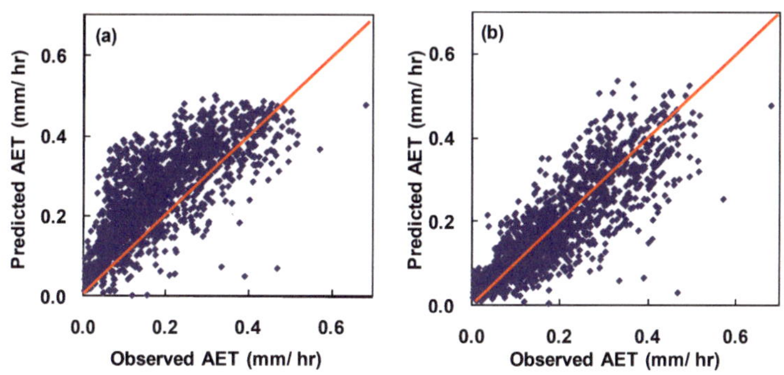

Figure 8. Scatter plots of predicted actual evapotranspiration (AET) versus observed AET by (a) ANN and (b) GP (Eq. 21) models using 2005 data.

Another point of interest, which was observed in this analysis, is the issue of representation of a set of optimum GP equations instead of representing only the best GP equation. Out of the best GP-evolved models, Eq. (20), as an example, performed better than Eq. (21) when they were applied to the testing dataset of 2006. However, Eq. (21) had better performance than that of Eq. (20) when 2005 data were tested. This indicates that no single GP equation can be adopted as the best GP model, and thus, representation of a set of GP equations as the optimum GP models is necessary (Parasuraman and Elshorbagy, 2008).

The proposed AET estimation models can also be compared in terms of their complexity and efficiency. The number and the data availability of the various inputs and the simplicity/ complexity of the model usage can also be investigated for better comparison of the proposed models. Table 7 provides the sum squared error (SSE) values of the proposed data driven models. Based on the SSE values, the ANN model has better fitness to the data than the GP models although it is more complex based on the number of input variables and optimized parameters. The GP model (Eq. 11) has also comparable SSE value, which indicates its good fitness though it is simple in terms of the number of inputs and estimated parameters.

Model	SSE*	No. of required input variables	No. of optimized parameters
ANN	2.16	5	24
GP			
- Equation (11)	2.83	3	4
- Equation (12)	3.50	2	3

* Sum squared error

Table 7. Akiak information criterion and sum squared error of the data driven models and their required inputs

The ANN technique provides an implicit model from which no explicit information about the AET process can be easily obtained. As a result, ANN models might be used when prediction of AET is the only concern of the modeling process. In other words, accurate estimation of AET is more important than the understanding of AET mechanism itself. Furthermore, the significant input variables that are employed for AET estimation in the ANN model cannot be explicitly/easily identified, since the associated information is stored in the network connection weights and cannot be easily interpreted. For the end user, application of ANN models is also not as easy as the equation-based models. The GP model is an equation-based model and is of interest for hydrologists and modellers because of its transparency and simplicity in application and usage. Explicit form of equation-based models, such as GP, makes it possible to extract some information about the physics of the process. The GP model has the advantage of using fewer input variables (Table 7) and also has simple and realistic structure. Consequently, the GP model becomes more applicable when a limited number of meteorological variables is available or can be measured. The simple structure of the GP models makes it easy for the users to understand how input variables are contributing to the AET process.

4.5.4. Wavelet analysis

As it was described in the previous chapter, the length of cone of influence (COI) is defined as a function of scale (e.g. $\sqrt{2}$ sfor Morlet wavelet) and increases with the scale. Since the studied range of scales mainly constitute of small scales (less than 16 hours), for all spectra shown hereafter, edge effects (COI) are negligible near both end regions of the wavelet transformation, and consequently, cannot be seen in the spectra. The thick black contour lines, seen in the wavelet spectra hereafter, enclose areas in which the values of wavelet powers are significantly greater than the background red noise at 95% confidence level. Black contour lines might be seen only as black areas in the spectra because the small-scale wavelet is narrow in time domain (high time resolution) and the peaks appear very sharp.

Daily variations are apparently the known cyclic pattern existed in the meteorological signals. Continuous wavelet transformation (CWT) conducted at scale range of 2 to 48 hours confirmed the presence of such diurnal cyclic variations. Figure 9 shows noticeably strong wavelet powers at the scale band of 16 to 32 hours, which is most likely due to the diurnal cyclic variations in the AET and R_n time series, as example. Similar spectra for other meteorological signals of T_g, T_a, RH, and W_s are provided in Appendix I. Strong wavelet powers at the band scale of 16 to 32 hours indicates that the larger-scale cyclic events are the dominant source of temporal variations in the studied time series. Consequently, small-scale cyclic events (e.g. less than 16 hours) may not play a considerable role in inducing the signals' temporal variations. In this study, the small-scale cyclic events were of interest to be investigated though they are not the main source of temporal variations. This is because the small-scale (hourly) variations and modeling of AET were the focus of this study, and the WA was examined for identifying the most important input variables in the estimation of small-scale AET values.

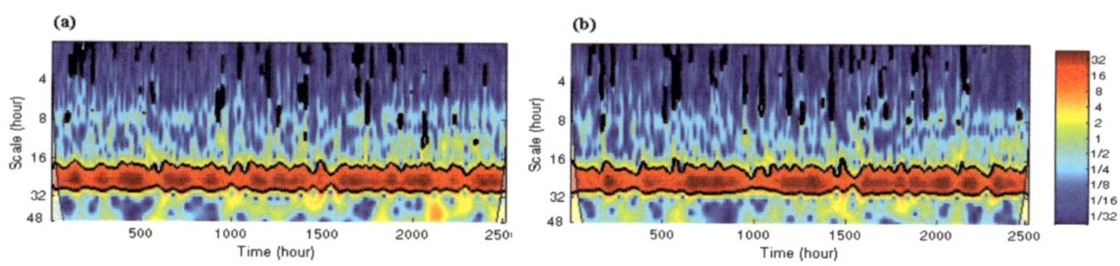

Figure 9. Continuous wavelet power spectrum of hourly time series for the scale range of 2 to 48 hours; (a) AET, (b) R_n.

Figure 10 shows continuous wavelet spectrum of the daytime hourly AET signal. Several wavelet peaks were found to be significantly different from the background red noise at scales of 2 to 8 hours, which were fairly observed along the studied period (growing season of 2006). The significant powers appeared at the scales of 2-4 hours are more frequent than those appeared at 4-8 hours showing that most of the short-time intermittent variations in the AET time series are probably produced by the 2-4 hours scale cyclic events.

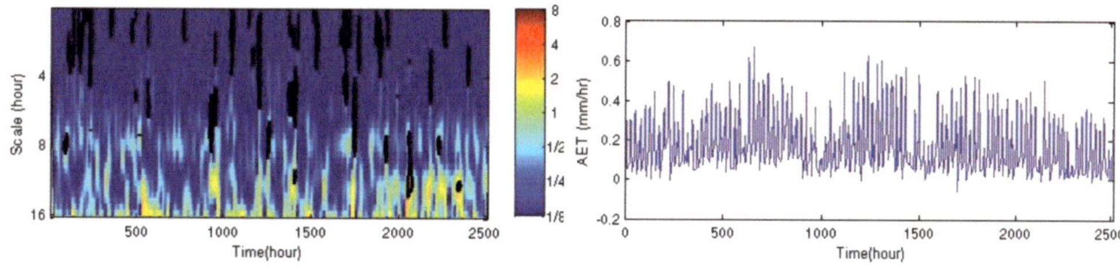

Figure 10. Continuous wavelet power spectrum (left) and time series of hourly AET (right).

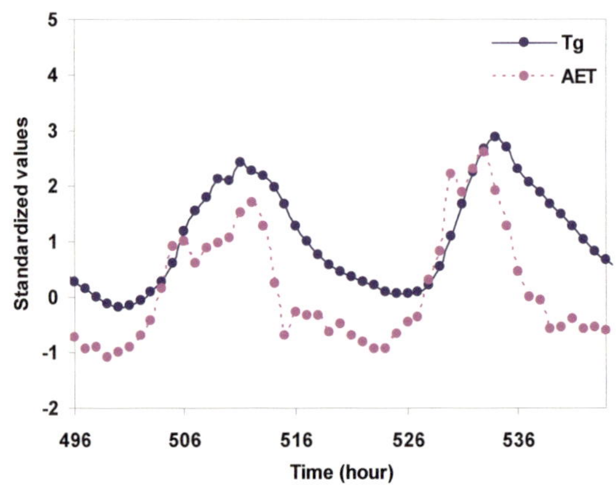

Figure 12. Time series of AET and T_g for a typical time-window of 48 hours.

Wavelet power spectrum of T_g signal is shown in Fig. 11 and exhibits no specific significant cyclic behaviour at scales less than 8 hour. The only cyclic features, which were identified to be significantly different from red noise, were at the scales of 8 to 16 hours. Detected features did not show high magnitude powers (mostly in green), which demonstrate the weak contribution of small-scale cyclic events in the temporal variations of T_g time series. This can also be observed in the zoomed time series of a typical 48-hour window of T_g signal (Figure 12). The time series of T_g does not exhibit much short-time cyclic variations, e.g., less than daily, compared to that of AET. This might be attributed to the physics of the T_g time series, which changes gradually over the short terms and is not immediately influenced by sudden fluctuations in the atmospheric condition.

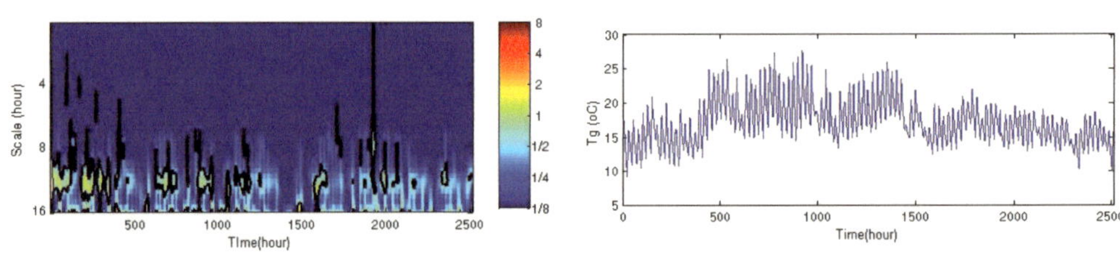

Figure 11. Continuous wavelet power spectrum (left) and time series of hourly T_g (right).

Figure 13 demonstrates limited detected cyclic features in the wavelet power spectrum of T_a, which are different from the background red noise at scales of 2 to 8 hours. The significant wavelet peaks that were identified at scales 8 to 16 hours do not contain large magnitude powers. Consequently, T_a signal might not constitute of considerable small-scale cyclic variations. Wavelet power spectrum of *RH* (Fig. 14) shows significant peaks at scales of 2-8 hours at several time locations along the studied period. Wavelet powers of *RH* spectrum at

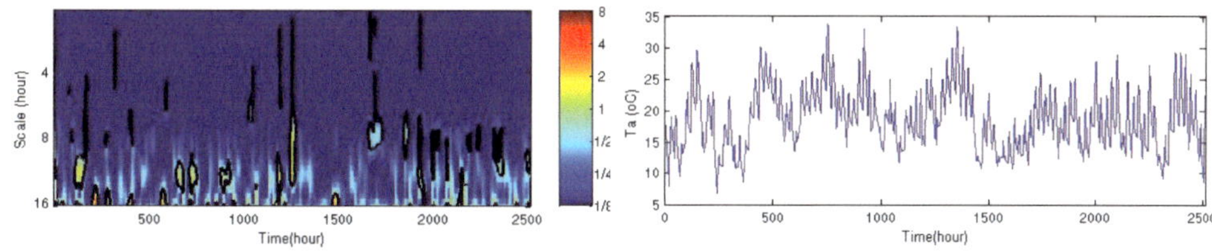

Figure 13. Continuous wavelet power spectrum (left) and time series of hourly T_a (right).

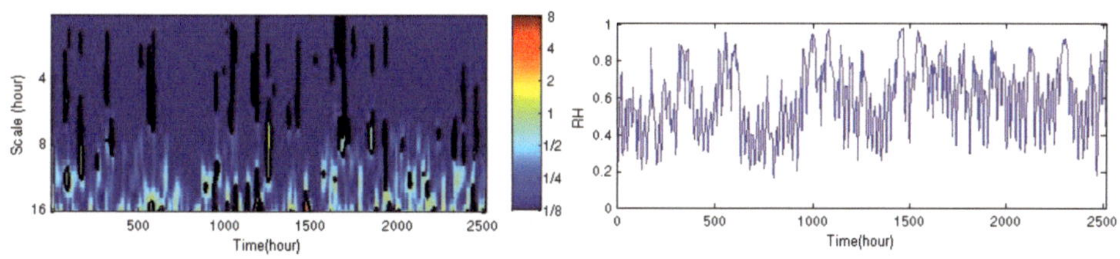

Figure 14. Continuous wavelet power spectrum (left) and time series of hourly *RH* (right).

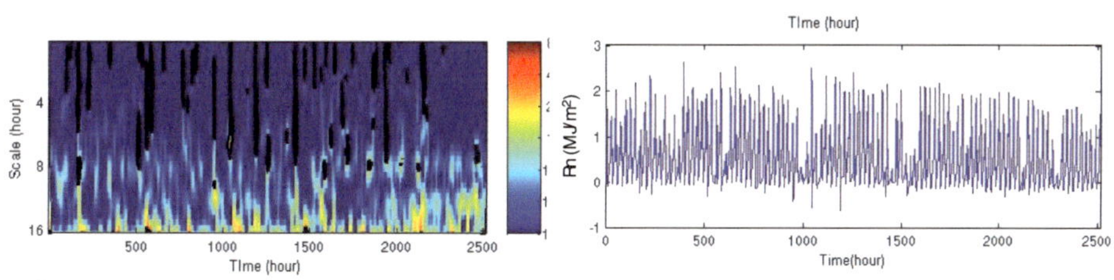

Figure 15. Continuous wavelet power spectrum (left) and time series of hourly R_n (right).

very small scales (around 2 hours) are not significantly different from the background red noise.

Cyclic temporal variations of R_n were identified to be significantly different from the red noise at small-scale band of 2 to 8 hours (Fig. 15). These significant cyclic features appeared quite frequently along the studied time duration especially at scales of 2-4 hours. Wavelet analysis of the W_s signal exhibited significant cyclic features at scales of 2 to almost 7 hours and 8 to 16 hours (Fig. 16). Small-scale cyclic features (2-4 hours) appeared more frequently than the larger-scale features (8-16 hours).

Out of the analyzed time series, the AET, R_n, *RH*, and W_s exhibited frequent small-scale cyclic features, which were found to be significantly different from the background red noise. No specific significant small-scale cyclic features were detected in the wavelet spectra of T_g and T_a, which could be attributed to two possible reasons. First, temporal variations of air and ground temperature signals do not involve considerable small-scale cyclic features and are mostly generated by larger-scale cyclic trends. Second, the likely existing small-scale cyclic

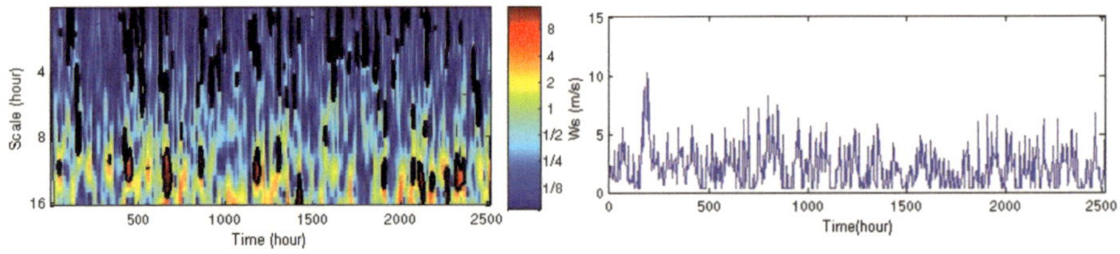

Figure 16. Continuous wavelet power spectrum (left) and time series of hourly W_s (right).

variations are not large enough, in magnitude (because of slight changes of these variables in small time scales), to be differentiated from the background red noise and consequently, cannot be detected as significant cyclic features in the wavelet power spectrum.

As it was discussed earlier in this section, although several small-scale cyclic features were found in the wavelet spectra of the time series, they might not substantially contribute to the temporal variations of the considered signals. The reason might be the existence of larger-scale features (e.g. scales of 16 to 32 hours), which induce the major temporal variations in most of the studied time series (Izadifar, 2010).

4.5.5. Cross wavelet analysis

It can be seen from the cross wavelet spectrum of AET-R_n (Fig. 17) that both time series have common significant powers at scales of 2 to 8 hours along the studied period. This demonstrates the significant linear correlation between AET and R_n signals at small 2scales at 95% confidence level. To be more specific, the significant AET-R_n correlations appeared at particular time locations but not continuously along the time axis. This means that the linear covariation of these two time series becomes significantly different from red noise only at some periods of time, which is more likely due to the low magnitude of variations at small scales compared to that of larger scales (e.g. diurnal). In other words, there might be non-noise small-scale covariances between the time series; however, since they are not major source of variations in the time series and are low in magnitude, associated cross wavelet powers cannot be distinguished from background noise. Significant powers of AET-R_n cross wavelet spectrum imply that small-scale variation of AET can be explained by R_n time series. Phase information is provided in the cross wavelet spectra using arrows. Arrows, pointing right and left, show in order in-phase and anti-phase relationship between the two time series. Fig. 17 indicates the in-phase relationship between AET and R_n at significant areas. By in-phase, it means that the two time series are positively correlated. The relationship between AET and R_n was observed to be not necessarily in-phase over all detected significant areas, since there are some cases when other involved factors affect the conventional cause and effect relationship between the two signals. Varying (not fixed) phase information was also observed in the cross wavelet spectra between AET and other considered meteorological signals, which might be attributed to the range of studied scales (small-scales). Small-scale cyclic events are not the dominant source of variation in the studied time series and consequently, might not carry solid phase

http://dx.doi.org/10.5772/52809

information. Larger-scale features, which have more contribution to the temporal variations, may contain less varying and more reliable phase information (Izadifar, 2010).

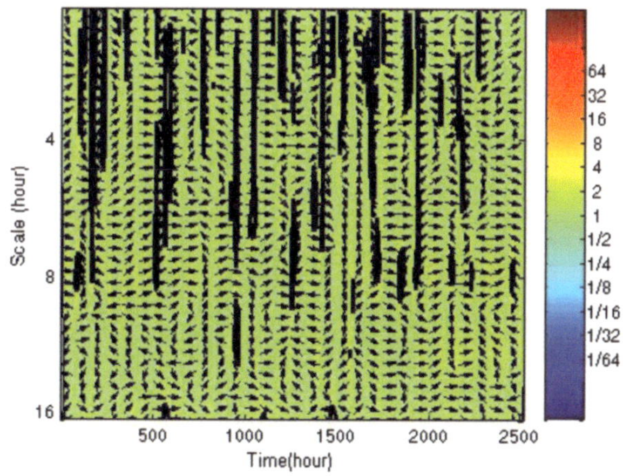

Figure 17. Cross wavelet transform of the AET-R_n time series.

Cross wavelet transform of AET and *RH* exhibited significant common features at the scale band of 2 to 8 hours (Fig. 18). Significant correlation between AET and *RH* indicates that the *RH* signal can describe some of the small-scale variations of AET. Although the magnitude of linear correlation between the two time series might not be identified quantitatively, it is seen that *RH* has a significant cause and effect relationship with the AET signal at small scales. Similar to the AET-R_n cross wavelet spectrum, phase relationship between AET and *RH* was not stable. However, some anti-phase relationship can be observed at specific time-scale locations. By anti-phase it means that the AET and *RH* are negatively correlated to each other. Unstable phase relationship, at low scales, also demonstrates the complexity that exists in the short-time variations of AET and its relationship with the involved meteorological factors, which cannot be easily explored by using the cross wavelet transformation.

W_s time series also exhibited significant covariances with AET signal at scales of 2 to 8 hours (Fig. 19), which were more frequent at scales less than 4 hours. All of the significant small-scale features found in individual wavelet transform of the W_s time series were not detected as common features between AET and W_s at 95% confidence level. It indicates that only specific numbers of short-time cyclic variations of W_s are linearly correlated to the small-scale cyclic variations of AET. Overall, the results of cross wavelet analysis of AET and W_s demonstrate the existence of linear correlation at small scales. Phase information of AET-W_s spectrum illustrates both in-phase and anti-phase relationship between the variations of the two analyzed signals.

As it was expected form the individual wavelet spectra of AET and T_g, no specific significant common power was found at small scales in the cross wavelet transform of AET-T_g (Fig. 20). This might be attributed to two possible reasons; first, the presence of non-linear correlation

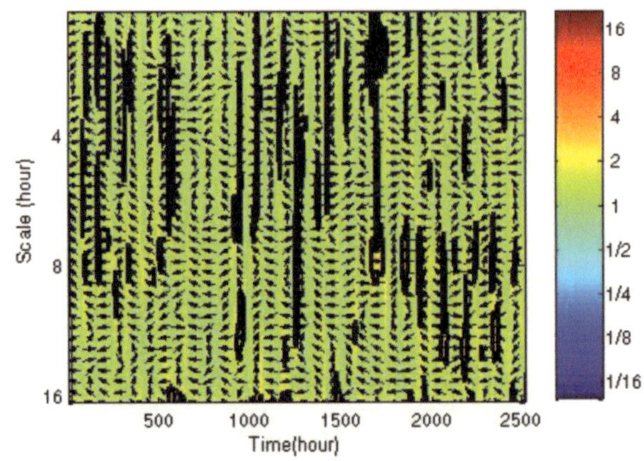

Figure 18. Cross wavelet transform of the AET-*RH* time series.

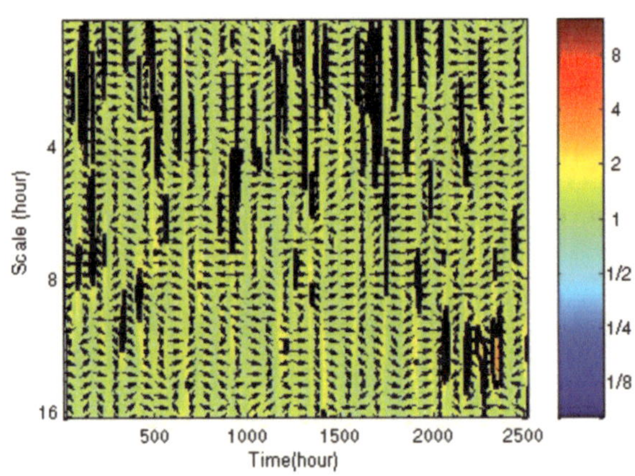

Figure 19. Cross wavelet transform of the AET-W_s time series.

between AET and T_g at small scales, which cannot be identified by the current cross wavelet analysis. Second, for a time series like T_g, which is not varying much over short time intervals (e.g. hourly), small-scale cyclic features do not have high powers at scales of 2-8 hours and result in low cross wavelet powers of AET-T_g spectrum that cannot be differentiated from background red noise.

Cross wavelet spectrum of AET-T_a shows limited detected features in the time-scale domain in which the two signals were linearly correlated and the power was significantly different from red noise at 95% confidence level compared to those of AET-R_n, AET-*RH*, and AET-W_s (Fig. 21). Considering the rare significant peaks detected in the band scales of 2 to 8 hours of T_a univariate power spectrum (Fig. 13), the identified powers in the cross wavelet spectrum

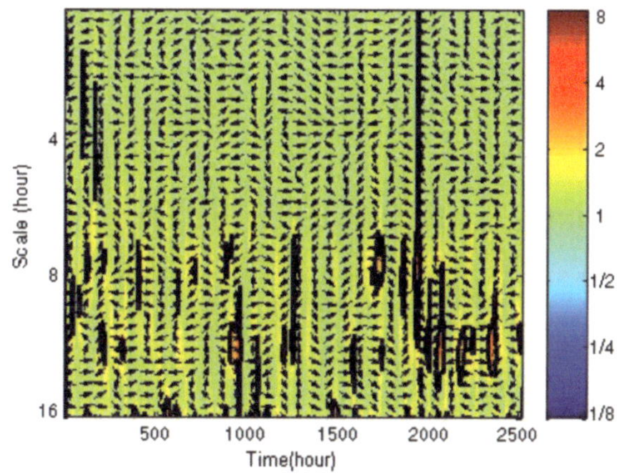

Figure 20. Cross wavelet transform of the AET-T_g time series.

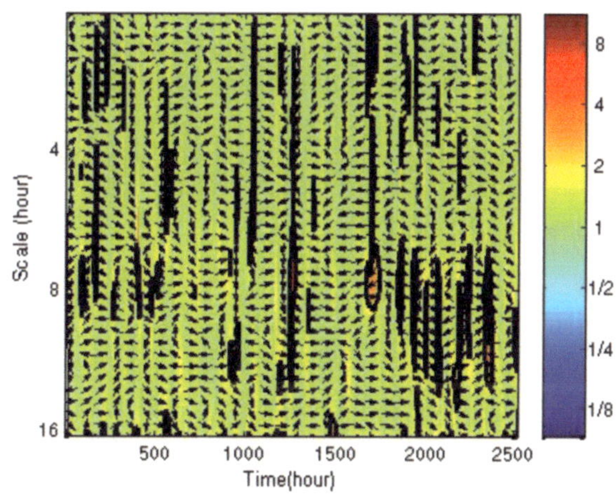

Figure 21. Cross wavelet transform of the AET-T_a time series.

might not indicate significant covariations between the two signals. The significant common powers in the cross wavelet spectrum of AET-T_a were most probably caused by the strong powers of univariate AET spectrum only, which were more frequent than those of T_a over the studied time period. Consequently, no specific and reliable cause and effect relationship can be perceived between AET and T_a time series at small scales. However, strong linear correlation, which was significantly different from background red noise, was observed between AET and T_a at about diurnal scale (scale band of 16 to 32 hours) when the range of studied scales was extended up to 48 hours, Fig. 22.

The results of the cross wavelet analysis determined, to some extent, the meteorological variables that have significant linear correlation with the AET signal at small scales at 95%

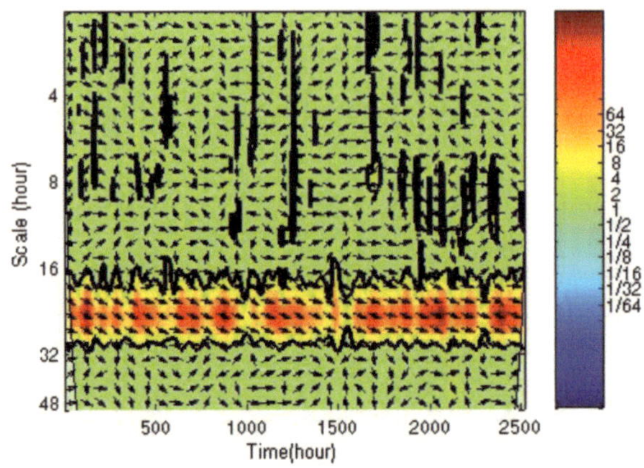

Figure 22. Cross wavelet transform of the AET-T_a time series for the scale range of 2 to 48 hours.

confidence level. Based on the cross wavelet analysis, R_n, *RH*, and W_s time series exhibited significant correlation with the AET signal and therefore, they are the important variables in the prediction of small-scale AET variations. Based on the values provided on the scales (color bars) of the cross wavelet spectra, R_n exhibited stronger correlation with AET than *RH*, which is stronger than W_s, with maximum cross wavelet power magnitudes of 256, 16, and 8, respectively. Unfortunately, the results of the cross wavelet analysis cannot be interpreted in a precise quantitative way to identify the importance of one variable over others in the prediction of AET at specific scales (or band scales). In addition, significant powers detected in the univariant wavelet spectra must always be considered when significant cross wavelet powers are interpreted. This is to avoid false common powers, which might be created by the large magnitude powers of one univariate spectrum only.

Based on the cross wavelet analysis, ground temperature has no important linear correlation with the AET time series at small scales. As a result, one might not select T_g as a predictor in the estimation of AET. However, the results of the data driven modeling demonstrated the importance of ground temperature in the prediction of AET. This inconsistency might be attributed to the previously mentioned ability of cross wavelet analysis in identifying only linear correlations between time series. As a result, any existing non-linear correlation between each pair of signals remains undiscovered using cross wavelet analysis. Another possible reason for insignificant common powers in AET-T_g cross wavelet spectrum is that the cross wavelet analysis can investigate the correlation between only two time series at a time (not multiple time series), which ignores the possible effect of other factors interacting with the considered signals. The importance of interaction effects, which exist among the variables involved in the AET process, was observed in the results of the data driven modeling. The above-mentioned limitations of cross wavelet analysis affect the accuracy of the correlation analysis of time series for determining the most important AET predictors.

The other issue, which most probably resulted in the inconsistency between the findings of wavelet analysis and data driven modeling is the time-scale at which the temporal variations of signals were investigated using wavelet analysis or modeled by data driven techniques. The results of the cross wavelet analysis at the larger-scales might be in agreement with the results of the data driven models, regarding the most effective variables in the prediction of AET. Cross wavelet analysis exhibited strong linear correlation between AET and T_g at about diurnal scale approximately over the whole studied period. As a result, it can be perceived that the proposed data driven models mainly characterized the larger-scale variations of AET (the dominant cyclic patterns) for which R_n and T_g were found to be the most effective predictors. The cross wavelet analysis, conducted in the present study, concerned more about the small-scale variations of AET. The results obtained using the WA at small scales might be useful in the development of AET models that aim to characterize the small-scale temporal variations.

The above-mentioned discussion highlights the importance of the time-scale of temporal variations, which might be of interest to be investigated, analyzed, and modeled. Depending on the specific time-scale of variations one is interested in, the employed modeling technique and/or, for instance, the range of studied scales in the wavelet analysis may vary. As a result, it is important to have better understanding of different types of temporal variation exist in the investigated time series prior to the signal analysis and/or modeling.

5. Conclusion

In conclusion, the investigated data driven modeling techniques were promising for the estimation of the hourly AET mechanism using the observed data, without assuming or applying significant knowledge of the physics of the process. The choice of the testing dataset was found to be important for realistically assessing the generalization ability of the proposed data driven models and also for the determination of the possible superiority of any of the modeling techniques over others. The genetic programming (GP) modeling technique was found to perform similarly and better than the ANN model with regard to generalization ability. The GP-evolved models also had the advantage of being structurally simple and requiring fewer input variables, which is of interest for many hydrological practitioners.

Furthermore, the proposed equation-based GP models showed that the AET process has the potential to be estimated by structurally simple (e.g. semi-linear) models. Equation-based AET models made it possible to extract some information about the physics of the process. It was observed that the meteorological variables of R_n and T_g have larger contribution, than other variables, to the estimation of AET. In addition, the interaction effects of the meteorological variables were found to be important and effective in the estimation of AET.

The results of wavelet analysis improved the understanding of the AET mechanism by revealing the importance and contributions of different time-scale cyclic variations exist in the AET time series. This highlights the issue of time-scale and the importance of its consideration in the modeling and prior-to-modeling input selection procedure. Although several small-scale cyclic features were detected in the AET signal, larger-scale variations were found to be the major frequency events at which the predictant-predictor (AET-meteorological variables) correlation analysis was more clear and reliable. GP models were noted to mainly model the larger-scale (dominant) temporal variations of AET, although short time-scale (hourly) data were employed for training and developing the models.

Consistency between the results of data driven modeling (especially GP) and wavelet analysis, regarding the most important predictor variables, at large time-scales (e.g. diurnal) indicated that wavelet analysis can be employed as a guide for identifying the most linearly correlated predictors for the modeling of AET. However, limitations of such signal analysis tool should be considered when it is used for input determination prior to modeling. Wavelet analysis helped to perceive the difference between the predictive abilities of various models from a new perspective, which is the time-scale of variations that the models characterize. For instance, when the performances of two prediction models are compared, it should be noted if both models are capturing the same time-scale of variations. Consideration of this point can make the models' comparative analysis more accurate and fair.

Author details

Zohreh Izadifar and Amin Elshorbagy

University of Saskatchewan, Canada

References

[1] Akaike, H. (1974). A new look at statistical model identification. IEEE Trans. Automat. Contr., AC, , 19, 716-722.

[2] Allen, M. R, & Smith, L. A. (1996). Monte Carlo SSA: Detecting irregular oscillations in the presence of coloured noise, J. Clim., , 9, 3373-3404.

[3] Anctil, F, & Tape, D. G. (2004). An exploration of artificial neural network rainfall-runoff forecasting combined with wavelet decomposition. J. Environ. Eng. Sci., 3, SS128., 121.

[4] ASCE(2000). Artificial Neural Networks in Hydrology. I. Preliminary Concepts. J. Hydrol. Eng., , 5(2), 115-123.

[5] Baldocchi, D. D, Hicks, B. B, & Meyers, T. P. (1988). Measuring biosphere-atmosphere exchanges of biologically related gases with micrometeorological methods. Ecology, 69, 1331–1340.

[6] Banzhaf, W, Nordin, P, Keller, R. E, & France, F. (1998). Genetic Programming: An Introduction. Morgan Kaufmann Publishers, Inc. San Francisco, California, USA.

[7] BhakarS.R; Oiha, S.; Singh, R.V. & Ansari, A. ((2006). Estimation of evapotranspiration for wheat crop using artificial neural network, Proceedigs of 4th World Congress on Computers in Agriculture, Orlando, FL, US.

[8] Box, G. E. P. (1957). Evolutionary operation: A method for increasing industrial productivity. Appl. Statistics, , 6(2), 81-101.

[9] Bremermann, H. J. (1962). Optimization Through Evolution and Recombination, In: Self-Organizing Systems, M.C. Yovits et al, (Ed.), Spartan Books, Washington, DC., 93-106.

[10] Cahill, A. T. (2002). Determination of changes in streamflow variance by means of a wavelet-based test. Water Resour. Res., , 38(6), 1065-1078.

[11] Chen, X, & Liu, D. (2008). Wavelet analysis on inter-annual variation of precipitation in Guangdong, China. IAHS publication, , 319, 3-9.

[12] Cheng, B, & Titterington, D. M. (1994). Neural networks: A review from a statistical perspective. Statistical Science, , 9(1), 2-54.

[13] Coulibaly, P. (2006). Spatial and temporal variability of Canadian seasonal precipitation (1900-2000). Advances in Water Resources, , 29, 1846-1865.

[14] Coulibaly, P, & Burn, D. H. (2004). Wavelet analysis of variability in annual Canadian streamflows. Water Resource Research, 40, W03105, doi:10.1029/2003WR002667.

[15] Cybenko, G. (1989). Approximation by superposition of a sigmoidal function. Math. Control Signals Syst., , 2, 303-314.

[16] Dai, X, Shi, H, Li, Y, Ouyang, Z, & Huo, Z. (2009). Artificial neural network models for estimating regional reference evapotranspiration based on climate factors. Hydrological processes, , 23(3), 442-450.

[17] Dingman, S. L. (2002). Physical Hydrology, Prentice-Hall, Inc., Upper Saddle River, NJ.

[18] El-Baroudy, I, Elshorbagy, A, Carey, S, Giustolisi, O, & Savic, D. (2009). Comparison of Three Data-driven Techniques in Modelling Evapotranspiration Process. Journal of Hydroinformatics, In press.

[19] Friedberg, R. M. (1958). A learning machine: Part I. IBM J., , 2(1), 2-13.

[20] Friedberg, R. M, Dunham, B, & North, J. H. (1959). A learning machine: Part II. IBM J., , 3(7), 282-287.

[21] Grinsted, A, Moor, J. C, & Jevrejeva, S. (2004). Application of the cross wavelet transform and wavelet coherence to geophysical time series. Nonlinear Processes in Geophysics, , 11, 561-566.

[22] Gupta, V. K, & Waymire, E. (1990). Multiscaling properties of spatial rainfall and river flow distribution. Journal of Geophysical Research, 95, 1999–2009.

[23] Hasselmann, K. (1976). Stochastic climate models: part I. theory. Tellus, , 28(6), 473-85.

[24] He, Y, Guo, R, & Si, B. C. (2007). Detecting grassland spatial variation by a wavelet approach. Int. J. Remote Sens., , 28, 1527-1545.

[25] Hornik, K, Stinchcombe, M, & White, H. (1989). Multilayer feedforward networks are universal approximators. Neural Networks, , 2(5), 359-366.

[26] Izadifar, Z. Modeling and analysis of actual evapotranspiration using data driven and wavelet techniques. M.Sc. thesis, University of Saskatchewan, Saskatoon, Saskatchewan, CA.

[27] Jain, S. K, Nayak, P. C, & Sudheer, K. P. (2008). Models for estimating evapotranspiration using artificial neural networks, and their physical interpretation. Hydrological Processes, , 22(13), 2225-2234.

[28] Kaheil, Y. H, Rosero, E, Gill, M. K, Mckee, M, & Bastidas, L. A. (2008). Downscaling and forecasting of evapotranspiration using a synthetic model of wavelets and support vector machines. IEEE Trans. Geosci.Remote Sens., , 46(9), 2692-2707.

[29] Kirkup, H, Pitman, A. J, Hogan, J, & Brierley, G. (2001). An initial analysis of river discharge and rainfall in Coastal New South Wales, Australia, using wavelet transforms. Australian Geographical Studies, , 39(3), 313-334.

[30] Koza, J. R. (1992). Genetic Programming: On the Programming of Computers by Means of Natural Selection, MIT Press, Cambridge, MA.

[31] Kumar, M, Raghuwanshi, N. S, Singh, R, Wallender, W. W, & Pruitt, W. O. (2002). Estimating evapotranspiration using artificial neural network. Journal of Irrigation and Drainage Engineering, , 128(4), 224-233.

[32] Kumar, P, & Foufoula-georgiou, E. (1993a). A multicomponent decomposition of spatial rainfall fields 1. Segregation of large-and small-scale features using wavelet transforms. Water Resources Research, 29, 8, 2515–2532.

[33] Kumar, P, & Foufoula-georgiou, E. (1993b). A multicomponent decomposition of spatial rainfall fields 1. Self-similarity in fluctuations. Water Resources Research, 29, 8, 2533–2544.

[34] Labat, D. (2006). Oscillations in land surface hydrological cycle. Earth and Planetary Science Letters, 242, 1-2, 143-154.

[35] Labat, D. (2008). Wavelet analysis of the annual discharge records of the world's largest rivers. Adv. Water Resour., , 31(1), 109-117.

[36] Labat, D, Ronchail, J, & Guyot, J. L. (2005). Recent advances in wavelet analyses: Part Amazon, Parana, Orinoco and Congo discharges time scale variability. J Hydrol., 314, 1-4, 289-311., 2.

[37] Lafreniere, M, & Sharp, M. (2003). Wavelet analysis of inter-annual variability in the runoff regimes of glacial and nival stream catchments, Bow Lake, Alberta. Hydrological Processes, , 17(6), 1093-1118.

[38] Landeras, G, Ortiz-barredo, A, & Lopez, J. J. (2008). Comparison of artificial neural network models and empirical and semi-empirical equations for daily reference evapotranspiration estimation in the Basque Country (Northern Spain). Agricultural Water Management, , 95(5), 553-565.

[39] Levenberg, K. (1944). A method for the solution of certain problems in least squares. Quart. Appl. Math. , 2, 164-168.

[40] MacKayD. ((1992). A practical bayesian framework for backproparation networks. Neural Computation, , 4(3), 448-472.

[41] Maier, H. D, & Dandy, G. C. (2000). Neural network for the prediction and forecasting of water resource variables: a review of modeling issues and applications. Environmental Modelling & Software, , 15, 101-124.

[42] Marquardt, D. (1963). An algorithm for least squares estimation of non-linear parameters. J. Ind. Appl. Math., , 11(2), 431-441.

[43] Marquardt, D. (1963). An algorithm for least squares estimation of non-linear parameters. J. Ind. Appl. Math., , 11(2), 431-441.

[44] MATLAB softwareVersion [6.5.1]. Copyright © (2009). MathWorks, Inc., 3 Apple Hill Drive, Natick, MA, USA., 1994-2009.

[45] Miao, C. Y, Wang, Y. F, & Zheng, Y. Z. (2007). Wavelet-based analysis of characteristics of summer rainfall in Nenjiang and Harbin. Journal of Ecology and Rural Environment, , 23(4), 29-32.

[46] Neural Network Toolbox User's Guide(2009). MathWorks, Inc., 3 Apple Hill Drive, Natick, MA, USA.

[47] Parasuraman, K, & Elshorbagy, A. (2008). Toward improving the reliability of hydrologic prediction: Model structure uncertainty and its quantification using ensemble-based genetic programming framework. Water Resour. Res., 44, W12406, doi: 10.1029/2007WR006451.

[48] Parasuraman, K, Elshorbagy, A, & Carey, S. K. (2006). Spiking modular neural networks: A neural network modeling approach for hydrological processes. Water Resour. Res., 42, W05412, doi:10.1029/2005WR004317.

[49] Parasuraman, K, Elshorbagy, A, & Carey, S. K. (2007). Modeling the dynamics of the evapotranspiration process using genetic programming, Hydrological Sciences Journal, , 52, 563-578.

[50] Polikar, P. (1996). The wavelet tutorial, http://www.site.uottawa.ca/~qingchen/wavelet/htm,accessed on July 2008.

[51] Russo, D, Bresler, E, Shani, U, & Parker, J. C. (1991). Analysis of infiltration events in relation to determining soil hydraulic properties by inverse problem, methodology. Water Resour. Res., , 27, 1361-1373.

[52] Saco, P, & Kumar, P. (2000). Coherent modes in multiscale variability of streamflow over the United States. Water Resour.Res., , 36(4), 1049-1067.

[53] SAS/STAT softwareVersion [9.1] of the SAS System for Windows. Copyright © SAS Institute Inc. SAS and all other SAS Institute Inc. product or service names are registered trademarks or trademarks of SAS Institute Inc., Cary, NC, USA., 2002-2003.

[54] Schaefli, B, Maraun, D, & Holschneider, M. (2007). What drives high flow events in the Swiss Alps? Recent developments in wavelet spectral analysis and their application to hydrology. Advances in Water Resources, , 30(12), 2511-2525.

[55] Sette, S, & Boullart, L. (2001). Genetic programming: principles and applications. Engineering Application of Artificial Intelligence, , 14, 727-736.

[56] Si, B. C. (2003). Spatial and scale-dependent soil hydraulic properties: A wavelet approach. In: Scaling methods in soil physics, Ya. Pachepsky et al., (Ed.), CRC Press, New York., 163-178.

[57] Si, B. C. (2008). Spatial Scaling Analyses of Soil Physical Properties: A Review of Spectral and Wavelet Methods. Vadose Zone Journal, , 7(2), 547-562.

[58] Si, B. C, & Zeleke, T. B. (2005). Wavelet coherency analysis to relate saturated hydraulic properties to soil physical properties. Water Resource Res., 41, 11, W11424. doi:10.1029/2005WR004118.

[59] Silva, S. (2005). GPLAB-a genetic programming toolbox for MATLAB. http://gplab.sourceforge.net.

[60] Sudheer, K. P, Gosain, A. K, & Ramasastri, K. S. (2003). Estimating actual evapotranspiration from limited climatic data using neural computing technique. Journal of Irrigation and Drainage Engineering, ASCE, , 129, 214-218.

[61] Swingler, K. (1996). Applying neural networks: A practical guide, Academic press, London.

[62] Torrence, C, & Compo, G. P. (1998). A practical guide to wavelet analysis. Bull. Am. Met. Soc., , 79, 61-78.

[63] Westra, S, & Sharma, A. (2006). Dominant modes of interannual variability in Australia rainfall analyzed using wavelets. Journal of Geophysical Research-part D: Atmospheres, 111, 5, D05102, doi:10.1029/2005JD005996.

[64] Zanetti, S. S, Sousa, E. F, Oliveira, V. P. S, Almeida, F. T, & Bernardo, S. (2007). Estimating evapotranspiration using artificial neural network and minimum climatological data. J. Irrig. and Drain. Engrg., , 133(2), 83-89.

Chapter 10

Influence of Soil Physical Properties and Terrain Relief on Actual Evapotranspiration in the Catchment with Prevailing Arable Land Determined by Energy Balance and Bowen Ratio

Renata Duffková

Additional information is available at the end of the chapter

http://dx.doi.org/10.5772/52810

1. Introduction

Actual evapotranspiration rate (ETa) represents a key element of landscape water balance. It plays an active role in the biomass production, establishes the cooling capacity of the region and, depending on soil properties, contributes to runoff formation in the catchment [1-3]. The rate of the process is determined by the gradient of water potential between soil, vegetation, and atmosphere and the prevailing aerodynamic and surface resistances. It integrates the effects of meteorological parameters (precipitation, radiation energy, water saturation deficit and wind speed), soil water content, soil hydraulic properties, vegetation density, height and roughness and the depth of the root system [4-8] on both the spatial and the temporal bases.

Physical properties of soils have a significant influence on their water regime and should be considered when selecting suitable agricultural crops for particular sites, taking into account the crop productivity and its water requirements. The impact of the soil on ETa depends upon the properties of its pore space, which are determined primarily by its grain size distribution and structure. Clay (fine-textured) soils tend to show higher porosity [8-9], higher soil water storage and ETa, but, on the other hand, lower hydraulic conductivity and subsurface runoff [2], compared to sandy (coarse-textured) soils. The highest available moisture-holding capacity is displayed by loamy soils, which, though possessing a somewhat lower field water capacity than the clay soils, exhibit a significantly lower wilting point than the latter. The movement of water in the soil can be extensively altered by the preferential (e.g., macropore) flow, which is 100 to 400 fold faster than water flow in the soil matrix [10], depending on rainfall and snowmelt patterns and, if applied, on irrigation management.

Physical properties of soils influence the selection of suitable agricultural crops with respect to their water consumption and their productivity. For example in [11] was showed that soil types with higher moisture-holding capacity are better suited to crops, such as corn, that are more sensitive to atmospheric drought than to less sensitive crops, such as wheat.

The spatial distribution of soil types and textures, and thus of the soil water storage, is determined by geology, terrain relief, climate and biotic factors. From a hydrogeologic viewpoint, the catchment can be divided into recharge zones, where precipitation infiltrates and then recharges the groundwater store, and discharge zones, where groundwater approaches the land surface or a surface water body [5]. The recharge zones are mainly located in the highest areas of the catchment, close to the catchment divide, peaks and ridges. The soils of these zones are typically shallow and stony, with high sand content and high infiltration capacity. The coarse-textured soils of the recharge zones are, with respect to groundwater resources, well suited to growing grass, which, beside water quality benefits, increases their field capacity and results in virtually complete infiltration of precipitation, including rainstorms [12-16]. The discharge zones can be found in the lowest parts of the slopes and along surface streams and lakes and are prone to surface waterlogging. The dominant soils in the discharge zones are generally deep, with higher clay content and a lower capacity for infiltration. A connection between the recharge zones and the discharge zones is provided by transient zones, where precipitation is mostly transformed to surface runoff and groundwater flows downslope in a quasi-steady way [12, 17]. The transient zones are located mainly in the middle sections of slopes. Groundwater in natural catchments flows from the recharge zones to the discharge zones. Actual spatial distribution of these zones depends on local geologic and geomorphologic conditions [18, 19].

2. The concept of the problem studied

The topic of the chapter is dealing with the influence of physical soil properties in the recharge, transient, and discharge zones on ETa in a cultivated catchment, where different crops are grown and some fields are tile-drained, in different periods and vegetation development phases. In addition to these comprehensive factors of the habitat, the results of ETa research are also connected with the method used (Bowen ratio method) which is dependent on air temperature and humidity coming over extensive plant cover from prevailing wind direction.

3. Description of the catchment and weather stations

The study was conducted in the experimental catchment Dehtáře, situated in the south-west Bohemo-Moravian Highland (Czech Republic), in the years 2004, 2006 and 2009. The outflow point of the catchment lies at 49°28′ N - 15° 12′ E.

The catchment Dehtáře (Fig. 1) has an area of 59.6 ha, with tile drained areas occupying 19 ha (~32%). The catchment area is mainly agricultural land (89.3%). Minor forested areas (3.3%) lie

http://dx.doi.org/10.5772/52810

at its north-western and northern borders. Grassland (20.3%) covers the southern part of the catchment, as well as the adjacent lowest lying south-western area, which is tile-drained. The remaining area (69%) is arable land, which is exploited mainly for cereals production. The catchment geomorphology belongs to the erosion-accumulation relief type [20]. The altitude ranges between 497.0 and 549.8 m. According to Quitt [21], the local climate is classified as moderately warm. According to Köppen [21], it belongs to the temperate broadleaf deciduous forest (Cfb) zone. The average annual total precipitation is 660 mm and the average air temperature is 7.0°C.

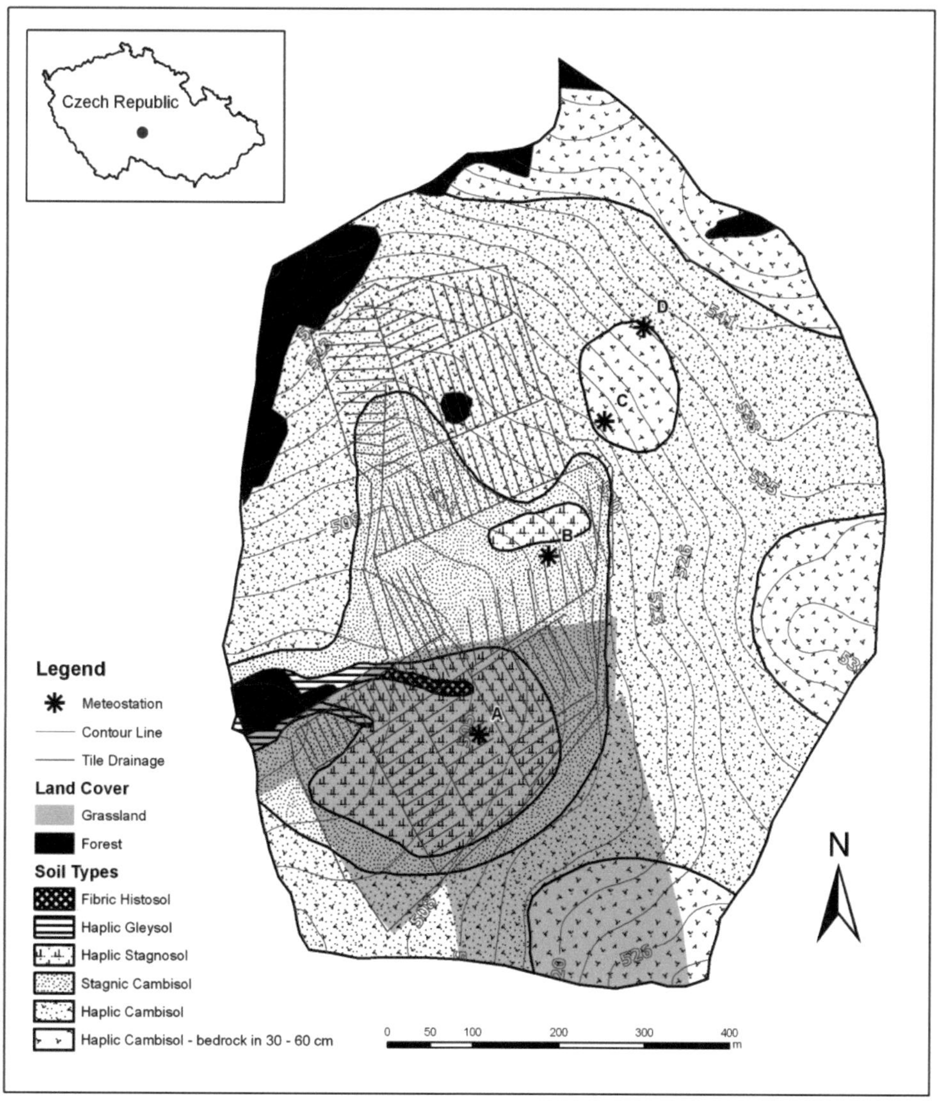

Figure 1. An overview map of the Dehtáře catchment and its soil types

There is no permanent surface drainage channel within the catchment. The catchment hydrogeology is characterised by shallow aquifers (with groundwater table in the discharge zone lying at 0.2–1.5 m), occurring in Quaternary deposits, in the weathered zone of the bedrock and in its fissures and faults. The bedrock is a partially migmatized paragneiss. Quaternary deposits are slope sands and bottom loams, reaching a thickness of 1–2 m. The bottom loams usually act as aquitards and have their own phreatic groundwater table, partially recharged from the atmosphere, so that the water from below mixes in them with the water from above. The dominant soil types according to [22] (see Fig. 1) are Haplic Cambisols (CMha) in the recharge and transient zones. These soils are light, shallow and stony (the thickness of the soil profile being only 30 cm in some parts) sandy loams and loamy sands (according to the USDA soil texture triangle, see [23]). Haplic Stagnosols (STha), Haplic Gleysols, Fibric Histosols and Stagnic Cambisols (CMst) are typical for the discharge and the discharge/ transient zones. Medium-deep sandy loams dominate in the lower parts of the slopes, while deep loams are most typical for the catchment bottom. However, the spatial variability of grain size distribution in the soils of the catchment bottom makes surveying difficult. This variability is due to natural erosion and accumulation processes as well as due to artificial mixing which occurred during the tile drainage installation. The clay (< 0.002 mm) content in the topsoil and subsoil varies from 10 to 13% and from 9 to 13%, respectively, near the water divide, and from 12 to 15% and from 13 to 25%, respectively, at the catchment bottom. A layer of clay loam was identified by geophysical survey [24, 25] at the catchment bottom at the depth 30 to 200 cm, in some places cropping up to the soil surface.

The tile drainage (still fully functioning) was laid in 1977 in the western, lower part of the catchment. The average slope of the drained land is about 5%. The spacing of lateral drains is either 13 or 20 m. The depth of the laterals is about 1.0 m, while the depth of the mains is about 1.1 m. Circumferential intercepting tile drains, provided with gravel filters, are placed at depths 1.1 to 1.8 m. The tile drainage system empties into a fire water reservoir. Seasonal ascending springs, either point springs or spring lines, emerged in middle parts of the catchment slopes before the tile drainage system installation, causing temporary waterlogging of the lands [26, 27]. Today, the soils in the tile-drained north-western part of the catchment (Fig. 1) can be classified as CMha. Before the tile drainage installation they mainly belonged to the STha type.

Four weather stations (A, B, C, D - Fig. 1 and Table 1) were placed in the experimental catchment to record conditions on sites with different soil types and textures and in different relief zones. The stations A and B were located in the discharge and the discharge/transient zones, respectively (Table 1) on texturally heavier STha and CMst soils, where the terrain slope was 2–4°. The station B was placed in concentrated flow paths. The stations C and D were situated in the transient and the transient/recharge zones, respectively, on the CMha soil and the terrain slope 5–7°. Permeable loamy-sand soils under stations C and D and between them are shallow; weathered bedrock can be found at 15–40 cm below the surface. Meteorological data of the station C, based on prevailing wind direction, was markedly influenced by mentioned drying shallow bedrock and deeper, but tile drained CMha (Fig. 1). In addition to

http://dx.doi.org/10.5772/52810

drying shallow bedrock the station D data was also significantly affected by deeper wetter soil without a drainage system.

Weather station	Terrain zone*	Altitude (m)	Soil type*	Textural class Topsoil/ subsoil*	Average clay content* < 0.002 mm (%)		Average sand content* 0.05-2.0 mm (%)		Maximum capillary water capacity** (% vol) at 20-30 cm	Land use*	Tile-drained area* (%)	Slope (°)*
					topsoil	subsoil	topsoil	subsoil				
A	Discharge zone	506	Haplic Stagnosol (STha)	Silt loam or loam or sandy loam / Sandy loam or loam or clay loam	12.3	16.1	48.3	48.4	40.8 (topsoil)	grassland	100	2-3
B	Discharge/ transient zone (in concentrated flow paths)	513	Stagnic Cambisol (CMst), Haplic Stagnosol (STha)	Loam / Sandy loam or loam (STha)	12.3 (CMSt) 16.5 (STha)	10.6 (CMSt) 14.9 (STha)	48.5 (CMSt) 44.4 (STha)	62.2 (CMSt) 53.5 (STha)	39.8 (subsoil)	arable land	69	3-4
C	Transient zone	523	Haplic Cambisol (CMha)	Sandy loam or loamy sand /Sandy loam or loamy sand	11.0	11.7	66.9	62.2	30.4 (subsoil)	arable land	29	5-6
D	Transient/ recharge zone	534	Haplic Cambisol (CMha)	Sandy loam or loamy sand /Sandy loam or loamy sand	12.6	10.7	62.5	66.2	30.3 (subsoil)	arable land	0	6-7

* within a 100 m radius circle around each weather station

** samples taken close to weather stations

Table 1. Location, soils, land use, tile-drainage and terrain conditions around the weather stations A to D in the Dehtáře catchment

Each station was equipped with an ETa-measuring system, which comprised a datalogger (MiniCube VV/VX, EMS Brno, CZ), two air temperature and two air relative humidity sensors (EMS 33, EMS Brno, CZ), selected pairwise to have similar characteristics and placed at 0.5–1.5 and 2.0–2.4 m above the ground, depending on the crop growth stage, a net radiometer (Schenk 8110, Philipp Schenk, AT, thermal principle, stability 3% per year), soil temperature sensors (PT 100/8, EMS Brno) at 0.1 and 0.2 m and a soil heat flux meter (HFP01, Hukseflux, NL). The stations B, C and D were operating during the growing season. The station A, operating year round, was in addition equipped with a pyranometer for measuring global radiation (EMS 11, EMS Brno, CZ, silicone diode sensor, calibration error under daylight condition max. 7%) and a wind sensors measuring wind speed and direction (Met One 034B, Met One, Oregon, U.S.A., 0.28 m s^{-1} starting threshold) placed at 2 m height. All stations recorded their data at one-minute intervals, while the dataloggers saved only 10 min averages.

The station A was surrounded by permanent grassland cut three times a year (end of May, second half of July and second half of October). Winter wheat was cultivated in the vicinity of the stations B, C and D in 2004 (sown 20 Sept. 2003, harvested 25 July 2004), winter rape was there in 2006 (sown 22 Aug. 2005, harvested 1 Aug. 2006) and spring barley in 2009 (sown 8 March, harvested 7 Aug. 2009); the latter was a cover crop for red clover.

4. Determination of actual evapotranspiration rate

ETa was determined from the latent heat flux (LE) in the simplified energy balance equation [28]:

$$Rn = G + LE + H \quad \left(\mathrm{W\ m^{-2}}\right) \tag{1}$$

where Rn (net radiation) and G (soil heat flux) could be directly measured with a sufficient accuracy and H (turbulent sensible-heat flux) was calculated from the Bowen ratio (β). The turbulent diffusion theory admits that, under some assumptions, the Bowen ratio can be calculated from the vertical air temperature and vapour pressure gradients. The basic assumptions are the equality of transport coefficients for vertical turbulent transport of heat and water vapour under conditions the neutral atmosphere stratification and a flat homogenous extensive plant cover over a certain distance upwind of the point of observation (fetch), ensuring that the gradient measurements can be made within the equilibrium sublayer, where the fluxes are assumed to be independent of height [29-32]. The thickness of the equilibrium sublayer for an aerodynamically smooth-to-rough transition is assumed to be 10% of the internal boundary layer thickness δ. The latter can be calculated using the Munro & Oke (1975) equation (cited in [29]):

$$\delta = x^{0.8} \cdot z_0^{\ 0.2} \left(\mathrm{m}\right) \tag{2}$$

where x is the fetch and z_0 is the momentum roughness length of the crop surface (it can be taken as 13% of the crop height).

The adequate fetch length for the internal boundary layer to be of sufficient thickness (2.0 to 2.5 m above the ground, depending on the crop height) can be then estimated as:

$$x = (\delta / z_0^{\ 0.2})^{1.25} \quad \left(\mathrm{m}\right) \tag{3}$$

The Bowen ratio β is defined as:

$$\beta = \frac{H}{LE} \tag{4}$$

http://dx.doi.org/10.5772/52810

After substitution from the equations of vertical turbulent heat and mass transport [28, 33] and after introduction of the psychrometric constant γ (kPa °C^{-1}) is obtained:

$$\beta = \gamma \frac{(T_2 - T_1)}{(e_2 - e_1)} \tag{5}$$

where $(T_2\text{-}T_1)/(e_2\text{-}e_1)$ is the ratio of the air temperature (°C) and vapour pressure (kPa) vertical gradients above the plant canopy.

ETa can be calculated by combining equations (1) and (4):

$$ETa = \frac{Rn - G}{L(1+\beta)} \cdot 3600 \left(\text{mm h}^{-1}\right) \tag{6}$$

where L is the latent heat of vaporization (J kg^{-1}) and 1 mm of water is taken as 1 kg m^{-2}.

Conditions needed to fulfil theoretical requirements for using this method usually cannot be achieved in the early morning, evening and night periods and sometimes even on cloudy or rainy days or on days with significant advection. Therefore, the cases when $|\text{Rn-G}| \leq 10$ W m^{-2} or $\beta < -0.1$ or $\beta > 4$ or LE = 0 or when simultaneously LE < 0 and H > 0 were excluded from further processing. In this way, it was ensured that the situations when the gradients of air temperature and vapour pressure had opposite or uncertain signs (due to insufficient resolution limits of the sensors or due to advection) or when the stratification was strongly instable (far from neutral) [34, 35] were not taken into account. Table 2 shows Bowen ratio data excluded, which comprises $\beta < -0.1$ or $\beta > 4$; Table 3 rejects ETa values with this "unfavourable" Bowen ratios in conjuction with all other cases mentioned above. During the periods of vapour condensation at the surface under conditions of nocturnal inversion and outgoing available energy (LE < 0 and H < 0), the equilibrium evaporation $\Delta(\text{Rn-G})/(\Delta+\gamma)$, which is negative under these conditions, was set as a lower limit of vapour condensation, i.e., the absolute value of the actual condensation could not be higher than the absolute value of this equilibrium evaporation [35]. The missing ETa values for the periods thus excluded were estimated based on linear regression between the valid ETa values and the equilibrium evaporation. Then the summation of the resulting uninterrupted series of 10-min ETa values gave the average daily ETa rates in mm h^{-1} and these were subsequently converted to daily totals of ETa in mm d^{-1}.

5. Measurement of precipitation and maximum capillary water capacity and the soil textural class estimation

A tipping-bucket rain gauge 276 mm in diameter (with the interception surface 0.06 m^2) was located near the centre of the catchment. Each tip corresponded to 0.1 mm increment of precipitation. The precipitation totals were recorded at 10-min intervals.

Both disturbed and undisturbed soil samples were taken at about 100 m distance from each weather station from both topsoil and subsoil. To assess the moisture-holding capacity of the soil, an empirical characteristic of the soil water retention capacity was determined in the laboratory according to Novák's procedure [36]. It is referred to as the maximum capillary water capacity (MCWC). The procedure consists of allowing an undisturbed soil sample (100 cm^3), previously fully soaked with water by capillarity from below, to drain by suction on a layer of filter paper over 2 h. Its moisture content at the end of the period is MCWC. [37] declares that MCWC corresponds approximately to the field capacity of the soil. The grain size distribution of the soil was determined according to [38]. It was expressed in percent by mass of individual particle size fractions (clay: < 0.002 mm, silt: 0.002–0.05 mm, sand: 0.05–2.00 mm). The soil texture was classified according to the USDA [23] soil texture triangle.

6. Statistical analysis

To assess the systematic effect of the categorical independent variable "weather station" (including soil type, crop, tile drainage and terrain position) upon the dependent quantitative variable (daily ETa) a paired t-test was used, in order to indicate whether or not the expected difference between two matching observations is zero (the null hypothesis), taking the probability of unwarranted rejection of the null hypothesis $p = 0.05$. The period of spring and early summer was separately tested, i.e., from May to mid-July (period 1, up to crop maturity), and the following period of summer from mid-July to mid-August (period 2, after crop maturity). The latter period ended shortly after the crop harvest (in the case of field crops) or in the middle of the interval between the second and the third grass cutting.

7. Bowen ratio and actual evapotranspiration values exlusion, FETCH determination

Table 2 shows the number of all Bowen ratio (β) values obtained in individual periods and years and the percentage of values excluded. Of all 10-min β values measured at individual weather stations, 19-27% were rejected (these and the following values being taken over the entire period of observation). Seventy-four to 84% of the rejected values were night measurements (between 19:10 and 05:50). Of the night-time β values, 36-43% were rejected, compared with 8-13% rejection rate of daytime β values (not shown in Table 2). These results agree with those by [32] who reported 29% and 9% of β values rejected at night and in daytime, respectively.

Altogether, 32-40% of 10-min ETa measurements were invalid (Table 3), according to the criteria set forth in the part 4. Seventy-one to 84% of the invalid data points occurred at night, when ETa values tended toward zero (being either very small positive or very small negative). Of the night-time ETa measurements, 62-70% were invalid, as opposed to the daytime ETa pattern, when 11.5-15% were invalid.

Period	Weather station A			Weather station B			Weather station C			Weather station D		
	All β data	**All β data excluded (%)**	**Night-time β data excluded (%)**	**All β data**	**All β data excluded (%)**	**Night-time β data excluded (%)**	**All β data**	**All β data excluded (%)**	**Night-time β data excluded (%)**	**All β data**	**β data excluded (%)**	**Night-time β data excluded (%)**
2004 29 May-12 Aug	1065 6	18.5	16.1* (87)**	1094 4	27.8	17.4* (62.8)**	1094 4	33.5	24.5* (73.1)**	1094 3	26.6	20.1* (75.6)**
2006 6 May-19 Aug	1526 4	16.7	13.6* (81.5)**	1526 4	24.2	17.0* (70.4)**	1526 4	25.0	19.1* (76.5)**	1518 7	25.4	22.7* (89.5)**
2009 1 May-23 Aug	1656 0	22.5	13.8* (61.5)**	1656 0	13.6	11.9* (87.5)**	1656 0	19.6	15.9* (81.0)**	1655 6	28.1	24.3* (86.5)**
Average		19.2	14.5* (76.7)**		21.9	15.4* (73.6)**		26.0	19.8* (76.9)**		26.7	22.4* (83.9)**

* related to all β values

** related to all β values excluded

Table 2. Total number of 10-min Bowen ratio (β) values and the percentage of values excluded in individual periods

Period	Weather station A			Weather station B			Weather station C			Weather station D		
	All ETa data	**All ETa data excluded (%)**	**Night-time ETa data excluded (%)**	**All ETa data**	**All ETa data excluded (%)**	**Night-time ETa data excluded (%)**	**All ETa data**	**All ETa data excluded (%)**	**Night-time ETa data excluded (%)**	**All ETa data**	**ETa data excluded (%)**	**Night-time ETa data excluded (%)**
2004 29 May-12 Aug	1065 6	27.6	14.6* (53)**	1094 4	36.5	25.4* (69.6)**	1094 4	41.7	31.4* (75.4)**	1094 3	38.9	30.9* (79.5)**
2006 6 May-19 Aug	1526 4	28.1	23.1* (82.3)**	1526 4	41.5	33.1* (79.8)**	1526 4	32.5	25.2* (77.4)**	1518 7	36.8	32.3* (87.8)**
2009 1 May-23 Aug	1656 0	41.5	31.8* (76.7)**	1656 0	40.6	35.5* (87.5)**	1656 0	31.7	25.7* (80.9)**	1655 6	38.8	32.7* (84.1)**
Average		32.4	23.2* (70.7)**		39.5	31,3* (79.0)**		35.3	27.4* (77.9)**		38.2	32.0* (83.8)**

* related to all ETa values

** related to all ETa values excluded

Table 3. Total number of 10-min ETa values and the percentage of values excluded in individual periods

The accuracy of the BREB method of ETa determination, provided that its theoretical assumptions are met, is approximately 10% [31, 35]. The thickness of the equilibrium sublayer is related to the fetch. It is recommended that the minimum fetch to upper measurement height ratio is at least 10:1 to 200:1, with 100:1 being considered adequate for most measurements. The BREB method is less sensitive to imperfect fetch conditions than other techniques, if the Bowen ratio is small (c. 0.3-0.4, [39]). According to [29], a significant boundary–layer adjustment occurs within the first 15 m of the fetch and, hence, when the Bowen ratio is small, the method can be used successfully at fetch-to-height ratios as low as 20:1, despite the fact that the measurements are not made strictly within the equilibrium sublayer. [32] show fetches ranging from 90 m to 360 m, [40] mention a sufficient fetch of 148–168 m.

In this case, considering the crop height of 0.4 to 1.5 m (cereals, rape) and an adequate equilibrium sublayer thickness of 2.0-2.5 m, a sufficient minimum fetch, according to Eq. 3, is 80-90 m. The prevailing wind directions observed were 60-120° and 210-300°. The actual wind direction remained within these two directions over 67.0% of the time in 2004, over 59.3% of the time in 2006 and over 64.6% of the time in 2009. The particular weather stations were far enough apart related to the minimum fetch, their mutual distances being 114 m (C-D), 175 m (B-C) and 204 m (A-B). The distances any of A, B, C and D from the upwind boundary of the crop stand were in most cases greater than 80-90 m. The minimum and maximum fetches were 95–300 m and 180–510 m, respectively, along the prevailing wind directions. Hence, no significant footprint overlapping of the weather stations occurred. The fetch of the station B from the permanent grassland boundary varied between 68 m and 90 m, if the wind direction varied between 125° and 215°. However, only 20-27% of wind directions measured lay within this interval. With the wind direction within this interval and considering only the Bowen ratios $\beta > 0.4$ (for which greater sensitivity to the perfectness of the fetch was expected), only another 10-14% of β values and 8-9% of ETa values would have been rejected. Based on this analysis, the fetches of all weather stations were considered sufficient and no data were rejected because of "wrong" wind directions. Of all ETa data considered valid and measured by the stations B and C, 70% and 57%, respectively, were influenced by the nearby tile-drained area. This happened when the wind direction was 25°–250° at station B and 160°–295° at station C.

8. Between-stations ETa and Bowen comparison

The soils around the stations A to D are characterized by markedly differing grain size distribution, which influenced their water retention capacity (Table 1). The Haplic Stagnosols and Stagnic Cambisols on which weather stations A and B were located display lower sand content in the topsoil as compared to the Haplic Cambisols around the stations C and D. However, all varieties of Cambisols (around the stations B, C and D) contain more sand in the subsoil than the Stagnosols. The actual soil water retention capacity was influenced not only by the sand and clay content but also by the presence of tile-drainage systems around the stations A, B, C and terrain relief (see Fig. 1 and Table 1).

The soils surrounding the station C, with the highest content of sand and partially influenced by the adjacent drainage system, manifested themselves in the lowest ETa values and the highest β over all three years (Table 4, Figs. 2-7). In contrast, the fine-textured soils with greater MCWC and affected more by the shallow groundwater table and the shallow lateral flow (STha, CMst, stations A and B), showed in most cases the highest ETa values. The findings by [2, 41-42] in this respect are similar.

Year and the crop around stations B to D	Station	Average daily ETa (mm)			Average daily Bowen ratio β (-)			Precipitation (mm): Total over period/ Daily average	
		May to mid-July (up to crop maturity)	Mid-July to mid-August (after crop maturity)	Entire growing season	May to mid-July (up to crop maturity)	Mid-July to mid-August (after crop maturity)	Entire growing season	May to mid-July (up to crop maturity)	Mid-July to mid-August (after crop maturity)
2004 Winter wheat	A*	2.97±1.18	3.46±0.76	3.15±1.06	0.32±0.11	0.40±0.15	0.35±0.13	109.8/2.29	6.4/0.22
	B	3.52±1.34	2.38±1.24	3.10±1.41	0.26±0.12	1.53±1.02	0.71±0.86		
	C	2.68±1.04	1.75±1.00	2.34±1.11	0.46±0.17	1.66±1.09	0.83±0.82		
	D	3.15±1.15	2.26±0.90	2.83±1.15	0.41±0.15	1.98±1.15	0.99±1.04		
2006 Winter rape	A*	3.43±1.39	2.96±1.58	3.27±1.47	0.47±0.25	0.47±0.27	0.47±0.25	222.3/3.18	126.3/3.41
	B	3.37±1.39	3.15±1.39	3.29±1.38	0.53±0.32	0.40±0.21	0.49±0.30		
	C	3.20±1.37	2.44±1.29	2.94±1.39	0.61±0.29	0.93±0.51	0.72±0.41		
	D	3.28±1.24	2.66±1.27	3.07±1.28	0.66±0.27	0.85±0.31	0.73±0.30		
2009 Spring barley as a cover crop for red clover	A*	3.06±1.31	3.65±1.33	3.26±1.34	0.35±0.27	0.27±0.30	0.32±0.28	207.2/2.73	109.1/2.73
	B	3.09±1.44	3.19±1.06	3.13±1.32	0.40±0.16	0.50±0.17	0.43±0.17		
	C	2.67±1.12	2.60±0.74	2.65±1.00	0.59±0.23	0.81±0.38	0.67±0.31		
	D	3.05±1.29	3.47±1.12	3.20±1.25	0.36±0.16	0.30±0.14	0.34±0.15		

*Station A is surrounded by grassland

Table 4. Average actual evapotranspiration (ETa) and Bowen ratio (β) values ± their standard deviations, precipitation totals and daily averages for the weather stations A to D in individual years and periods

8.1. Dry weather conditions

The reported differences in daily ETa were related to the periods of limited transpiration either due to the onset of crop maturity (when the plant water consumption was already low and the excessive precipitation was absorbed by the soil) and/or due to drought (when the soil water supply to plants was limited). Hence, in some cases (in period 2 but, in 2004, also in the second half of period 1 - from late June to mid-July), the statistical tests signalled systematic ETa differences between different soil types under different land use (mostly A vs. C) and even between different soil types under the same land use (always B vs. C, mostly D vs. C, Table 5). It means that the fine-textured soils in the discharge and discharge/transient zones, having higher MCWC and affected more by shallow groundwater table or/and shallow subsurface flow due to concentrated flow paths (STha, CMst, stations A and B) were marked in most cases

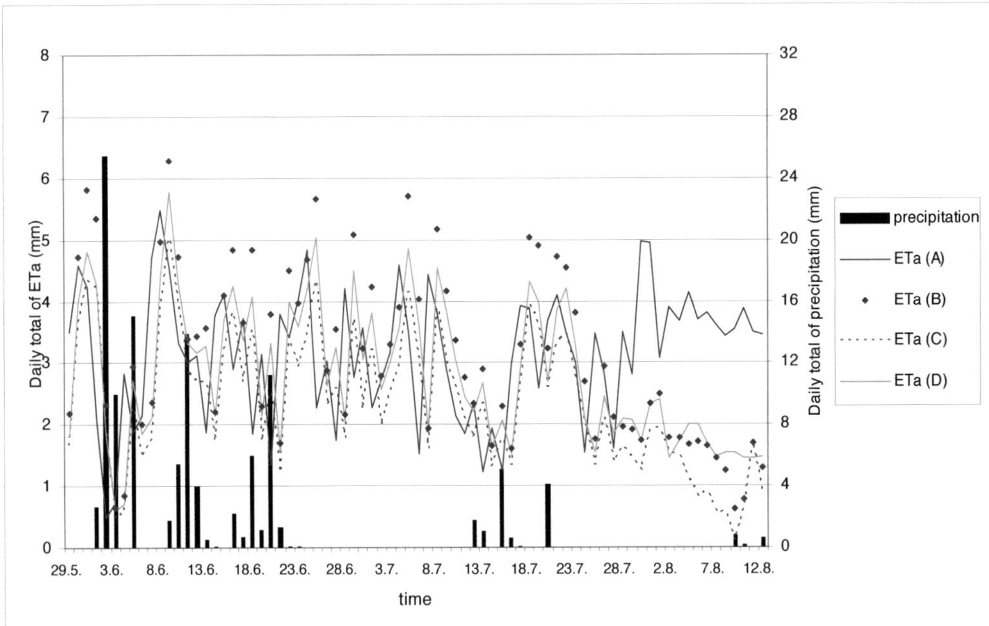

Figure 2. Average daily values of actual evapotranspiration for the Dehtáře catchment, 2004

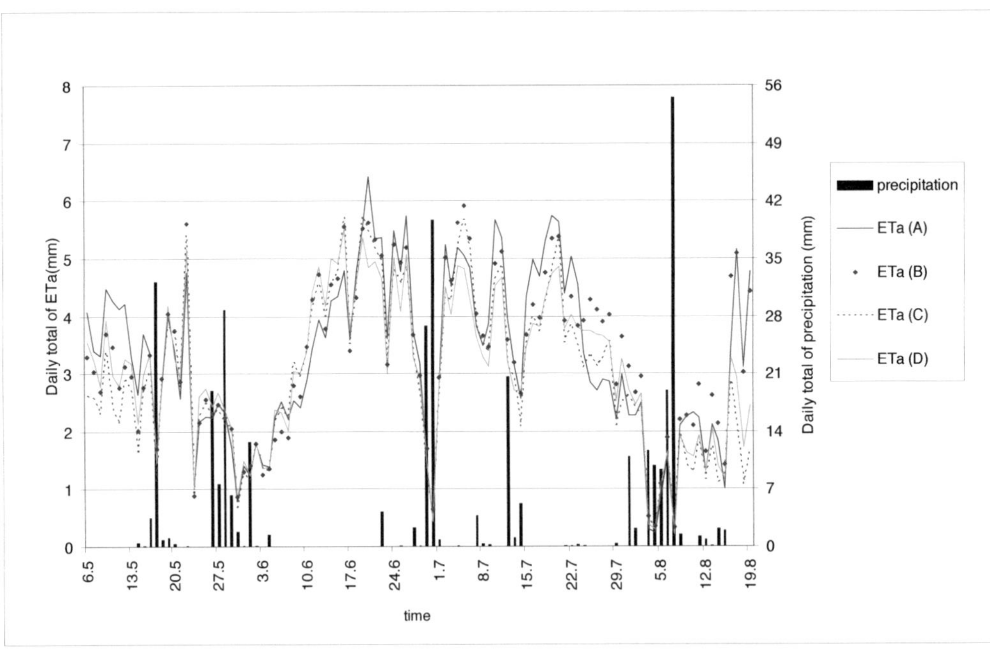

Figure 3. Average daily values of actual evapotranspiration for the Dehtáře catchment, 2006

http://dx.doi.org/10.5772/52810

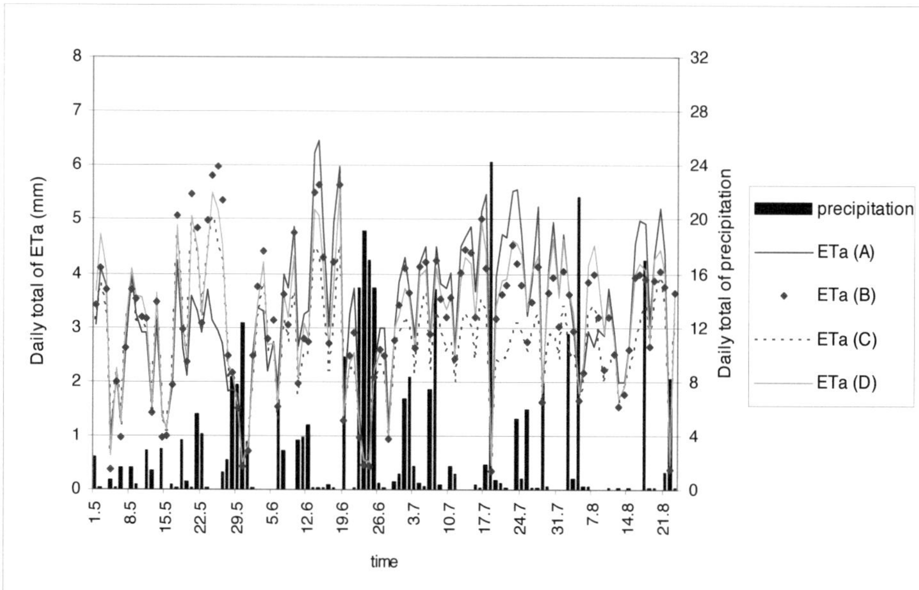

Figure 4. Average daily values of actual evapotranspiration for the Dehtáře catchment, 2009

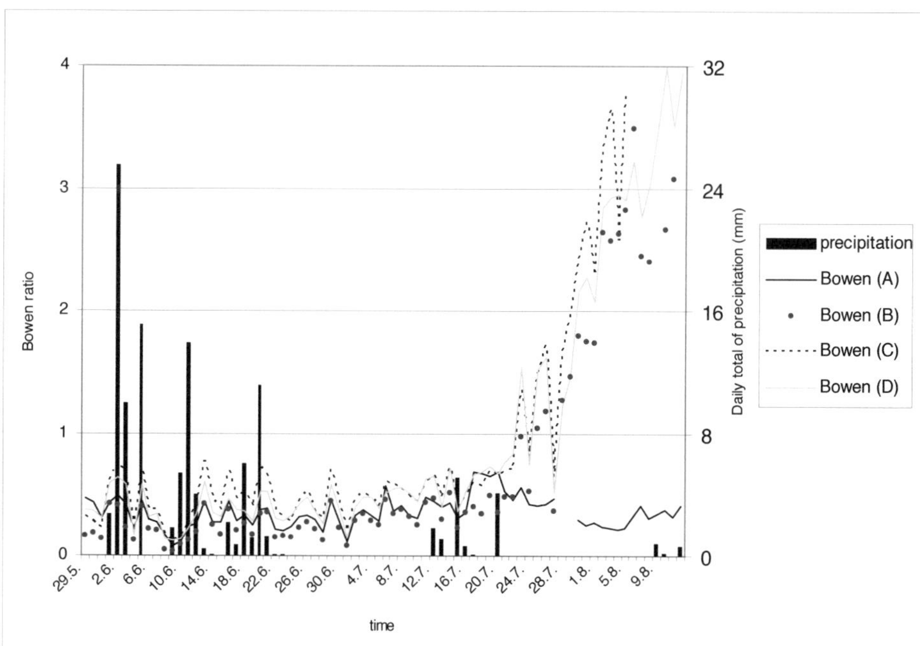

Figure 5. Average daily values of the Bowen ratio for the Dehtáře catchment, 2004

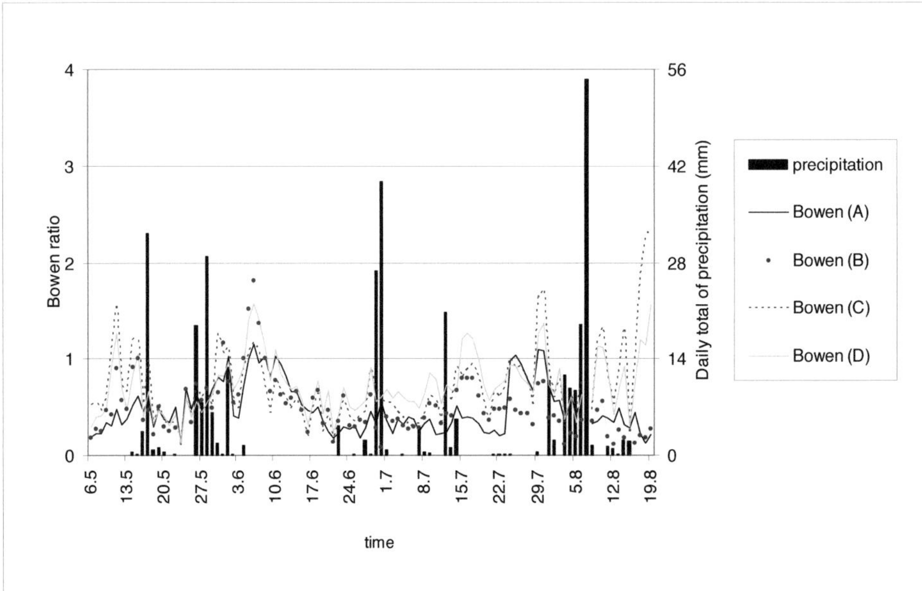

Figure 6. Average daily values of the Bowen ratio for the Dehtáře catchment, 2006

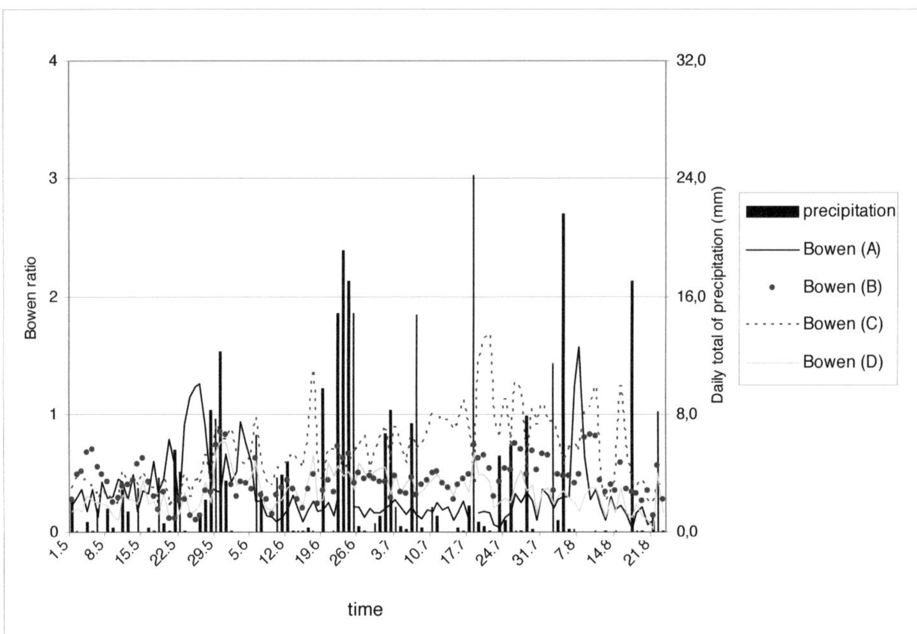

Figure 7. Average daily values of the Bowen ratio for the Dehtáře catchment, 2009

by the highest ETa and the lowest Bowen ratio values. The daily averages of ETa at the station C on soils with the highest content of sand, shallow and stony (CMha), influenced by air mass coming from drying shallow CMha and deeper, but tile drained CMha were always the lowest ones. As a result, the average daily ETa was significantly lower and the corresponding β significantly higher over coarser-textured soils (shallow CMha, C), namely (year–ETa(mm d^{-1})/ β): 2004 – 1.75/1.66; 2006 – 2.44/0.93; 2009 – 2.60/0.81, than over finer-textured soils (CMst and STha), namely: 2004 – 2.92/0.97; 2006 – 3.06/0.44; 2009 – 3.42/0.39.

Subjects of comparison	Compared pair of stations	Year	*p* From May to mid-July	*p* From mid-July to mid-August
Soils affected by hydromorphism (STha, CMst) under different land use (A – permanent grassland vs. B – field crops) and also the discharge zone vs. the discharge/transient zone, both being tile-drained	A x B	2004	0.03448*	0.00024*
		2006	N.S.	N.S.
		2009	N.S.	N.S.
Different Cambisol varieties in different terrain zones under the same land use and different intensity of tile-drainage: B (CMst/CMha, discharge/transient zone, tile-drained) C (CMha, transient zone, partially tile-drained) D (CMha, transient/recharge zone, not tile-drained)	B x C	2004	0.00096*	0.03982*
		2006	N.S.	0.02649*
		2009	0.04805*	0.00546*
	B x D	2004	N.S.	N.S.
		2006	N.S.	N.S.
		2009	N.S.	N.S.
	C x D	2004	0.03808*	N.S.
		2006	N.S.	N.S.
		2009	N.S.	0.00011*
Permanent grass cover on Stagnosol, discharge zone, tile-drained (A) vs. field crops on Haplic Cambisol, transient (C) or recharge/transient (D) zone, partially tile-drained (C) or not tile-drained (D)	A x C	2004	N.S.	$<10^{-5}$*
		2006	N.S.	N.S.
		2009	N.S.	0.00004*
	A x D	2004	N.S.	$<10^{-5}$*
		2006	N.S.	N.S.
		2009	N.S.	N.S.

* the null hypothesis rejected

N.S. = not significant

Table 5. The significance levels (*p*) for paired t-tests comparing average daily ETa values

With regard to the soil water regime and ETa, the CMst and STha of the station B with a drainage system and being in concentrated water flow acted similarly to the CMha of the station D without a drainage system (D), i.e. no differences were found between stations B and D.

Differences in ETa between stations C and D (i.e. higher values of the station D), both being at the interface shallow and deeper CMha, could be explained by the share of wind coming from drying shallow and tile drained CMha. Within valid ETa values, the station C was influenced of 80-85 % by the wind coming from drying shallow CMha and tile drained areas, while to the station D the wind went from shallow CMha only from 33-47%. Thus the significance of prevailing wind direction in ETa determination in soil heterogeneous conditions was evident.

Under extremely dry conditions (period 2, 2004), there were significant differences between the station A and the other stations, because of the dried-out soil profile under arable land (stations B to D), while the lowest part of the catchment around the station A was still relatively wet. The differences in terms of Bowen ratios (the average values between 10:00 and 18:00 of each day when $Rn > 70\ W\ m^{-2}$) were also visible. The β values pertaining to the stations B to D rose in July and August 2004 very sharply while those measured at A remained low (Fig. 5). Additionally, the soil water regime of the station A was markedly influenced by the reduction of ETa after grass cutting (i.e. after the above-ground biomass removal). Taken separately over the periods 1 and 2 as well as over the entire growing seasons in particular years, the stations C and D typically gave the highest Bowen ratios (Figs. 6, 7).

8.2. Wet weather conditions

In the period 1 the crop transpiration, a critical component of evapotranspiration, was in most cases not limited by the (non-existing) soil water deficit. The water supply to plants was sufficient and uninterrupted, while the crop stand was already fully developed. Under these conditions, both soil evaporation and plant transpiration were affected by weather factors in a similar way. The uplift of water through the plant tissues is markedly more efficient than the soil water upward movement during physical evaporation only [43]. After precipitation, the water that has infiltrated into the soil is mainly utilized for transpiration of the fully developed stand, that is, the soil physical properties have only a limited effect on its upward movement. Thus, the prevalence of the transpiration component of evapotranspiration acted as an equalizing factor on ETa from heterogeneous soil areas. The effect of physical properties of the soil was thereby masked. As a result, the daily ETa values were statistically the same across various soil types and crop species (a similar conclusion was made by [44]), except in the dry period of 2004. [45] arrived at similar conclusions, stating that the values of cumulative physical evaporation showed a more pronounced change with alterations in soil texture than did the values of cumulative transpiration. Figs. 3-4 and 6-7 support this conclusion by showing that, in the period 1, the differences in ETa and β among individual weather stations were negligible.

9. Conclusion

Different soil physical properties of the catchment, interacting with the tile-drainage system effects, terrain relief, manifested themselves in corresponding daily evapotranspiration differences during the periods of limited transpiration, either at the onset of crop maturity and/or during the soil drought. The fine-textured soils in the discharge and discharge/transient zones, affected more by shallow groundwater table and shallow subsurface flow were marked in most cases by the highest ETa and the lowest Bowen ratio values. The daily averages of ETa referring to soils with the highest content of sand, shallow and stony, were always the lowest ones.

The transpiration of a green plant cover, which took up a major portion of the soil water storage through its root suction force and was not, in most cases, limited by the soil water deficit, acted as an equalizing factor of evapotranspiration from heterogeneous soil areas. The transpiration also mitigated the differences in evapotranspiration among different soils, even when these were carrying different crops. The vegetation canopy thus minimized runoff in any form and reduced the infiltration and the groundwater recharge in the recharge zones.

In addition to soil conditions, the results of ETa research were also connected with the method used (Bowen ratio method) which is dependent on air temperature and humidity coming over extensive plant cover from prevailing wind direction.

Acknowledgements

This study was supported by the grants from the Ministry of Agriculture of the CR within the research programmes MZE 0002704902-03-01, MZE 0002704902-01-02 and the project no. QH92034.

Author details

Renata Duffková*

Address all correspondence to: duffkova.renata@vumop.cz

Research Institute for Soil and Water Conservation, Praha, Czech Republic

References

[1] Monteith, J. L. Vegetation and the Atmosphere. Principles. London, New York, San Francisco: Academic Press; (1976). , 1

[2] Ward, A. D, & Elliot, W. J. Editors. Environmental hydrology. Boca Raton, New York: Lewis Publishers; (1995).

[3] Allen, R. G, Pereira, L. S, Raes, D, & Smith, M. crop evapotranspiration, Guidelines for computing crop water requirements. FAO Irrigation and Drainage Paper 56. Rome: Food and Agriculture Organization of the United Nations; (1998).

[4] Dunn, S. M, & Mackay, R. Spatial variation in evapotranspiration and the influence of land use on catchment hydrology. Journal of Hydrology (1995). , 171, 49-73.

[5] Serrano, S. E. Hydrology for Engineers, Geologists and Environmental Professionals. Kentucky: HydroScience Inc. Lexington; (1997).

[6] Mengelkamp, H. T, Warrach, K, & Raschke, E. SEWAB- a parameterization of the Surface Energy and Water Balance for atmospheric and hydrologic models. Advances in Water Resources (1999). , 23, 165-175.

[7] Jhorar, R. K. Bastiaanssen, WGM., Feddes RA., Van Dam JC. Inversely estimating soil hydraulic functions using evapotranspiration fluxes. Journal of Hydrology (2002). , 258, 198-213.

[8] Brutsaert, W. Hydrology- an Introduction. Cambridge: Cambridge University Press; (2005).

[9] Luxmoore, R. J, & Sharma, L. M. Evapotranspiration and soil heterogeneity. Agricultural Water Management (1984). , 8, 279-289.

[10] Bronstert, A, & Plate, E. J. Modelling of runoff generation and soil moisture dynamics for hillslopes and micro-catchments. Journal of Hydrology (1997). , 198, 177-195.

[11] Popova, Z, & Kercheva, M. CERES model application for increasing preparedness to climate variability in agricultural planning-risk analyses. Physics and Chemistry of the Earth (2005). , 30, 117-124.

[12] Doležal, F, & Kvítek, T. The role of recharge zones, discharge zones, springs and tile drainage systems in peneplains of Central European highlands with regard to water quality generation processes. Physics and Chemistry of the Earth, Parts A/B/C (2004).

[13] Fucík, P, Kvítek, T, Lexa, M, Novák, P, & Bílková, A. Assessing the Stream Water Quality Dynamics in Connection with Land Use in Agricultural Catchments of Different Scales. Soil & Water Research (2008). , 3, 98-112.

[14] Kvítek, T, et al. Grassing of arable land with high infiltration hazard- a tool for reducing nitrate load in waters. Methodics of RISWC. Praha; (2007). in Czech)

[15] Lexa, M. The evaluation of nitrate concentratins of small streams in Želivka catchment and its analysis. Ph.D. thesis. Charles University, Faculty of Science, Praha; (2006). in Czech)

[16] Rychnovská, M, Balátová-tulácková, E, Úlehlová, B, & Pelikán, J. Meadows ecology. Praha: Academia; (1985). in Czech)

[17] Zheng, F. L. Huang ChH., Norton LD. Effects of Near-Surface Hydraulic Gradients on Nitrate and Phosphorus Losses in Surface Runoff. Journal of Environmental Quality (2004)., 33, 2174-2182.

[18] Barrett, M. E, & Charbeneau, R. J. A parsimonious model for simulating flow in a karst aquifer. Journal of Hydrology (1997)., 196, 47-65.

[19] Minár, J, & Evans, S. Elementary forms for land surface segmentation: The theoretical basis of terrain analysis and geomorphological mapping. Geomorphology (2008)., 95, 236-259.

[20] Demek, J, et al. Geography Lexicon of the Czech Socialist Republic- Mountains and Basins. Praha: Academia; (1987). in Czech)

[21] Tolasz, R, et al. Climate Atlas of Czechia. Praha-Olomouc: Czech Hydrometeorological Institute, Palacký University Olomouc, 1st edition; (2007).

[22] World reference base for soil resources (2006). Rome: World Soil Resources Reports 103. Food and Agriculture Organization of the United Nations; 2006.

[23] USDA-NRSC Soil taxonomyAgricultural Handbook Second Edition; (1999). (436)

[24] Karous, M, & Chalupník, T. Geophysical research of soil characteristics in a non-saturated zone at the Dehtáře locality. Praha; (2006). in Czech)

[25] Karous, M, & Chalupník, T. Geophysical research of soil characteristics in a non-saturated zone at the Dehtáře locality phase (2007). Praha; 2007. (in Czech)

[26] Haken, D, & Kvítek, T. Dynamics of the water regime in the drained meadow soil. Praha: Scientific works of the Institute of Agricultural Land Improvement (1982). in Czech), 1, 23-35.

[27] Haken, D, & Kvítek, T. The effectiveness of full-scale reclamation of waterlogged meadow sites in the potato vegetation region. Collection of the Institute of Scientific and Technical Information for Agriculture. Meliorations (1984). in Czech), 2, 121-132.

[28] Monteith, J. L. Principles of Environmental Physics. London: Edward Arnold (Publishers) Limited; (1973).

[29] Heilman, J. L, & Brittin, C. L. Fetch requirements for Bowen ratio measurements of latent and sensible heat fluxes. Agricultural and Forest Meteorology (1989). , 44, 261-273.

[30] Pauwels VRNSamson R. Comparison of different methods to measure and model actual evapotranspiration rates for a wet sloping grassland. Agricultural Water Management (2006)., 82, 1-24.

[31] Tattari, S, Ikonen, J. P, & Sucksdorff, Y. A comparison of evapotranspiration above a barley field based on quality tested Bowen ratio data and Deardorff modelling. Journal of Hydrology (1995)., 170, 1-14.

[32] Todd, R. W, Evett, S. R, & Howell, T. A. The Bowen ratio-energy balance method for estimating latent heat flux of irrigated alfalfa evaluated in a semi-arid, advective environment. Agricultural Forest Meteorology (2000). , 103, 335-348.

[33] Grace, J. Plant-Atmosphere Relationships. London, New York: Chapman and Hall; (1983).

[34] Inman-bamber, N. G, & Mc Glinchey, M. G. Crop coefficients and water-use estimates for sugarcane based on long-term Bowen ratio energy balance measurements. Field Crops Research (2003). , 83, 125-138.

[35] Perez, P. J, Castellvi, F, Ibanez, M, & Rosell, J. I. Assessment of reliability of Bowen ratio method for partitioning fluxes. Agricultural and Forest Meteorology (1999). , 97, 141-150.

[36] Klika, J, Novák, V, & Gregor, A. Manual of phytocenology, ecology, climatology and pedology. Praha: Publishing House of the Czechoslovak Academy of Sciences; (1954). in Czech)

[37] Kutílek, M. Water Management Pedology. Praha: Státní nakladatelství technické literatury; (1966). in Czech)

[38] ISO 11277(1998). Cor 1. Determination of particle size distribution in mineral soil material- Method by sieving and sedimentation; 2002.

[39] Yeh, G. T, & Brutsaert, W. H. A solution for simultaneous turbulent heat and vapour transfer between a water surface and the atmosphere. Boundary-Layer Meteorology (1971). , 2, 64-82.

[40] Steduto, P, & Hsiao, T. C. Maize canopies under two soil water regimes. I. Diurnal patterns of energy balance, carbon dioxide flux, and canopy conductance. Agricultural and Forest Meteorology (1998). , 89, 169-184.

[41] Salvucci, G. D, & Entekhabi, D. Hillslope and climatic controls on hydrologic fluxes. Water Resources Research (1995). , 31, 1725-1739.

[42] Yokoo, Y, Sivapalan, M, & Oki, T. Investigating the roles of climate seasonality and landscape characteristics on mean annual and monthly water balances. Journal of Hydrology (2008). , 357, 255-269.

[43] Novák, V. Water evaporation in nature and methods for its determination. Bratislava: VEDA. Publishing House of the Slovak Academy of Sciences; (1995). In Slovak)

[44] Mahmood, R, & Hubbard, K. G. Simulating sensitivity of soil moisture and evapotranspiration under heterogeneous soils and land uses. Journal of Hydrology (2003). , 280, 72-90.

[45] Kozak, A. J, Ahuja, L. R, Ma, L, & Green, T. R. Scaling and estimation of evaporation and transpiration of water across soil textures. Vadose Zone Journal (2005). , 4, 418-427.

Chapter 11

Uncertainty Evaluation of Water Budget Model Parameters for Different Environmental Conditions

Zoubeida Kebaili Bargaoui, Ahmed Houcine and
Asma Foughali

Additional information is available at the end of the chapter

http://dx.doi.org/10.5772/55236

1. Introduction

Modelling time distribution of soil moisture is a key issue for evapotranspiration and biomass evaluation and is often adopted for deriving drought awareness indices. Water budget models help computing time evolution of soil moisture provided hydroclimatological and soil information. While runoff time series are often used to drive water budget models calibration, it may conduct to false conclusions about the other model outputs such as percolation and evapotranspiration fluxes, in the absence of vegetation response observations. Thus a lot of uncertainty is attached to the calibrated model parameters and may constitute a handicap against model application. The aim of this study is to propose a methodology to cope with vegetation information inside the calibration process of a water balance model using a qualitative approach. A review of evapotranspiration estimation through water balance modelling is reported in Kebaili Bargaoui (2011). In section 2, we present the data used to apply this methodology. In section 3, we present the methodology of uncertainty quantification using kernel distribution of model parameters. In section 4, resuling kernels are provided as well as a sensitivity study of results to the choice of soil parameters evaluation method.

2. Data

Two watersheds are studied: the Wadi Sejnane watershed (North Tunisia) and Wadi Chaffar watershed (South Tunisia). They are of comparable moderate sizes (respectively 376 km^2 and 250 km^2). They have distinguishable occupation and climate. Sejnane basin is a forest basin under subhumid climate. Comparatively, for the Chaffar basin, vegetation cover comprises mainly olives under an arid climate. The soil type is principally sandy for Chaffar basin while a dominance of clay soils is outlined for Sejnane basin.

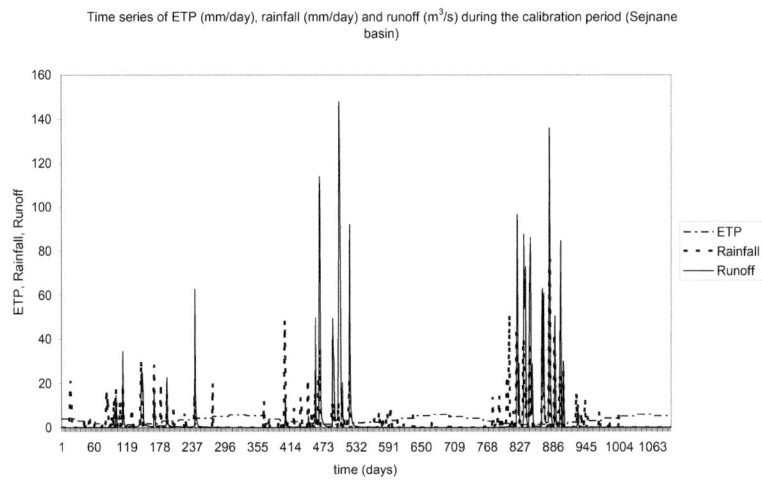

(a)

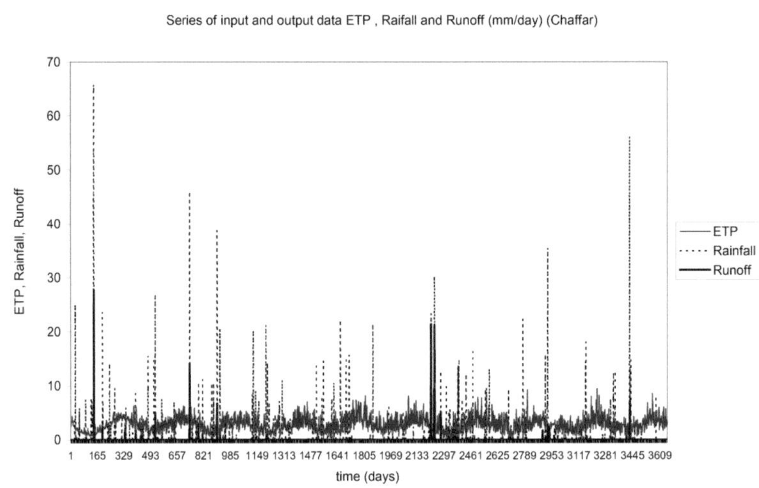

(b)

Figure 1. Time series of observed ETP (mm/day), rainfall (mm/day) and river discharges (m3/s) for (a) Sejnane basin (b) Chaffar basin

Potential evapotranspiration series are computed using the Turc formula based on monthly solar radiation and mean air temperature observed series at surrounding meteorological stations. A mean daily value is obtained for each month. Runoff (mm) series are estimated using observed daily stream discharges at the basin outlets with standard gauging methods. A ten year calibration period from September 1989 to August 1999 is considered for Chaffar basin including daily basin average rainfall evaluating using Thiessen method based on a network of 10 raingauges. A three year calibration period from September 1988 to August

http://dx.doi.org/10.5772/55236

1990 is available for Sejnane basin based on a rainfall network of 14 stations. Both basins have water tables. However piezometric data are not included in the study. Fig 1a and Fig. 1b report the time series of ETP, rainfall and runoff during the calibration periods.

3. Methodology

The water budget lumped BBH model presented by Kobayashi et al. (2001) is performed at daily time scale. Table 1 reports model equations. Mean daily rainfall and mean daily potential evapotranspiration are model inputs. Soil moisture content W (mm) and actual evapotranspiration ETR (mm) are model results of interest as well as runoff Rs (mm), percolation (Gd >0 mm) and capillary rise (Gd <0 mm). Seven parameters control model input-output transformations: thickness of active soil layer (*D* mm), effective soil porosity *p*; parameter related to the field capacity (*a* mm); parameter representing the decay of soil moisture (*b* mm); parameter representing the daily maximal capillary rise (*c* mm); parameter representing the moisture retaining capacity ($0< \eta <1$); parameter representing the stomatal resistance of vegetation to evapotranspiration ($0< \sigma <1$). The parameter W_{max} (mm) which represents the total water-holding capacity is a key parameter of the model.

According to Kobayachi and al. (2001) a/W_{max} is "nearly equal to or somewhat smaller than the field capacity". After Teshima et al., (2006), *b* is a measure of soil moisture recession that depends on hydraulic conductivity and active soil layer depth *D*. In Iwanaga et al. (2005) a sensitivity analysis of BBH model applied to an irrigated area in semi-arid region suggest that soil moisture RMSE is most sensitive to σ, η and c. All parameters are subject to calibration using soil, vegetation as well as climatic and hydrologic information.

Water balance equation	$\Delta W = W(t+1) - W(t) = P(t) - ETR(t) - Rs(t) - Gd(t)$
	t: time (day)
	W(t) : soil moisture content (mm)
	P: daily precipitation (mm)
	ETR: daily actual evapotranspiration (mm)
	Rs: daily surface runoff (mm)
	Gd: daily percolation (if Gd >0) or capillary rise (if Gd <0) (mm)
Daily actual evapotranspiration	$ETR(t) = M(t)\ ETP(t)$
	ETP: daily potential evapotranspiration (mm)
	$M(t) = Min(1, W(t)/(\sigma \times W_{max}))$
	σ : parameter representing the resistance of vegetation to evapotranspiration
	$W_{max} = pD$
	W_{max}: total water-holding capacity (mm)
	D: thickness of active soil layer (mm)
	p: effective soil porosity
Daily percolation and capillary rise	$Gd(t) = \exp((W(t)-a)/b)-c$
	a: parameter related to the field capacity (mm)
	b: parameter representing the decay of soil moisture (mm)
	c: parameter representing the daily maximal capillary rise (mm)

Water balance equation	$\Delta W = W(t+1) - W(t) = P(t) - ETR(t) - Rs(t) - Gd(t)$
	t: time (day)
	W(t) : soil moisture content (mm)
	P: daily precipitation (mm)
	ETR: daily actual evapotranspiration (mm)
	Rs: daily surface runoff (mm)
	Gd: daily percolation (if Gd >0) or capillary rise (if Gd <0) (mm)
Daily surface runoff	$Rs(t) = \max\left[P(t) - (W_{BC} - W(t)) - ETR(t) - Gd(t),\ 0\right]$
	$W_{BC} = \eta W_{max}$
	η : parameter representing the moisture retaining capacity ($0 < \eta < 1$).

Table 1. Equations and parameters of the BBH model

Moreover, we have introduced pedo transfer functions in the model in order to reduce the number of parameters to be calibrated on the basis of hydrometeorological series (Bargaoui and Houcine, 2010). It is worth noting that Kobayachi et al. (2001) adjusted soil humidity profiles measurements for BBH model calibration. As such observations are not often available; it seems an important task to adapt the original model using pedotransfer submodels especially when dealing with ungauged or partially gauged basins. To that purpose, three key soil characteristics are considered: saturated hydraulic conductivity K_s, soil water retention curve shape parameter B and field capacity S_{FC}. Assuming the percolation function as an exponential decay function, the leakage $L(s)$ is identified according to Guswa et al. (2002) model as reported in (Eq. 1) where s is the ratio W/W_{max}. Consequently, parameters a, b, c of the original BBH model are obtained by identification (Eq. 2, 3, 4) using the three soil parameters K_s, S_{FC}, and B.

$$L(s) = K_s \frac{e^{B(s-S_{FC})} - 1}{e^{B(1-S_{FC})} - 1} \tag{1}$$

$$a = W_{max}\left[S_{FC} - \frac{1}{B} Ln\left(Ks \frac{1}{e^{B(1-S_{FC})} - 1}\right)\right] \tag{2}$$

$$b = W_{max}\frac{1}{B} \tag{3}$$

$$c = \left(\frac{1}{e^{B(1-S_{FC})} - 1}\right) Ks \tag{4}$$

Rawls et al. (1982) model is adopted for estimating K_s while S_{FC} is derived according to two different models: Cosby and Saxton model which was recently adopted by Zhan et al., (2008) and Cosby et al. (1984) model. Effectively, this is suggested as a way to take into account uncertainty related to soil parameters. For the basin of Chaffar, because of lack of detailed information, we assume that p as well as K_s and S_{FC} parameters are those corresponding to the dominant soil class. For the Sejnane basin, a spatial mean of soil class properties is adopted using the spatial repartition of soil types as well as the area they cover within the basin. On the other hand, for the two cases, B =9 is adopted according to Rodriguez-Iturbe and al. (1999).

3.1. Model calibration

Finally, only the set of parameters (D, σ, η) remains subject to calibration through fitting observed and predicted runoff time series. The daily time step is adopted to run the model while annual, monthly and decadal time steps are adopted for its fitting. Many trials are firstly performed to adjust D choosing simply between three alternatives: D =1000 mm; D = 500 mm; D =300 mm which represent common values adopted in water balance models. Then, once D is fixed, the set of parameters (σ, η) is selected according to annual absolute relative runoff error AARE. Based on the idea of equifinality (Beven, 1993), a threshold value $AARE_s$ related to AARE is adopted for eliminating poor solutions using a grid of candidate solutions with $\Delta\sigma = \Delta\eta = 0.01$. Hence, only those pairs for which AARE > $AARE_s$ are selected and analyzed in the following.

Eq. (5) reports the objective function. It quantifies the absolute relative runoff bias during the calibration period:

$$AARE(\sigma,\eta) = \frac{1}{N}\sum_{i=1}^{N}\left|(y_{si} - y_{oi})/y_{oi}\right| \tag{5}$$

where y_{oi} is the annual observed runoff (mm) for year i; y_{si} is the annual computed runoff (mm) for year i; N is the number of years of the calibration period. Additionally, for the selected solutions Nash coefficient $R_{N,M}$ evaluated on the monthly basis as well as Nash coefficient $R_{N,D}$ evaluated on the decadal basis are reported. The assumption of existence of capillary rise response is tested through the calibration process. It is further believed that if performance criteria (AARE, $R_{N,M}$, $R_{N,D}$) are better in presence (or in absence) of capillary rise assumption, then the assumption is retained.

3.2. Uncertainty quantification

Thus, the adoption of a fixed value for the threshold $AARE_s$ will give rise to a number of acceptable solutions (σ, η). Here, the marginal kernel which represents a non parametric estimation of the statistical distribution of a given random variable (here the parameters σ and η) is adopted to represent parameter uncertainty. Similarly, the kernels of resulting outputs are computed in order to analysis the effect on model outputs especially evapotranspiration which is the variable of interest. A Gaussian kernel is adopted to perform the analysis.

3.3. Including vegetation information

It is now proposed to accurate (σ, η) kernel distribution by introducing the ratio Kv of mean annual actual evapotranspiration to mean annual potential evapotranspiration. In effect, as noticed by Eagleson (1994) after works of Ehleringer (1985), ecologists recognize three types of vegetation selection and adaptation in response to environmental stress due to water shortage (Type 1: desert annual grasses and humid climate trees; Type 2: semi-arid and sub-humid trees and shrubs; Type 3: perennial desert plants). Considering actual evapotranspiration as surrogate of vegetation productivity, three typical curves of Kv versus the inverse of the soil moisture are drawn by Eagleson (1994). Here, we assume the interval $0.45 < Kv < 0.55$ (mean Kv = 0.5) for type 2 (Sejnane basin) and $0.15 < Kv < 0.25$ (mean Kv = 0.2) for type 3 (Chaffar basin) which correspond to the values reported into the graph of Eagleson (1994) in case of weak environmental stress. Effectively, such an hypothesis is justified by the fact that the calibration periods represent mean water conditions for the two basins.

So, kernels of parameters and evapotranspiration conditional to the above conditions will also be drawn in order to evaluate the effect of including vegetation information supplementary to runoff observations on model results.

4. Results

Table 2 reports the 5 parameters which are not subject to fitting on basis of hydroclimatological series.

	Thickness of active soil layer D (mm)	**Effective soil porosity p**	**Saturated hydraulic conductivity K_s (mm/day)**	**Field capacity S_{FC}**	**Soil water retention curve shape parameter B**
Sejnane basin	1000	0.48	213.4	0.37(after [1]) and 0.45 (after[2])	9
Chaffar basin	500	0.34	3634	0.166 (after [1]) and 0.108 (after[2])	9

Table 2. Soil type model parameters (with 1: Cosby and Saxton model and 2: Cosby model for estimating S_{FC})

The thresholds $AARE_s = 5\%$ and $AARE_s = 20\ \%$ have been applied respectively to Sejnane basin and Chaffar basin. Effectively, it was assumed that owing the more important time variability of rainfall and runoff for Chaffar basin series, it was more indicated to enlarge the threshold $AARE_s$ for this basin. The analysis of simulation results and runoff performance criteria relatively to the hypothesis of taking account or not for capillary rise (CR) results in not taking it into account for Chaffar basin (CR=0) while taking it into account for Sejnane basin (CR≠0).

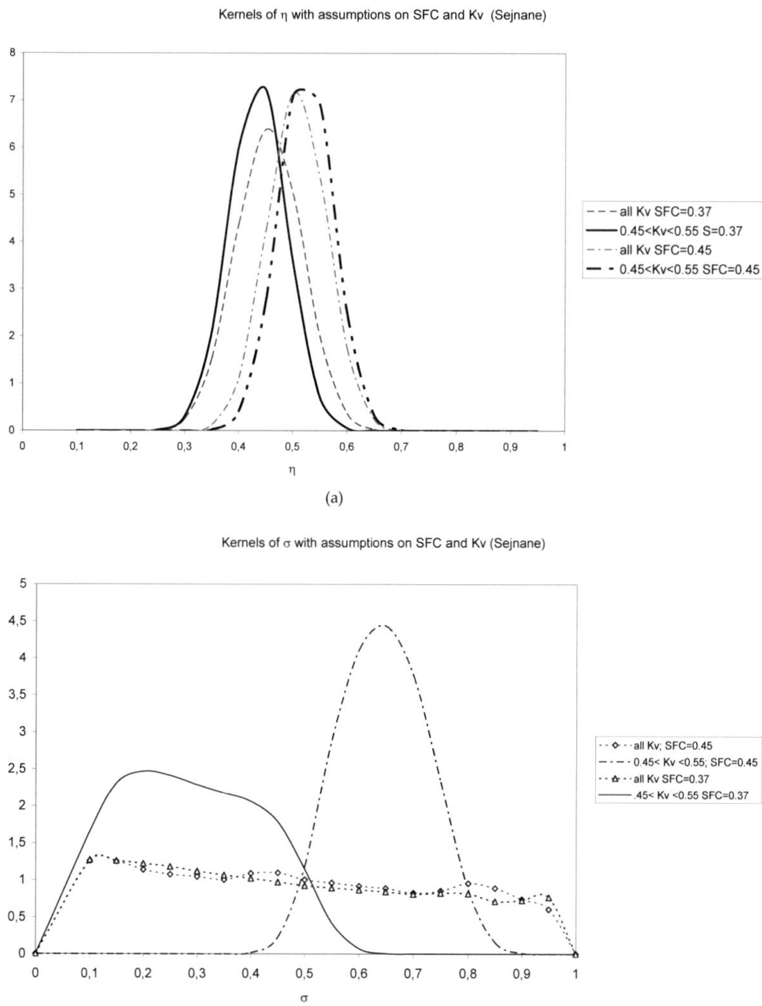

Figure 2. Parameters kernels (*all Kv* signifies that all selected solutions are considered in the Kernel estimation; *0.45 < Kv < 0.55* signifies that only solutions corresponding to this range of Kv are considered in the Kernel estimation)

4.1. Sejnane basin results

Fig. 2a reports the kernels corresponding to η in case of Sejnane basin. The result is sensitive to the choice of the parameter S_{FC} while kernels do not change with the change of the class of Kv in both assumptions on S_{FC}. Fig. 2b reports the resulting kernels of σ. It is worth noting that in both assumptions on S_{FC}, kernels are of uniform type reflecting the importance of uncertainty about σ. Conversely, the conditioning of results to the appropriate class of Kv ($0.45 < Kv < 0.55$) in relation with the vegetation and climate conditions of Sejnane watershed, reduces the uncertainty on σ and leads to two different intervals of variability for σ (smaller value of σ under the assumption of smaller value of S_{FC}).

More generally, the comparison of model error variances on monthly and decadal time scales suggests that the assumption of S_{FC} = 0.45 is more suitable for this basin. In effect, Fig. 3 which reports variances corresponding to the selected (σ,η) sets with $AARE_s$= 5% under the two assumptions on S_{FC}, shows that smaller variance values are achieved in the case where S_{FC} = 0.45.

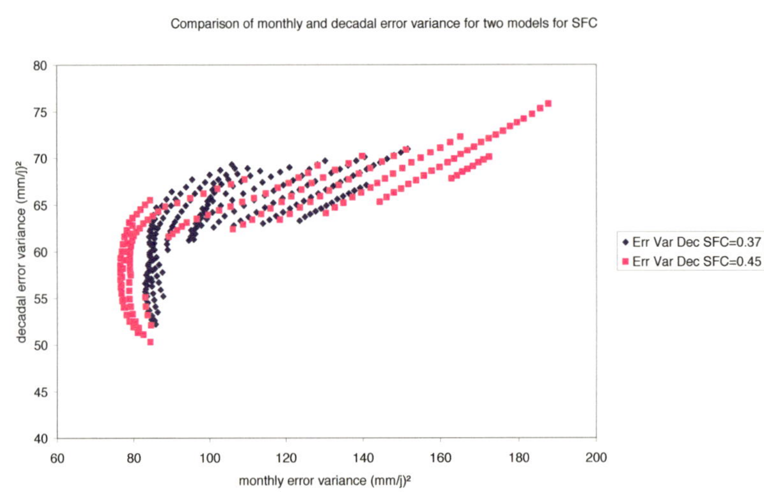

Figure 3. Comparison of monthly and decadal error variances

Fig. 4 reports AARE values as well as the values achieved by the other performance criteria ($R_{N,M}$ and $R_{N,D}$). Results are sorted according to Kv and dispatched by class of Kv. Four classes are considered: $0.35 < Kv < 0.45$; $0.45 < Kv < 0.55$; $0.55 < Kv < 0.65$; $Kv > 0.65$. It is worth noting that the parameter sets (σ,η) which result in $0.45 < Kv < 0.55$ exhibit the best AARE performance criteria while the other criteria are less sensitive to Kv conditions. Such a result might constitute a justification of adopting Eagleson (1994) Kv versus environmental stress condition variable within model fitting. Fig. 5 which reports the kernels of predicted actual evapotranspiration shows a clear reduction in uncertainty due to the inclusion of the constraint about vegetation and climate type. As well, it is noticeable that the kernel is less sensitive to the choice of S_{FC} when including such a constraint.

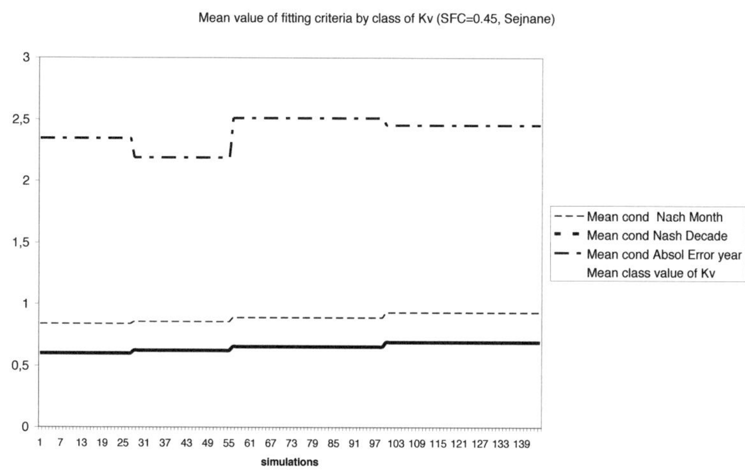

Figure 4. Values of fitting criteria when solutions are sorted according to the range of Kv

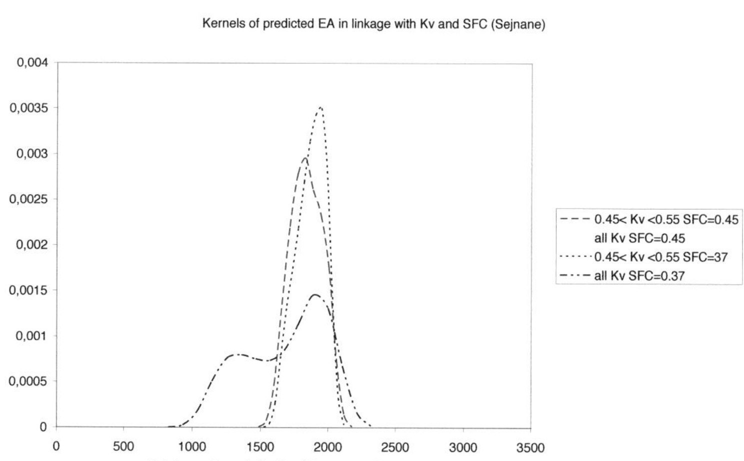

Figure 5. Kernels of the total calibration period (3 years) evapotranspiration (Sejnane basin)

4.2. Chaffar basin results

Fig. 6a andFig. 6b report respectively $R_{N,M}$ and $R_{N,D}$ values obtained in case where S_{FC} =0.166 and CR≠0. They are reported according to the corresponding Kv. It is noticeable that Kv values with 0< Kv <1 result from such simulations. Negative values of $R_{N,M}$ are often encountered suggesting very poor performances. Also, values of $R_{N,D}$ are sometimes very low. Better results are obtained when assuming CR=0 (Fig. 7a and Fig. 7b) with Kv values lying only in the interval (0.1< Kv < 0.2) which is more coherent with vegetation and climate information (type 3 curve).

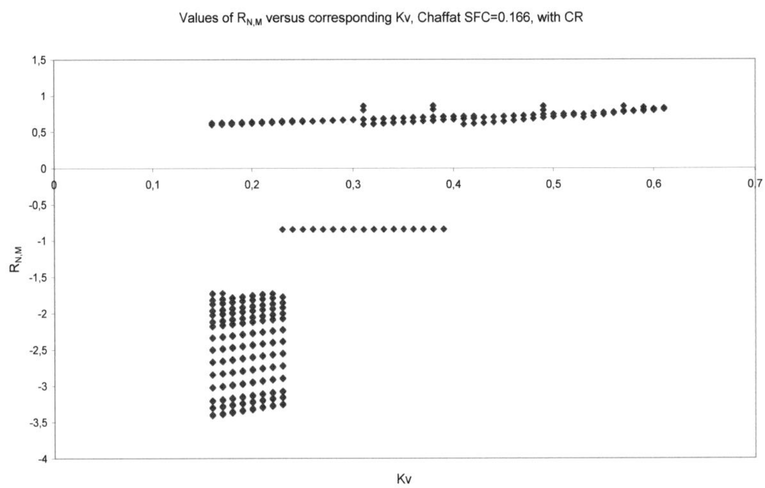

(a)

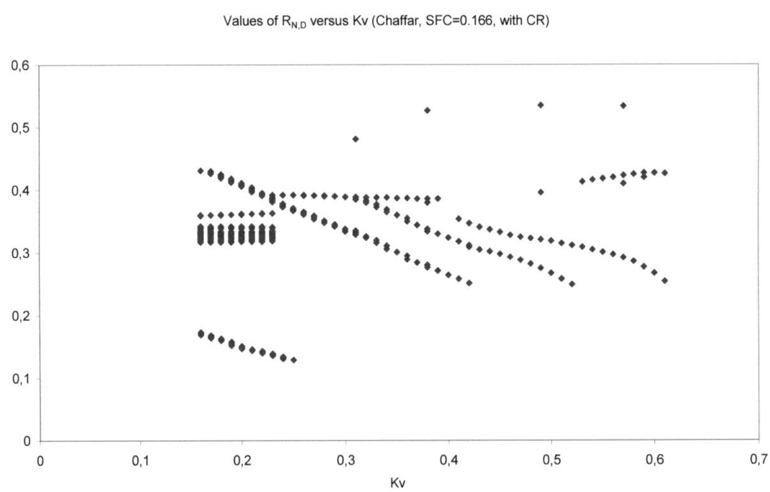

(b)

Figure 6. Values of the criterion RN,D according to the range of Kv (a) with CR=0 assumption (b) with CR ≠ 0 assumption

Finally Fig. 8 reports η kernels in the two cases (CR=0 and CR≠0). It is noticed that the distribution of resulting evapotranspiration is sensitive to the model assumption about CR. The introduction of the constraint about Kv reduces a little the spread of the kernel distribution. The kernels of σ are reported in Fig. 9. It is noticeable that they are of uniform type in the interval (0,1) : U(0,1) in the case where CR=0 and U(0.5, 1) in the case where CR≠0. For the case CR=0, the constraint about Kv reduces the uncertainty and results in a uniform distribution U(0, 0.5). Fig. 10 reports the kernel distribution of evapotranspiration in the case CR=0. The constraint about Kv highly reduces the uncertainty about this output.

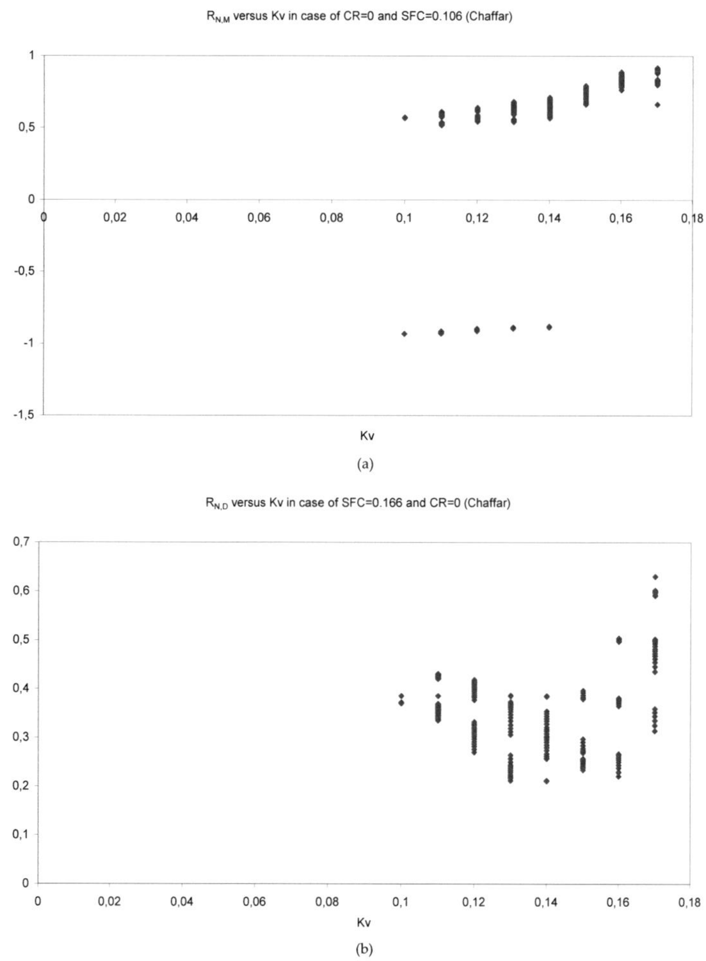

Figure 7. Values of the criteria according to Kv (a) criterion RN,M (monthly basis) and (b) criterion RN,D (decade basis)

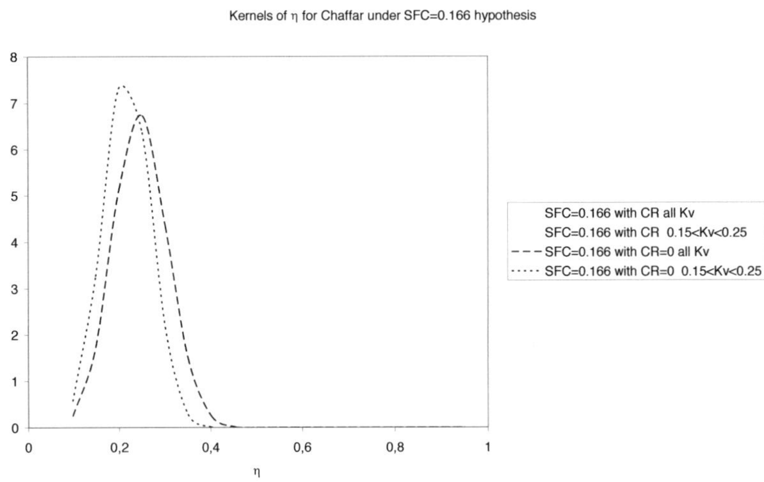

Figure 8. Kernels for the parameter η under various assumptionsKernels for the parameter

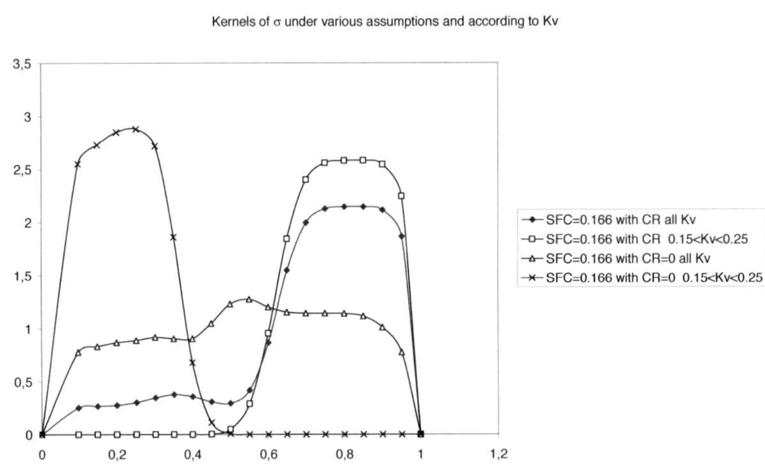

Figure 9. Kernels for the parameter ▪ under various assumptions

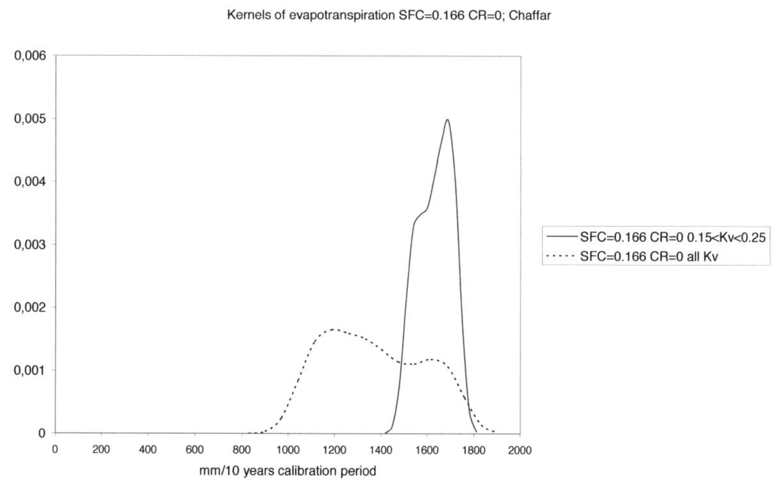

Figure 10. Kernels of the total calibration period (10 years) evapotranspiration (Chaffar basin)

5. Conclusions

The methodology developed herein aimed to integrate the type of vegetation response within the calibration process of a water budget model at basin scale and daily time step. From developments using two different watershed of moderate size under two different climatic and vegetation conditions, it results in reducing the uncertainty about the parameters σ representing the resistance of vegetation to evapotranspiration and the parameter η representing the moisture retaining capacity. Hence, the uncertainty about actual evapotranspiration predictions has been also reduced due to such an analysis. This methodology is easily transferable to other water balance models as well as vegetation and climate situations.

Author details

Zoubeida Kebaili Bargaoui, Ahmed Houcine and Asma Foughali

Université de Tunis El Manar, Ecole Nationale d'ingénieurs de Tunis, ENIT, Laboratoire LMHE, Tunis, Tunisia

References

[1] Bargaoui, Z, & Houcine, A. (2010). sensitivity to calibration data of simulated soil moisture related drought indices. *Revue Sécheresse*, , 21

[2] Beven, K. J. Prophecy, reality and uncertainty in distributed hydrological modeling. *Adv. in Water Resour.* (1993). , 6, 41-51.

[3] Cosby, B. J, Hornberger, G. M, Clapp, R. B, & Ginn, T. R. A Statistical Exploration of the Relationships of Soil Moisture Characteristics to the Physical Properties of Soils. *Water Resour. Res.* (1984). , 20(6), 682-690.

[4] Eagleson, P. S. (1994). the evolution of modern hydrology (from watershed to continent in 30 years), *Advances in water resources* 17 (1994), 3-18.

[5] Guswa, A. G, Celia, A, & Rodriguez-iturbe, I. Models of soil moisture dynamics in ecohydrology: A comparative study. *Water Resour. Res.* (2002). , 38(9), 1166-1180.

[6] Ehleringer, J. Annuals and perennials of warm deserts. In *Physiological ecology of North American plant communities*. Ed: B. F. Chabot and H. A. Mooney. Chapman et Hall. London. (1985). , 171.

[7] Iwanaga, R, Kobayashi, T, Wang, W, He, W, Teshima, J, & Cho, H. Evaluating the Irrigation requirement at a Cornfield in the Yellow River Basin Based on the "Dynamic Field Capacity " *J. Japan Soc. Hydrol. & Water resour.* (2005). , 18, 664-674.

[8] Kebaili Bargaoui Z(2011). Estimation of evapotranspiration using soil water balance modelling. In Evapotranspiration / Book 1. Intech. 979-9-53307-009-3http://www.intechweb.org/books

[9] Kobayashi, T, Matsuda, S, Nagai, H, & Teshima, J. (2001). A bucket with a bottom hole (BBH) model of soil hydrology. In: Soil-Vegetation-Atmosphere Transfer Schemes and Large-Scale Hydrological Models (ed. by H. Dolman et al.), IAHS Publ. 270. IAHS Press, Wallingford, UK., 41-75.

[10] Teshima, J, Hirayama, Y, Kobayashi, T, & Cho, H. Estimating evapotranspiration from a small area on a grass-covered slope using the BBH model of soil hydrology. *J. Agric. Meteorol.* (2006). , 62(2), 65-74.

[11] Rawls, W, Brakensiek, D, & Saxton, K. (1982) Estimation of soil water properties. Trans. Am. Soc. Agric. Engrs 25, 51-66.

[12] Rodriguez-Iturbe, I, Proporato, A, Ridolfi, L, Isham, V, & Cox, D. (1999) Probabilistic modelling of water balance at a point: the role of climate, soil and vegetation. Proc. R. Soc. London Ser. A, 455, 3789-3805.

[13] Zhan, Ch.-S, Xia, J, Chen, Z, & Zuo, Q.-T. (2008) An integrated hydrological and meteorological approach for the simulation of terrestrial evapotranspiration. Hydrol. Sci. J. 53(6), 1151–1164, doi :10.1623/hysj.53.6.1151.

Chapter 12

Evapotranspiration of Succulent Plant *(Sedum aizoonvar.floibundum)*

A. Al-Busaidi, T. Yamamoto, S. Tanak and S. Moritani

Additional information is available at the end of the chapter

http://dx.doi.org/10.5772/53213

1. Introduction

Fresh water resources available for agriculture are declining quantitatively and qualitatively. Therefore, the use of less water or lower-quality supplies will inevitably be practiced for irrigation purposes to maintain economically viable agriculture. Globally arid and semiarid areas are facing salinization of soils along with the acute shortage of water resources. The utilization of marginal waters for agriculture is getting considerable importance in such regions. In hot and dry climate, one of the most successful ground covers is Sedum. It is perennial plant, which grows by natural moisture even if there is a little soil [1]. As their common name of stonecrop suggests, they do very well in rocky areas, surviving on little soil and storing water in their thick leaves. While some do well in very sunny areas, others thrive in shade and they all tend to like good drainage. Sedums are suitable plants for rock gardens and flower borders. They are very easy to propagate as almost any tiny leaf or piece of stem that touches the ground will root. Some types become rather invasive but are easy to control since the roots are never very deep [1].

Sedum is one of the promising plants in dry areas. It has the characteristics of fire prevention and dry resistance. It has low transpiration value in the daytime compared to other plants. It uses latent heat transmission to control water loss. Generally, succulents, such as Sedum, have been the most studied and used plants for green roofs [2-5]. Greenroofs are increasingly being used as a source control measure for urban storm water management as they detain and slowly release rainwater. Their implementation is also recognized as having other benefits, including: habitat creation for birds and insects [6] filtering of aerosols; energy conservation by providing thermal insulation [7, 8]; improvement of local microclimate through evaporation; reduction of rooftop temperatures [8]. One of the main reasons Sedums seem ideally suited to green roof cultivation is the fact that many possess Crassulacean acid metabolism (CAM). During periods of soil moisture deficit, CAM plants keep their stomata closed during the day when transpiration rates are normally high and open them at night when transpiration rates are significantly lower. This is in contrast to C3 and C4 plants, which do not keep their stomata closed during the day and therefore have higher water use rates than CAM plants.

Research that examines the growth obstruction moisture point in Sedum is little, and its growth is confirmed as for the amount of pF 3.0 or lower moisture content [9, 10]. Sedum has the characteristic of doing shutting transpiring control and when plant under water stress conditions, carbon dioxide is absorbed at nighttime which also common in some Crassulaceous plants that have Crassulaceous Acid Metabolism (CAM). For terrestrial plant species CAM is generally considered to be an adaptation to growth in dry environments [11, 12]. CAM species generally have high water use efficiencies and slow growth rates and are most abundant in arid regions and dry microhabitats. The degree of CAM expression (the proportion of nighttime CO_2 assimilation by PEP carboxylase) potentially may vary from CO_2 uptake only at night, to CO_2 uptake both at night and during daylight, to CO_2 uptake only during the day. A greater proportion of nighttime CO_2 uptake has been associated with greater water use efficiency. This phenomenon helps Sedum to save much water and keep it for longer time. Sedum is a drought tolerant plant and its growth and survival under very dry conditions, still not well known. Therefore, objective of this study was to evaluate the ability of Sedum plant to grow under different soils condition where evapotranspiration was the main indicator for plant interaction with dry conditions. In addition, the studies on the growth and survival of Sedum under saline water conditions are scanty and not well documented. Therefore, the other objective of the study was to evaluate Sedum growth under saline water irrigation either by surface or shower method.

2. Materials and methods

Glasshouse study

Plot experiment was carried out in a glasshouse at Arid Land Research Center of Tottori University, Japan. The plots were made in two directions (North & South) with a slope of 20 and 30 degree, respectively. Twelve plots were filled up to 10 cm thickness with five types of soils (Table 1). The used soils have different criteria in which four of them were artificial and the other two were sandy and clayey soils. Sedum (*Sedum aizoonvar.floibundum*) plants were transplanted uniformly in all types of soils. Plants were irrigated by sprinkler with intensity of 20 mm/h. Air temperature, relative humidity and solar radiation were measured continuously day and night by Hobo (Pro series, onset, USA) meter. Evaporation and evapotranspiration were measured by using micro-lysimeters and evaporation pan (class A) following pan evaporation method ($ET_o = E_{pan} * K_{pan}$, $ET_p = ET_o * K_c$) where ET_o: reference evapotranspiration, E_{pan}: pan evaporation, K_{pan}: pan coefficient, ET_p: potential evapotranspiration, K_c: crop factor [13].

http://dx.doi.org/10.5772/53213

Soil properties	Sand (%)	Silt (%)	Clay (%)	Bulk density (gcm^{-3})	S. Hydraulic Conductivity (cm/s)	CEC ($cmol_{(+)}$/kg)	C (%)	N (%)	C/N
KS	89.6	5.8	4.6	0.48	3.6×10^{-2}	40.4	12.0	0.96	12.4
VS	92.9	1.7	5.4	0.67	9.2×10^{-1}	12.5	1.0	0.09	10.7
Powder Pearlite	98.6	1.4	0	0.19	3.2×10^{-1}	0.8	-	-	-
Coarse Pearlite	100	0	0	0.10	5.6×10^{-1}	-	-	-	-
Sand	96.1	0.4	3.5	1.42	3.4×10^{-2}	1.5	0.12	0.02	5.9
Touhaku	53.9	17.8	28.3	1.11	5.8×10^{-4}	9.5	0.50	0.09	5.4

Table 1. Soil physicochemical properties

Growth chamber study

Plastic containers of 15 cm height and 22 cm diameter were filled up to 10 cm height with different types of soils. The physicochemical properties of the used soils were same as the soils used in glasshouse study (Table 1). Sedum plant was transplanted in each pot with intensity of 30 plants/pot. All soils were irrigated until the field point of pF 1.8. After 24 hours, the evaporation process was inhibited by covering soil surface with plastic sheet. All pots were transferred to growth chamber with a day time temperature of 40 °C, 60 % relative humidity and light of 10000 Lux. Whereas, at the night time, the temperature and relative humidity were 20 °C and 60 %, respectively. All pots were placed in weighing balance scale so water lost by transpiration process was monitored (Figure 1).

Figure 1. Weighing balance for transpiration measurements

Physicochemical analysis

All used soils were air dried and passed through 2 mm sieve. Soil texture was determined by pipette method. Cation exchange capacity (CEC) was determined by atomic absorption spectrophotometer (Model Z-2300 Hitachi corp, Japan) after leaching with ammonium acetate solution and using sodium acetate as an index cation. Saturated hydrolyic conductivity was measured by constant head method. Whereas, percentage N and C were measured by C/N coda (MT700, Yanagimoto, Japan). The pF values for soil moisture characteristic curve were measured by suction and centrifuge methods for pF values of 0 - 4.2 and Saicromatar method for 4.2 - 6.0 (Figure 2). The selected properties of the soils are given in Table 1.

Salinity study

Pot experiment was carried out in same glasshouse at Arid Land Research Center, Tottori University, Japan. Sand dune soil was placed in 4 L pots. Sedum *(Sedum aizoon var. floibundum)* was planted in 24 pots at the planting density of 4 plants per pot. One group of the pots was irrigated with the saline water directly on the surface of the soil and the other group of pots was showered by the same water treatments. Irrigation with saline water was started after 14 days of planting. Saline water treatments were consisted of four levels:

i. fresh water (0.7 dS m^{-1}),

ii. saline (15 dS m^{-1}),

iii. highly saline (30 dS m^{-1}), and iv) sea water (46 dS m^{-1}).

Sea water was diluted by tap water to achieve these EC_w levels of irrigation water. Four saline water treatments were combined with two types of irrigation methods e. g., surface or normal irrigation (N) and shower or sprinkler irrigation (S). These treatments are denoted as 0.7(N), 0.7(S), 15(N), 15(S), 30(N), 30(S), 46(N), and 46(S) respectively. Plants were irrigated twice a week depending on the loss of evapotranspiration (ET_c) which was estimated by gravitational measurement. Extra water at the rate of 10% was added for leaching purpose. Evaporation was measured by using evaporation pan (class A). Air temperature and relative humidity were measured during the day as well as night by Hobo meter (Pro series, onset, USA). Prior to the harvesting of the plants for their fresh and dry weight, plant height and leaf area (by portable area meter LI-3000A) were also measured. Post-harvest soil samples were collected from each pot at a depth of 0-20 cm. Soil electrical conductivity (EC) was measured in the 1: 5 soil-water suspensions. Data were analyzed statistically for analysis of variance (ANOVA) and the means were compared at the probability level of 5% using least significant difference (LSD) test [14].

3. Results and discussion

Glasshouse study and Soil characteristics

Table 1 details the properties of the studied soils. All of them contain high percentage of sand particles with varied values of bulk density and saturated hydraulic conductivity. By checking cation exchange capacity (CEC) and C/N values, it seems that KS, VS and Touhaku soils are the most fertile soils. Generally, the physiochemical properties of these soils support plant growth but survival time or drought effect dependson how much moisture the soil can keep and the plant can take which usually related to soil water plant interactions. From Figure 2, it can be seen that Touhaku soil got the highest values for water content followed by K soil. That usually related to clay and silt contents in the soil. Moreover, as sand particles increase, volumetric water content decrease and that was the case with perlite and sand dune soils.

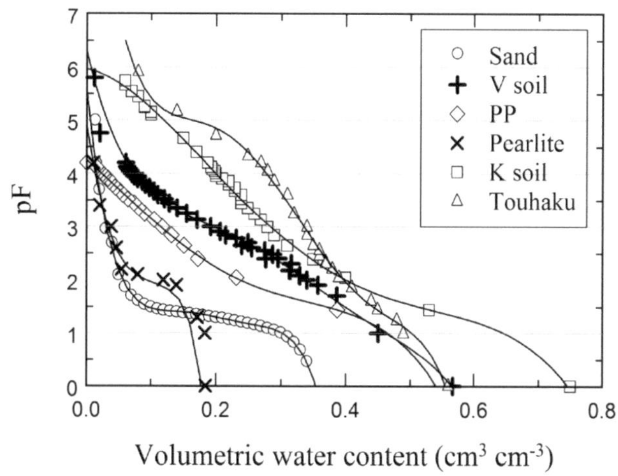

Figure 2. Soil water retention curve of studied soils

During the study period, the weather was changing with the average glasshouse temperature and humidity of 29 °C and 74 %, respectively. Average value of temperature seems to be suitable for Sedum growth but high value of 35 °C could enhance water loss through evapotranspiration (ET) process (Figure 3). Comparing ET values for both slopes of glasshouse study, it wasn't a big difference in the water lost between both slopes. However, there was a big difference between soil types (Figures 3-4) and that mainly related to the physiochemical properties of each soil. Generally, it can be noticed that after irrigation the water loss was high and gradually decreased with time. It is ranking in the following order: V soil > K soil > Sand >Touhaku> Pearlite. Moreover, it can be seen that Touhakusoil got the highest value for mean ET (Figure 4). Whereas, Pearlite soil got the lowest values. This can be related to the soil physical properties in which Touhakusoil has much clay content that can keep much water that could be subjected to ET losses. In same time, it was encouraging plant growth and plant lost much water through transpiration process compared to other soils. For Pearlite soil, since it has coarse particles so most water was lost through drainage and the rest of water was used for plant growing mechanisms. However, water lost from different slopes and directions (North and South) was inconsistence and was changing with time.

Since ET is one of the growth indicators, it seems that plant growing in K, V andTouhaku soils was growing very good by giving high values for ET. Whereas, plant growing in Pearlite soil was saving the water and reduce ET process (Figure 3 & 4).

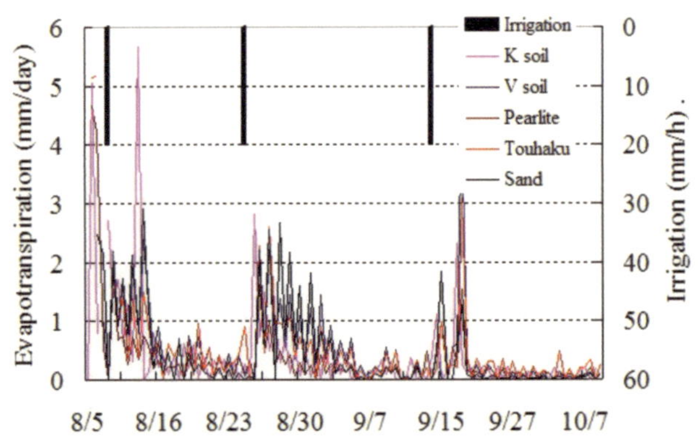

Figure 3. Irrigation and evapotranspiration values of Sedum in glasshouse study

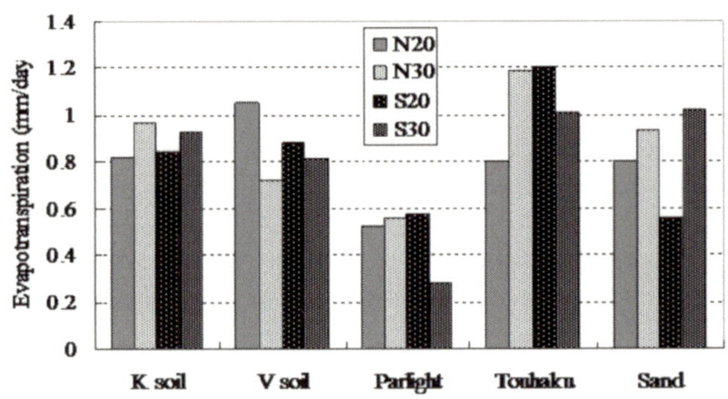

Figure 4. Mean values for Sedum evapotranspiration in glasshouse study

Growth chamber study

Growth chamber is a controlled environment and what happened inside the chamber can be more understandable than outside environment. In this study, Sedum transpiration ratio was continuously monitored by weighing scale (Figures 5-6). Day time was considered from 0-12 o'clock and night time from 12-24 o'clock. Figure 5 represent transpiration ratio at the first day of study. It can be seen that plants grown in V, K and sand soils got the highest transpiration values among others. Whereas, plants in Pearlite and PP soils got the lowest values in both intervals. This can be related to the soil physical properties which supply water to plants. Pearlite and PP are coarse soils with particle size of 3 and 1 mm, respectively, in which most water in the soil was in vapor form and not directly available for the plant. Whereas, K and V soils have high values of clay and silt contents and that increasedwater holding capacity compared to others.

http://dx.doi.org/10.5772/53213

In this study, plants were irrigated at the beginning of the study and soil surface was sealed so the only way to loss water was through transpiration process. Amount of transpiration water usually depend on plant growth and available soil moisture content. In CAM plants the transpiration process usually increase with day time and decrease at night. In this study the plant was under water stress condition so soil water content was decreasing with time. In the first day, since there was much water, plant was losing much water in the day time compared to night time (Figure 5). Whereas, at the last day of the study (Figure 6), plant was under stress and was losing less water in the day compared to the night time. This phenomenon usually happen in CAM plants when they are under stress condition. Under heat and water stress condition plants were trying to save much water by closing their stomata and opening them at normal conditions. This phenomenon can be seen very clear in Figure 6.

Generally Sedum plant is keeping much water in their leaves and since the plant was getting much water in the first day so it was losing much water in the day and night time (Figure 5) but in the last day (Figure 6) water lost was low and transpiration rate was almost constant in which the plant was keeping constant water potential value [10].

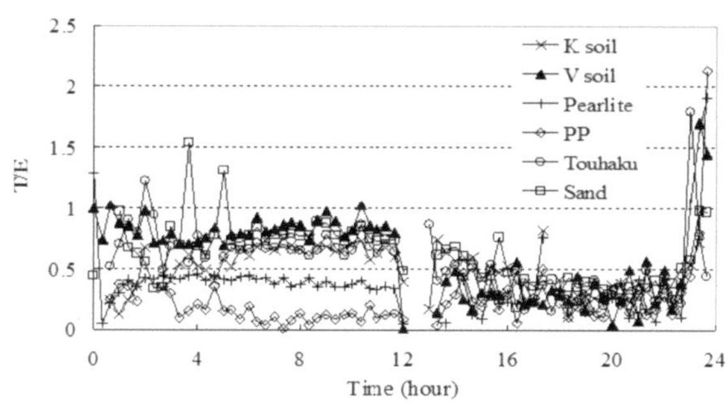

Figure 5. Transpiration ratio at early stage of study

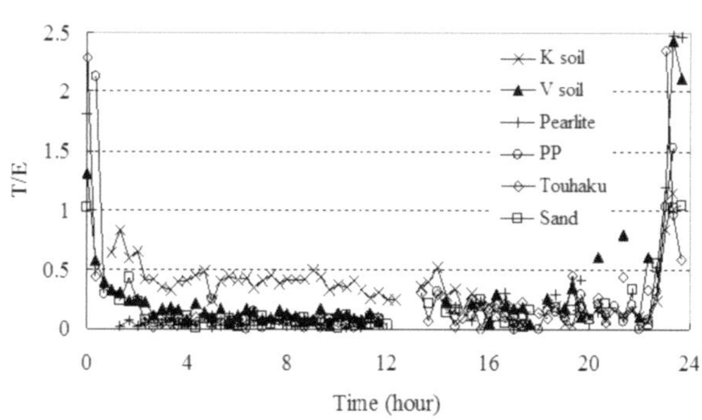

Figure 6. Transpiration ratio at late stage of study

Transpiration ratio assumption

In case of water stress, evapotranspiration could be a good indicator for plant survival. In this study cumulative transpiration and ET were related to square roots of elapse time (Figure 7). Both studies in glasshouse and growth chamber were agreed with starting point but with the time, each study gave different patterns especially in the last days of measurements. This can be related to the environmental conditions of both studies. Glasshouse conditions were varied and both evaporation and transpiration were counted. In same time plant was growing under normal condition without any heat or drought stresses. Whereas, in growth chamber, the growth conditions were almost constant and transpiration was the main factor that was measured. In addition plant was growing under stress condition in which transpiration was decreasing with time. However, all values in growth chamber were highly correlated with R^2 value of 0.99. This finding was also confirmed by Moritaniet al. [9] and Iijima [10].

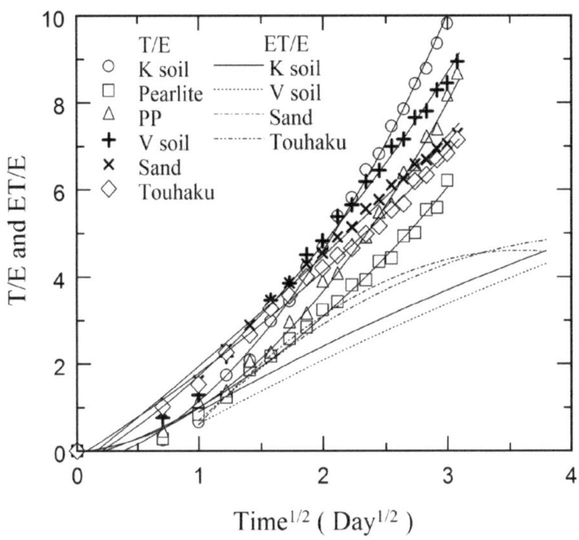

Figure 7. Relationship between cumulative transpiration ratio and square root elapse time

Transpiration ratio and soil water potential

Figure 8 shows soil water potential with transpiration ratio. The transpiring ratio for the study period decreased gradually with the increase in soil water potential (pF). The dotes shown in Figure 8 are the observation points. Whereas, the line was the predicted values found by equation 1.

$$y = \alpha / \{1 + \beta * \exp(-\gamma * x)\} + \delta \qquad (1)$$

whereα, β, γ, and δ are fitting parameters. Root mean square error (RMSE) was calculated between dotes and plotted line using equation 2.

$$RMSE = \sqrt{\frac{\sum_{i=1}^{n} d_i^2}{N}} \tag{2}$$

where d is the difference between observed and calculated values and N is the total number of the data. The average value for RMSE is 0.0097 which mean there is a good fit between observed and calculated data. It can be seen that K and V soils were giving different graphs depending on physicochemical properties of each soil. Point 1 and point 2 are the changing points. A decrease of 10 % from highest point in the graph is a representation of point 1 and an increase by 10 % from lowest point is a representation for point 2. Many points were checked and 10 % of increase or decrease was giving the best data and matching with value of changing point. The best values for point 1 were found in K soil followed by Touhaku > V > Pearlite > Sand > PP soils. For point 2 the order is Touhaku > K > V > Pearlite > Sand > PP soils (Table 2). Soils of K, V, and Touhaku types gave the highest values for point 1 and 2. The starting point for V, Touhaku and Sand soils is 0.8 and 1.6 for K soil. Whereas, Pearlite and PP soils got 0.3. From table 2 and within evapotranspiration ratio, it can be seen that K, V and Touhaku soils had almost similar values for point 1 and 2. Whereas, Sand, Pearlite and PP soils got the lowest values for transpiration. This reflects the poor structure of the soils and disability of the soil to hold water. It is also mean that soil was losing much water in short time and plant got stressed within short period. Since different soil has different properties of sand, silt and clay so each one was storing different amount of water and that was reflected in plant growth.

Sedum evapotranspiration ratio showed same results as found by many researchers [9, 10]. However, using different soil with different properties will give different values for point 1 and 2 (Table 2).

	T/E		**ET/E**	
	Point 1	**Point 2**	**Point 1**	**Point 2**
K soil	3.1	4.6	3.0	4.2
V soil	2.2	3.2	3.0	4.2
Sand	1.2	1.5	2.2	4.2
Touhaku	2.6	4.7	3.0	4.2
Pearlite	2.0	2.7	-	-
PP	0.8	1.3	-	-

Table 2. The pF value of point 1and 2

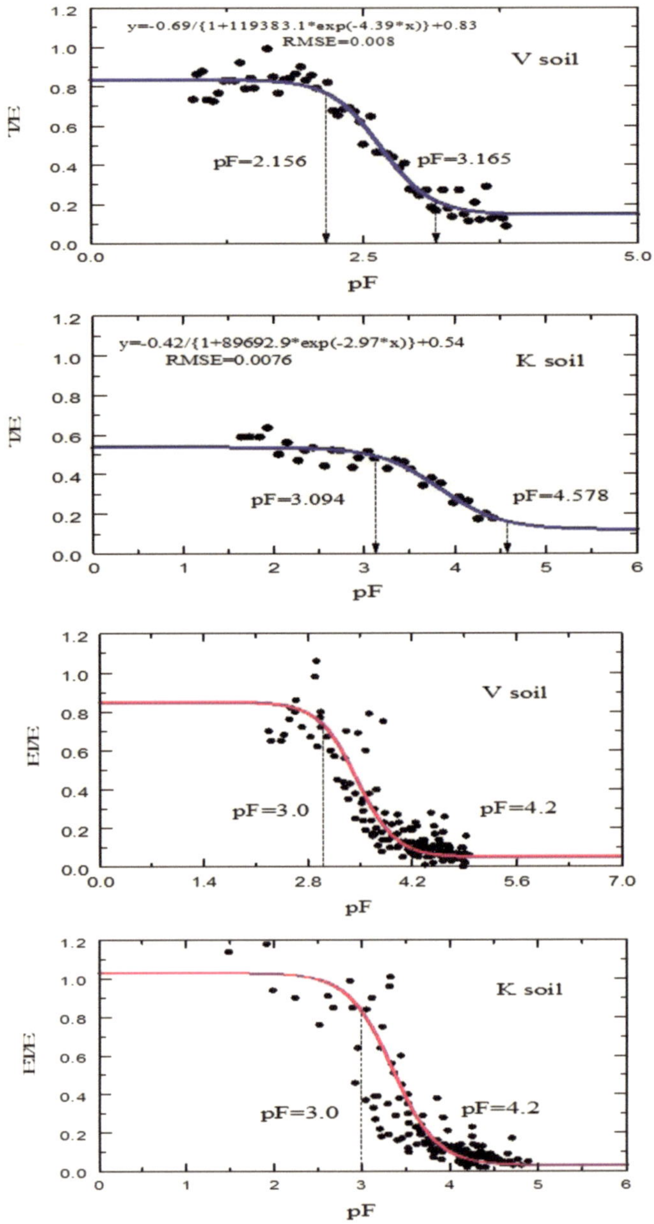

Figure 8. Relationship between transpiration ratio, evapotranspiration ratio and water potential

Salinity study

During the experiment, weather fluctuated with the average glasshouse temperature of 29 °C and humidity of 74 %. Changes in the temperature and humidity during the experiment are shown in Figure 9. Under fresh water treatment the plants exhibited the highest values of evapotranspiration as compared to saline water treatments (Figure 10). In general the higher level of evapotranspiration and accumulation of salts on the soil surface was caused by the variations in the temperature over time.

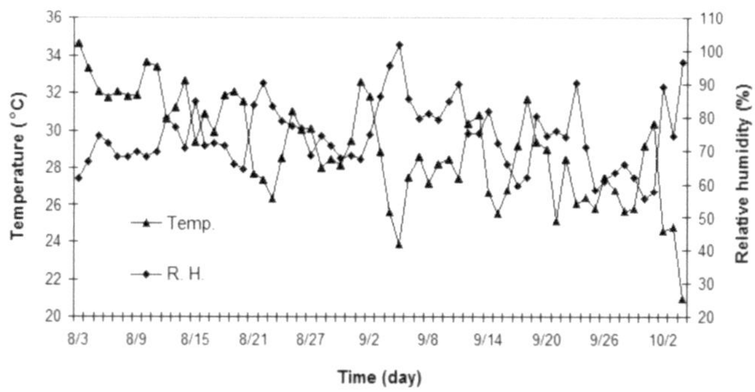

Figure 9. Variations in temperature and humidity during study period

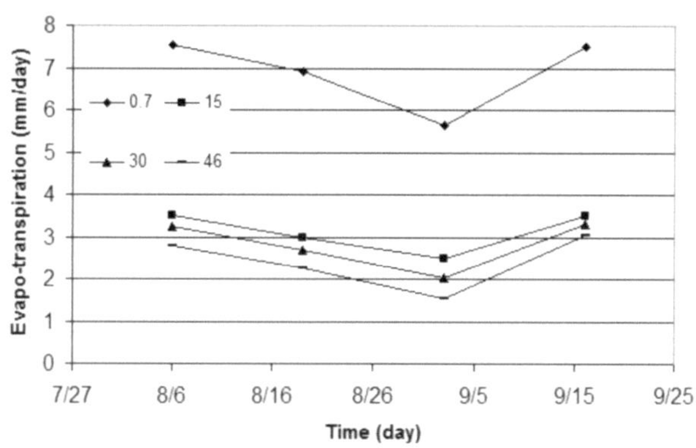

Figure 10. Variability in the evapotranspiration as affected by saline treatments

Fresh water encouraged evaporation process more than saline water. Maximum evapotranspiration occurred with good quality water. Since the plants absorb water in saline conditions with higher pressure therefore the water losses through transpiration were retarded. Thus the magnitude of the evapotranspiration was inversely related to the amount of salts in the irrigation water. Reduced bioavailability of water and retarded plant growth under saline irrigation produced poor evapotranspiration in the system. On the other hand presence of salts in the saline irrigation inhibits evapotranspiration and reduces water consumption. Water density, viscosity and formation of salt crust are factors that could reduce evaporation and maintain higher water in the soils. Al-Busaidi and Cookson [15] reported salt crust formation on the soil surface due to saline irrigation inhibited evaporation and reduced leaching efficiency. It has been reported elsewhere that salt accumulation in root zone causes the development of osmotic stress and reduces plant development [16, 17].

Application of irrigation water with certain level of salts results the deposition of soluble salts in the soils. Evaporation and transpiration of irrigation water eventually accumulate excessive amounts of salts in the soils unless an adequate leaching and drainage systems are not practiced [18]. During the study, a low electrical conductivity of soil was noted under normal water whereas sea water irrigation largely increased the salinity level of soil (Figure 11). The saline water accumulated salts in the soil in spite of the leaching process. Petersen [19] reported that the accumulation and release of salts could depend on the quality and quantity of irrigation water, soil type and plant response. Abu-Awwad [20] reported high salt concentration on the soil surface due to evaporation.

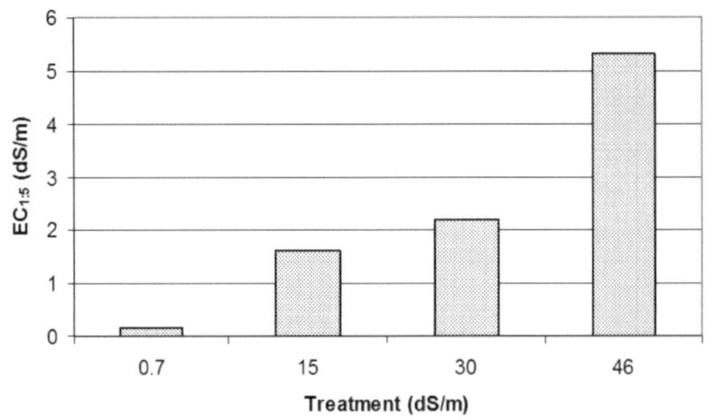

Figure 11. Soil salinity under different saline irrigation treatments

Plant growth

Plant parameters were the function of irrigation water treatments. Sedum plant grew well under non-saline conditions. Highest plant fresh and dry biomass, plant height and leaf area were noticed with normal irrigation water. While, sea water treatment gave the lowest values of the plant parameters (Table 3). Soil salinity was the main reason behind the lower plant growth whereas the effects of irrigation methods were statistically found insignificant. Sedum plants accumulated more salts and leaf injuries were seen especially under high saline treatments. The physiological thickness of the Sedum leaves with higher water absorbing potential could possibly facilitate Sedum plants to survive under high saline conditions. Usually, CAM plants are capable of transporting water very effectively to those tissues necessary for survival. This also happened in Sedum when the plants were dry, water was transported from the older leaflets to the younger parts of the shoots, which were thus kept turgescent, whereas the older leaflets died. Closure of the stomata helped further to maintain a sufficiently high water potential [21].

There is a general consensus that higher salinity profoundly impaired plant growth parameters. The response of crops to salinity could depend upon plant species, soil texture, water

http://dx.doi.org/10.5772/53213

holding capacity and composition of the salts. Abu-Awwad [20] reported that saline soils with considerable soluble salts interfered the growth of crop species. Heakal et al. [16] reported that dry matter yield of plants decreased with increasing salinity of irrigation water.

Treatment	Plant height (cm)	Leaf area (cm^2)	Fresh weight (g)	Dry weight (g)
0.7 (N)	31	11	355	44
0.7 (S)	25	10	314	39
15 (N)	20	5	69	22
15 (S)	20	4	57	16
30 (N)	17	3	39	17
30 (S)	16	3	43	18
46 (N)	14	2	42	17
46 (S)	14	2	35	20

Table 3. Plant parameters as affected by saline water irrigation

In general the plant biomass is dependent absolutely on the growth of plants. Differences were found in the fresh and dry weights among the irrigation treatments. Water deficit level increased with the increasing salinity (Figure12). The ratio of dry weight to fresh plant weight increased significantly with the increasing level of salinity treatments. The stress caused by the ion concentrations allows the water gradient to decrease, making it more difficult for water and nutrients to move through the root membrane [22]. Accumulation of salts in the root zone affects plant performance through creation of water deficit and disruption of ion homeostasis [23] which in turn cause metabolic dysfunctions. The differences in the water content of the plants between the irrigation methods could reflect the efficiency of surface irrigation which can provide enough water to the plant without physically touching the leaves. Sprinkler or shower irrigation adds salts directly on the leaves and may disturb its normal functions.

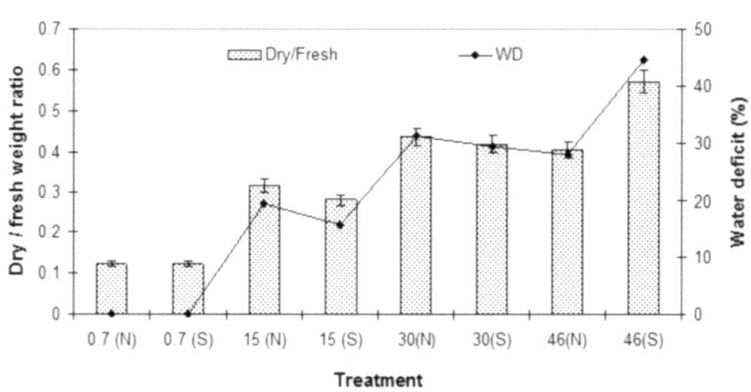

Figure 12. The ratio of dry to fresh weight and water deficit (WD) as affected by the saline treatments

Water salinity and irrigation method can affect plant growth. Moreover, the interaction effect of both independent parameters was affecting plant height and biomass (Table 4). However, it can be seen that all dependent parameters were significantly affected by applied treatments. Volkmar et al. [22] reported that plants grown in saline soils have diverse ionic compositions and concentrations of salts. The fluctuations in the salts concentrations could be related to the changes in the water source, drainage, evapotranspiration, and solute availability. The two major environmental factors that currently reduce plant productivity are drought and salinity and these stresses cause similar reactions in plants due to water stress [24].

Parameter	Saline water (S)	Irrigation method (I)	S x I
		P-value	
Plant height	0.0001*	0.0001*	0.0002*
Leaf area	0.0001*	NS	NS
Fresh weight	0.0001*	0.0001*	0.0001*
Dry weight	0.0001*	0.0006*	0.0001*

Table 4. Summary of two-way analysis of variance on the effects of saline water and irrigation method on plant parameters* denotes the level of significance at P value < 0.05 and NS denotes non-significance.

Most studies of CAM plants have focused on the physiology and ecology of individual plant performance. It has generally been assumed that the expression of CAM is associated with adaptive success in arid environments because traits related to water use efficiency and tolerance of low water availability are genetically correlated with CAM. Water loss in CAM plants is reduced as a result of low stomatal frequency and high cuticular resistance [11, 25-27].

It is known that malate accumulation in at least some CAM plants can result in a substantial level of osmotic adjustment [28]. The maintenance of negative leaf potential during drought appears to be related to the continued ability of the Amistad plants to take up CO_2 and to accumulate dry weight. Both Troughton et al. [29] and Mooney et al. [30] have found that reproductive tissue in a variety of leaf succulents may have carbon isotope ratios that are considerably less negative than those of vegetative tissue from the same plant. While this may relate to the concurrence of drought and reproduction, it does illustrate that CAM activity may make an important contribution to reproductive carbon sinks.

4. Conclusion

All plants are subjected to a multitude of stresses throughout their life cycle. Depending on the plant species and stress source, the plant will respond in different ways. Sedum is a drought tolerant plant. Its ability to grow with different soil types and under water stress condition was investigated. The main reason for the difference in water loss between treatments was differences in soil types. At the end of growth chamber study, it was found that transpiring ratio at the day and night was almost equal. However, transpiration value can be predicted from the relationship of the square root time and measured values for plant transpiration. Low transpiration values in the day time were related to the photosynthesis characteristic of Sedum with leaf water potential. In the night time where low temperature was observed, the stomata was open and plant was exchanging CO_2 with more transpiration rate compared to the day time. In salinity study, the experiment could confirm that Sedum plants can tolerate salinity stress and can survive with water deficit conditions, which was related to its ability to store water for long time. However, the saline waters remarkably affected the evapotranspiration rate, salts accumulation in the soils and plant biomass production. Water deficit increased with the increase in salinity level. The salinity of the soil significantly increased with higher saline water. The plant growth was not affected by surface or sprinkler irrigation methods. The use of sea water up to certain dilution could be an option for Sedum production in water scarce areas.

Author details

A. Al-Busaidi[1*], T. Yamamoto[2], S. Tanak[2] and S. Moritani[2]

*Address all correspondence to: albusaidiahmed@yahoo.com

1 College of Agricultural & Marine Sciences, Department of Soils, Water and Agricultural Engineering, Sultan Qaboos University, Oman

2 Arid Land Research Center, Tottori University, Japan

References

[1] Stephenson, R. (1994). Sedum Cultivated Stonecrops. *Timber press, Inc., Oregon, USA.*

[2] Berghage, R., Beattie, D., Jarrett, A., & Rezaei, F. (2007). Green Roof Plant Water Use. *BerghageR., et al. Quantifying Evaporation and Transpirational Water Losses from Green Roofs and Green Roof Media Capacity for Neutralizing Acid Rain.The Pennsylvania State University, State College, PA.*, 18-38.

[3] Durhman, A. K., Rowe, D. B., & Rugh, C. L. (2007). Effect of Substrate Depth on Initial Growth, Coverage, and Survival of 25 Succulent Green Roof Plant Taxa. *HortScience*, 42, 588-595.

[4] Latocha, P., & Batorska, A. (2007). The Influence of Irrigation System on Growth Rate and Frost Resistance of Chosen Ground Cover Plants on Extensive Green Roofs. *Annals of Warsaw University of Life Sciences-SGGW.Horticulture and Landscape Architecture.Warsaw University of Life Sciences Press, Warsaw, Poland*, 131-137.

[5] Sendo, T., Inagaki, N., Kanechi, M., & Uno, Y. (2007). What Kind of Plant Species Are the Best for Urban Rooftop Gardening. *Acta Hort*, 762, 333-339.

[6] Scholz-Barth, K. (2001). Green on Top. *Urban Land*, 83-97.

[7] Wong, N. H., Cheong, D. K. W., Yang, H., Soh, J., Ong, C. L., & Sia, A. (2003). The Effects of Roof Top Garden on Energy Consumption of A commercial Building in Singapore. *Energy Build*, 35, 353-364.

[8] Köhler, M., Schmidt, M., Grimme, F. W., Laar, M., de Assunção, Paiva. V. L., & Tavares, S. (2002). Green Roofs in Temperate Climates and in the Hot-humid Tropics- Far Beyond the Aesthetics. *Environ. Management and Health*, 13, 382-391.

[9] Moritani, S., Yamamoto, T., Andry, H., Inoue, M., & Kaneuchi, T. (2010). Using Digital Photogrammetry to Monitor Soil Erosion under Conditions of Simulated Rainfall and Wind. *Soil Research*, 48(1), 36-42.

[10] Iijima, K. (1995). About the Water Stress Reaction of the Sedum Belonging When Put under A dry Condition. *Landscape Research*, 58(5), 69-72.

[11] Kluge, M., & Ting, I. P. (1978). Crassulacean Acid Metabolism: Analysis of An ecological Adaptation. *Billings WD, Golley F, Lange OL, Olson JS. Ecological Studies: Analysis and Synthesis.Springer, Berlin, Heidelberg, New York.*

[12] Ting, I. P. (1985). CrassulaceanAcid Metabolism. *Ann Rev Plant Physiol*, 36, 595-622.

[13] FAO. (1998). Crop Evapotranspiration. *FAO Irrigation and Drainage Papers* [56], Rome.

[14] Kinnear, P. R., & Gray, C. D. (1997). SPSS for windows made simple. *Psychology Press.UK.*

[15] Al-Busaidi, A., & Cookson, P. (2005). Leaching Potential of Sea Water. *Journal for Scientific Research: Agricultural and Marine Sciences (SQU)*, 9, 27-30.

[16] Heakal, M. S., Modaihsh, A. S., Mashhady, A. S., & Metwally, A. I. (1990). Combined Effects of Leaching Fraction Salinity and Potassium Content of Waters on Growth and Water Use Efficiency of Wheat and Barley. *Plant and Soil*, 125, 177-184.

[17] Abdul, K. S., Alkam, F. M., & Jamal, M. A. (1988). Effects of Different Salinity Levels on Vegetative Growth, Yield and its Components in Barley. *ZANCO*, 1, 21-32.

[18] U.S. Salinity Laboratory Staff. (1954). Diagnosis and Improvement of Saline and Alkali Soils. *USDA Handbook. 60. U.S. Gov. Print. Office, Washington, DC.*

[19] Petersen, F. H. (1996). Water Testing and Interpretation. *Reed, D.W. (ed.) Water, Media, and Nutrition for Greenhouse Crops. Batavia: Ball.*

[20] Abu-Awwad, A. M. (2001). Influence of Different Water Quantities and Qualities on Limon Trees and Soil Salts Distribution at the Jordan Valley. *Agricultural Water Management*, 52, 53-71.

[21] Steudle, E., Smith, J. A. C., & Lfittge, U. (1980). Water-Relation Parameters of Individual Mesophyll Cells of the Crassulacean Acid Metabolism Plant Kalaneho6 Daigremontiana. *Plant Physiol*, 66, 1155-1163.

[22] Volkmar, K. M., Hu, Y., & Steppuhn, H. (1998). Physiological Responses of Plants to Salinity: A review. *Can. J. Plant Sci*, 78, 19-27.

[23] Munns, R. (2002). Comparative Physiology of Salt and Water Stress. *Plant Cell Environ*, 25, 239-250.

[24] Serrano, R. (1999). A glimpse of the Mechanisms of Ion Homeostasis DuringSalt Stress. *Journal of Experimental Botany*, 50, 1023-1036.

[25] Nobel, P. S. (1976). Water Relations and Photosynthesis of A desert CAM Plant, Agave deserti. *Plant Physiol*, 58, 576-582.

[26] Nobel, P. S. (1977). Water Relations and Photosynthesis of A barrel Cactus, FerocactusAcanthodes. *Oecologia(Berlin)*, 27, 117-13.

[27] Osmond, C. B. (1978). CrassulaceanAcid Metabolism: A curiosity in Context. *Ann Rev Plant Physiol*, 29, 379-414.

[28] Lfittge, U., & Nobel, P. S. (1984). Day-night Variations in Malate Concentration, Osmotic Pressure, and Hydrostatic Pressure in Cereus Validus. *Plant Physiol*, 75, 804-807.

[29] Troughton, J. H., Mooney, H. A., Berry, J. H., & Verity, D. (1977). Variable Carbon Isotope Ratios of DudleyaSpecies Growing in Natural Environments. *Oecologia (Berl)*, 30, 307-311.

[30] Mooney, H. A., Troughton, J. H., & Berry, J. A. (1977). Carbon Isotope Ratio Measurements of Succulent Plants in Southern Africa. *Oecologia (Berl)*, 30, 295-306.

Chapter 13

Reference Evapotranspiration (ETo) in North Fluminense, Rio de Janeiro, Brazil: A Review of Methodologies of the Calibration for Different Periods of Analysis

José Carlos Mendonça,
Barbara dos Santos Esteves and
Elias Fernandes de Sousa

Additional information is available at the end of the chapter

http://dx.doi.org/10.5772/52278

1. Introduction

Water is the essential element for life on Earth planet, where currently different regions suffer from shortages due to the large population growth and depletion of natural sources. The agricultural sector is the human activity that consumes the most water in the world (about 70% of drinking water sources) and one of the main problems of irrigated agriculture is the correct quantification of crop water requirements. In this sense, there is a constant search to implement sustainable practices for the management of water resources, one of the more efficient determination of evapotranspiration (ET), which is the term used to describe the amount of water effectively ceded the land surface to atmosphere and an important component of the hydrological cycle and used for quantifying the calculation of water balance in soil, detection of water stress conditions and use as input for quantitative models of harvesting or other applications (Ferreira et al., 2011.)

With the objective to standardize the definition of evapotranspiration given by various authors, as Penman (1948) and Thornthwaite (1948), it became necessary to define the reference evapotranspiration (ETo), which according to Allen et al. (1998) can be defined as the rate of evapotranspiration from a hypothetical crop with an assumed height of 0.12 m, with a surface resistance of 70 sec/m and an albedo of 0.23, closely resembling the evaporation from an extensive surface of green grass of uniform height, actively growing and adequately watered.

Several researchers have developed methods for estimating and measuring evapotranspiration. Burman et al. (1983) did a review of these methods in different parts of the world and commented that many methods have been proposed and the methods may be broadly classified as those based on combination theory, humidity data, radiation data, temperature data, and miscellaneous methods which usually involve multiple correlations of ET and various climate data. Usually the reference evapotranspiration methods are classified in Combination methods, Radiation method, Temperature methods, pan evapotranspiration, etc.

Allen et al. (1998) mentioning that evapotranspiration is not easy to measure. Specific devices and accurate measurements of various physical parameters or the soil water balance in lysimeters are required to determine evapotranspiration. The methods are often expensive, demanding in terms of accuracy of measurement and can only be fully exploited by well-trained research personnel. Although the methods are inappropriate for routine measurements, they remain important for the evaluation of ET estimates obtained by more indirect methods.

Since the 1930s there are several methods for estimating ETo. However, whatever the method is detailed and rigorous, there will always be the needs of local or regional calibrations if you are being adopted outside the region where it was developed. Burman et al (1983) argue that several equations to estimate reference evapotranspiration developed around the world use the grass and alfalfa as a standard surface. This situation creates difficulties as the proposal for an empirical equation bears a strong dependence on the standard surface, causing undesirable and significant errors in estimation. Based on these discussions is that the Penman-Monteith equation was parameterized by Allen et al. (1998).

The surface resistance is defined as the resistance of water vapor through the openings of stomata and drag as that of the upper plant, involving the friction of the air flow over the surface vegetated. The aerodynamic resistance is a parameter dependent on the local weather and its demonstration of layers depends on the roughness governing the processes of transport of momentum and heat, and the offset and zero plane. This displacement of the zero plane refers to the height to which the speed is zero. Thereafter the profile starts log wind speed. However, the aerodynamic resistance scheme used in the formulation Penman-Monteith (FAO, 56) is restricted to the condition parameter neutral atmosphere, ie when the air temperature, atmospheric pressure and wind speed field close to the adiabatic condition. One can also be noted that the displacement of the zero plane and the layers of roughness, the processes that govern the amount of heat transport is correlated with the height of culture and as regards the parameter of surface resistance, it is directly proportional stomatal resistance and inversely proportional to the active leaf area index, stomatal resistance being directly affected by atmospheric conditions and the availability of water for the crop.

Allen et al. (1998) clain that the Penman-Monteith(FAO-56) for estimating reference evapotranspiration does not allow controversy and provides consistent and reliable information in different weather conditions and location is recommended by the FAO, the lack of lysimeters as calibration standard in the world.

Zanetti et al. (2007) tested an artificial neural network (ANN) for estimating the reference evepotranspiration (ETo) as a function of the maximum and minimum air temperatures in the Campos dos Goytacazes, Rio de Janeiro State. The data used in the network training were obtained from a historical series (September 1996 to August 2002) of daily climatic data collected in Campos dos Goytacazes. When testing the artificial neural network, two historical series were used (September 2002 to August 2003) relative to Campos dos Goytacazes, Rio de Janeiro and Viçosa, Minas Gerais State. The ANNs (multilayer perceptron type) were trained to estimate ETo as a function of the maximum and minimum air temperatures, extraterrestrial radiation, and the daylight hours; and the last two were previously calculated as a function of either the local latitude or the Julian date. According to the results obtained in this ANN testing phase, it is concluded that when taking into account just the maximum and minimum air temperatures, it is possible to estimate ETo in Campos dos Goytacazes, RJ.

One work was performed with the aim of proposing an artificial neural network (ANN) to estimate the reference evapotranspiration (ETo) as a function of geographic position coordinates and air temperature in the State of Rio de Janeiro (Zanetti et al., 2008). Data used for the network training were collected from 17 historical time series of climatic elements located in the State of Rio de Janeiro. The daily ETo calculated by Penman-Monteith (FAO-56) method was used as a reference for network training. ANNs of multilayer perceptron type were trained to estimate ETo as a function of latitude, longitude, altitude, mean air temperature, thermal daily amplitude and day of the year. After training with different network configurations, the one showing best performance was selected, and was composed by only one intermediary layer (with twenty neurons and sigmoid logistic activation function) and one output layer (with one neuron and linear activation function). According to the results obtained it can be concluded that, considering only geographical positioning coordinates and air temperature, it is possible to estimate daily ETo in 17 places of Rio de Janeiro State by using an ANN.

Another method of estimating the ETo are evaporimeters, which measure the evaporation of water, the most common Class "A" Pan developed by the U.S. Weather Service (USWB) and widespread use. According to Pereira et al. (1997) Class "A" Pan (TCA) is influenced by solar radiation, wind speed, temperature and relative humidity and thus, different researchers have questioned the methodology of choice of the pan coefficient (Kt) and should be determined by results of scientific research estimates that there are no wrong.

It is observed that the choice of methodology to be adopted should be based on the availability of climate data, the necessary precision, convenience and cost. In irrigation projects are required for short periods, ranging from daily to a maximum of fortnightly research is needed to evaluate the efficiency of the methodologies in these conditions. Using a series of ten years of daily average data collected at Evapotranspirometric Station of Universidade Estadual do Norte Fluminense Darcy Ribeiro, this study aimed to evaluate the performance of indirect methods for estimating reference evapotranspiration (ETo) proposed by Hargreaves-Samani (1985), FAO-24 Radiation Solar (1977), Jensen-Haise (1963), Linacre (1977), Makkink

(1957), Penman Simplified (2006) and Pan Class "A" estimated using four equations for determining the coefficient of the Pan - Kt: Allen (1998), Bernardo et al. (1996), Cuenca (1989) and Snyder (1992) for periods of 1, 5 and 10 days, with the Penman-Monteith FAO-parameterized, in the North Fluminense, Rio de Janeiro, Brazil.

2. Materials and methods

2.1. Study area

The city of Campos dos Goytacazes located in the North Fluminense occupies an area of 4.027 km^2. The downtown area is located in the following geographical coordinates: 21° 45″ 23′ south latitude, 41° 19″ 40′ west longitude and 14 m above sea level. In Figure 1 is presented the study area contained in the North Fluminense, in reference to the state of Rio de Janeiro and Brazil. According Köeppen climate, this region's clime is classified as Aw, that is, tropical humid, with rainy summer, dry winter and the temperature average above 18°C during the coolest months. The annual average temperature stands at around 24 º C and the small temperature range; The climatological normal rainfall is 1055.3 mm (Ramos et al., 2009).

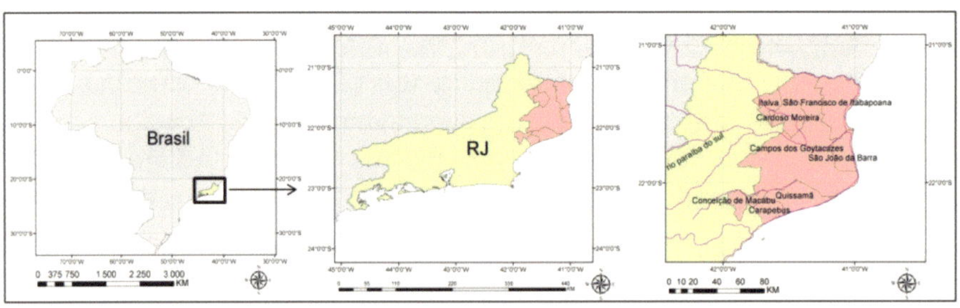

Figure 1. Study area localization in reference to the Rio de Janeiro State and Brazil.

In this study were used weather data daily, a period of 10 years (1996-2006), collected by an automatic station, model Thies Clima, installed at the Experimental Station of Pesagro-Rio (geographical coordinates: 21°18'47" south; 41º18'24" west and altitude of 11 meters).

The Thies automatic weather station is equipped with sensors for measuring meteorological data the following:

- Wind Speed: a sensor to detect the wind speed in the range 0.3 to 50 m / s;
- Atmospheric Pressure: A barometer can measure values in the range 946 to 1053 hPa;
- Temperature and Relative Humidity: A thermo-hygrometer allows register values of relative humidity in the range 1-100% and a temperature range of -35 to + 70 °C;

- Global Solar Radiation: A pyranometer can measure values in the range 0 to 1400 W/m2;
- Precipitation: A rain gauge measuring rainfall intensity of up to 7 mm/min.

All sensors are connected to a datalogger model DL-15 - V. 2:00 – Thies Clima, with total capacity of 256 Kbytes of memory storage, recording daily averages between 24-h. The sensor values recorded every minute, and a stored mean value every 6 minutes

Observations the conventional meteorological station (Class A pan and weighing lysimeter) for such work were performed at 9 h.

The lysimeter tank with dimensions of 3.0 x 2.0 x 1.5 m, made of sheet metal had their weight carried by a set of four load cells manufactured by J-Star Electronics, Wisconsin, and installed at the tank base, and determining the lysimeter blade evapotranspired obtained by variation in weight observed in the period divided by evaporating surface area (6 m2).The station area is covered with grass Batatais (Paspalum notatun Fluegge).

2.2. Methods for obtaining the reference evapotranspiration (ETo)

2.2.1. FAO Penman-Monteith method (FAO-PM - 1998)

The Penman-Monteith parameterized by Allen et al.(1998) was selected as a benchmark method for comparation can be derived (Equation 1).

$$ETo = \frac{0,408\Delta\left(Rn - G\right) + \gamma\frac{900}{T + 273}u_2\left(e_s - e_a\right)}{\Delta + \gamma\left(1 + 0,34u_2\right)} \quad (1)$$

Where:

ETo is the reference evapotranspiration, in mm/day;

Rn - net radiation at crop surface, in MJ m^{-2}/day;

G -soil heat flux density, in MJ/m^2/day;

T - air temperature at 2 m height, in °C;

γ- psychrometric constant, kPa/C;

Δ -slope vapor pressure curve, in kPa;

u_2 - wind speed at 2 m height, in m/s;

e_s - saturation vapor pressure, in kPa;

e_a - actual vapor pressure, kPa;

e_s - e_a-the saturation vapor pressure deficit, in kPa.

2.2.2. FAO – 24 Radiation method (1977)

The estimate ETo by the FAO - 24 Radiation method was described for Doorenbos and Pruitt (1977) andcan be derived (equation 02)

$$ETo = c_o + cL\ W\,Rs \tag{2}$$

Where

C_0 =-0,3; a0 = 1,0656 ; a1 = -0,0012795 ; a2 = 0,044953 ; a3 = -0,00020033 ; a4 = -0,000031508 ; a5 = -0,0011026

cL =- a0 + a1 UR + a2 Vd + a3 UR Vd+ a4 UR^2 + a5 Vd^2

UR = Relative Humidity (%);

Vd = Average wind speed during the day to 2 m in height, in m/s (considered to Vd = 70% of the average wind speed within 24 h);

Rs - Radiation at the surface, expressed as equivalent evaporation (Rs, mm/day);

W - Weight factor dependent on the temperature (Tair).

The weighting factor (W) can be obtained by the following equations:

$$W = 0,407 + 0,0145T_{air}\ if\ 0 < T_{air} < 16^{º}C \tag{3}$$

$$W = 0,483 + 0,01T_{air}\ if\ 16,1 < T_{air} < 32^{º}C \tag{4}$$

Where T_u is the average daily temperature of air, to take $T_u = T_{ar}$.

2.2.3. Hargreaves –Samani method (1985)

The Hargreaves - Samani method data requires only air temperature and extraterrestrial radiation to estimate ETo. For applicationused thefollowing equation:

$$ETo = 0,0023\,Ra\left(Tmáx - T\min\right)^{0,5}\left(T + 17,8\right) \tag{5}$$

Where: Ra extraterrestrial radiation, in mm day^{-1}, T max is the maximum temperature in °C;Tmin is the minimum temperature, in °C.

2.2.4. Jansen-Haise method (1963)

For the estimation of ETo by Jensen-Haise method used the equation 06:

$$ETo = Rs\,(0{,}0252\,T + 0{,}078) \tag{6}$$

2.2.5. Makkink method (1957)

For the estimation of ETo by the Makkink method used the equation 07:

$$ETo = Rs\,(\frac{\Delta}{\Delta + \gamma}) + 0{,}12 \tag{7}$$

Where

Rs - Radiation at the surface, expressed as equivalent evaporation (Rs, mm/day);

Δ -slope vapor pressure curve, in kPa/°C;
γ- psychrometric constant, kPa/°C.

2.2.6. Linacre method (1977)

For the estimation of ETo by the Linacre method used the equation 08:

$$ETo = \frac{\dfrac{J\left(T + 0{,}006\ h\right)}{100 - \varphi} + 15\left(T - T_{dew}\right)}{80 - T} \tag{8}$$

Where J is a dimensionless constant equal to 700; h is the local altitude in meters; ϕ is the local latitude degrees and T_o is the temperature of dew point. The dew point temperature (T_{dew}) can be estimated by equation 09:

$$T_{dew} = \frac{237{,}3\log\left(e_a\right) - 156{,}8}{8{,}16 - \log\left(e_a\right)} \tag{9}$$

Where ea is the vapor pressure of water in kPa, determined by equation 10:

$$e_a = e_s\left(T\right)0{,}01\,RU(\%) \tag{10}$$

Where RU (%) - relative humidity and e_s - saturation vapour pressure, inkPa/°C.

2.2.7. Simplified Penman's Method (2006)

A simplified estimation method to calculate the potential evapotranspiration was developed to Villa Nova et al. (2006) based on the Penman approach, considering only the diurnal values of evapotranspiration rates thatare more representative of the water vapor transfer process to the atmosphere for a givenagricultural ecosystem. In addition, the classical expression of the Bowen ratio (b) was modifiedherein by considering the sensible heat flux (H) emergent from the evaporative surface inconjunction with the air turbulent flux, which transports also latent heat flux (LE). Such procedureresults in a similarity between the aerodynamic resistances of sensible heat and latent heat fluxes soas to allow for a considerable simplification without impairing the estimates.

ETo estimated by the SPM proposed by Villa Nova et al. (2006) was obtained from equation 11:

$$ETo = 0,408\ \frac{(Rn - G)}{(2 - W)} \tag{11}$$

Where:

ETo - Evapotranspiration from the wet surface (mm/day) during the sunshine period;

G - heat flux in the soil (MJ/m^2 /day) during the diurnal period;

Rn - diurnal net radiation at a vegetated surface (MJ/m^2 /day), and

W - tangent of water the vapor saturation pressure curve at the point of diurnal daily mean airtemperature (Tair).

2.2.8. Class A Pan Method (TCA)

The ETo is estimated by the Class "A" Pan by using the following equation:

$$ETo^{TCA} = EV.\,Kt \tag{12}$$

Where: Ev - Evaporation of Class A Pan, in mm dia^{-1} and Kt, the Pan coefficient (dimensionless).

2.2.8.1. Methods to estimate the Pan coefficient – Kt

To estimate the Pan coefficient - Kt method used in the Class "A" Pan were evaluated four methodologies are described below:

2.2.8.1.1. Methodology proposed by Cuenca (1989)

$$\begin{aligned} Kt = {} & 0,475 - 2,4.10^{-4}\,U_2 + 5,16.10^{-3}.\,H + 1,8.10^{-3}.\ F - 1,6.10^{-5}.\,RU^2 \\ & -1,01.10^{-6}.\,F^2 - 8,0.10^{-9}\,RU^2.\,U_2 - 1,0.10^{-8}.\,RU^2.\,F \end{aligned} \tag{13}$$

2.2.8.1.2. Methodology proposed by Snyder (1992)

$$Kt = 0{,}482 + 0{,}024\ln(F) - 0{,}000376\,U_2 + 0{,}0045\,RU \tag{14}$$

2.2.8.1.3. Methodology proposed by Bernardo et al. (1996)

$$Kt = 0{,}69 \tag{15}$$

2.2.8.1.4. Methodology proposed byAllen et al. (1998)

$$Kt = 0{,}108 - 0{,}0286\,U_2 + 0{,}0422\ln(F) + 1434\ln(RU) - 0{,}000631\left[\ln(F)\right]^2\ln(RU) \tag{16}$$

Where U_2 is the wind speed at 2 m height in km/day, RU is the average relative humidity (%), and F is the boundary of the green crop area, considered in this study equal to 15 m.

2.3. Evaluation of methods

To evaluate the performance of the methods we proceeded to linear regression analysis, considering the linear model y = bx (regression through the origin), in which the independent variable was the Penman-Monteith (ETo_{PM}), and the dependent variable, the other methods. Was also used the Index of agreement of Willmott (D) (Willmott, 1981), the mean absolute error (MAE), the maximum error (EMAX) and the efficiency of the method (EF), based on the equations 17, 18, 19 and 20:

$$D = 1 - \frac{\sum_i^n (Pi - Oi)^2}{\sum_i^n (\mid Pi - \bar{O} \mid + (Oi - \bar{O} \mid)^2} \tag{17}$$

$$MAE = \frac{1}{n}\sum_i^n (Oi - Pi) \tag{18}$$

$$EMAX = MAX(|Oi - Pi|)_{in}^{n} \tag{19}$$

$$EF = \frac{\sum(O - \bar{O})^2 + \sum(O - Pi)^2}{\sum(O - \bar{O})^2} \tag{20}$$

Where: O = estimated values by ETo_{PM}, Pi = estimated by other methods; $\bar{O}$ = mean value ETo_{PM}.

3. Results and discussion

3.1. Comparison of Penman-Monteith with other methods

Table 1 shows the monthly averages of air temperature, the relative humidity, wind speed and solar radiation for the ten years of data analyzed.

Months	Tair (ºC)	RU (%)	U_2 ($m\ s^{-1}$)	Rs ($W\ m^{-2}$)
January	26.47	74.35	2.36	305
February	26.54	73.55	2.12	296
March	25.69	75.75	1.73	247
April	24.41	76.70	1.54	207
May	21.85	75.16	1.50	176
June	20.98	77.41	1.51	159
July	20.28	76.35	1.68	160
August	21.60	75.90	2.24	194
September	22.16	75.80	2.43	209
October	23.07	75.90	2.31	227
November	24,29	76,19	2,34	251
December	25,79	75,99	2,21	281

Table 1. Average monthly values of air temperature (Tar), relative humidity (RU), wind speed at 2m (U_2) and solar radiation (Rs) for 1996 to 2006 period.

These meteorological variables are required as input data to the standard method and estimation of other variables, as well as entries for the other methods tested. Evapotranspiration is a complex phenomenon and non-linear, because it is dependent on the interaction between various climatological elements (Kumar et al, 2002).

Table 2 shows the parameters for statistical analysis: correlation coefficient (r^2), index of agreement of Wilmott (D), mean absolute error (MAE), efficiency of the method (EF), maximum error (EMAX) and the slope (b) comparing the parameterized Penman-Monteith method (FAO PM) with other methodologies and Figure 2, the graphs of correlation in respect to the line 1:1.

Periods (days)	**Methods**	**R^2**	**D**	MAE (mm d^{-1})	EF (mm d^{-1})	EMAX (mm d^{-1})	**b**
1	H-S	0.84	0.95	0.46	0.83	2.18	1.02
5	H-S	0.91	0.97	0.29	0.90	1.42	1.02
10	H-S	0.93	0.80	0.24	0.93	1.24	1.02
1	RS-FAO	0.92	0.98	0.38	0.89	1.56	0.94
5	RS-FAO	0.95	0.99	0.26	0.93	1.06	0.95
10	RS-FAO	0.96	0.99	0.22	0.95	0.90	0.96
1	MAK	0.93	0.89	0.82	0.60	2.37	1.23
5	MAK	0.96	0.87	0.82	0.51	1.81	1.23
10	MAK	0.97	0.87	0.82	0.48	1.59	1.24
1	J-H	0.92	0.89	1.00	0.34	2.80	0.80
5	J-H	0.94	0.88	0.95	0.23	2.39	0.81
10	J-H	0.95	0.87	0.95	0.19	2.28	0.81
1	LIN	0.54	0,78	0,84	0,51	3,08	0.95
5	LIN	0.64	0,83	0,68	0,60	2,05	0.94
10	LIN	0.67	0,85	0,63	0,62	1,79	0.94
1	MSP	0.89	0.93	0.66	0.72	2.45	1.16
5	MSP	0.91	0.93	0.66	0.68	1.85	1.16
10	MSP	0.92	0.93	0.66	0.67	1.45	1.16

Table 2. Analysis of statistical methods for estimating evapotranspiration for averaging periods of 1, 5 and 10 days.

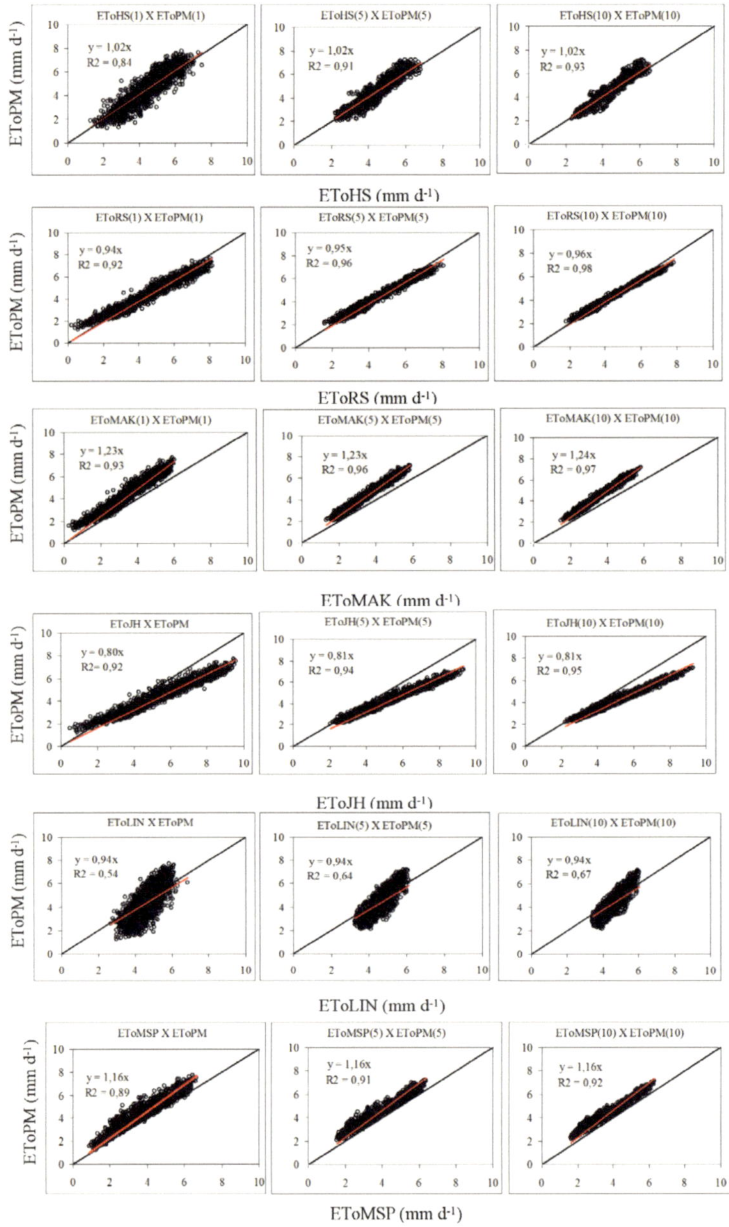

Figure 2. Graphs of correction in respect to the line 1:1.

Observing the values in Table 2 and Figure 2, it appears that all methods were evaluated value of the correction coefficient (r^2) larger than 0.80, with the exception of the method of Linacre, where r^2 values to the two periods varied from 0.54 (for 1 day) to 0.67 (for 10 days). This adjustment increased r^2, a measure that increased the periods studied was common to

http://dx.doi.org/10.5772/52278

all methods. This finding agrees with Mendonça (2001, 2003) justified by smoothing the averages of the sampled values. Another observation concerning the increment of the remaining days is that on the Hargreaves-Samani, Solar Radiation-FAO and Linacre, the mean absolute error (MAE) suffered a decrement. For the method of Hargreaves-Samani, this range was 0.46 mm d^{-1} (for the period of 1 day) going to 0.24 mm dia^{-1} (for 10 days). For the FAO-24 solar radiation method, the variation was of 0.38 mm d^{-1} (within 1 day) to 0.22 mm d^{-1} at 10 days. To Linacre method, the same variation was from 0.84 to 0.63 mm d^{-1}.

Makkink method and Simplified Penman method showed no variation in the mean absolute error, keeping them constant in d-1 0.82 mm and 0.66 mm d^{-1} for all periods, respectively. Since the Jensen-Haise method presented the MAE of 1.00, 0.96 and 0.95 mm d^{-1}, respectively, for periods of 1, 5 and 10 days.

Analyzing the slopes of the methods evaluated was observed that Makkink and Simplified Penman Method showed values above 1 for all periods, with the group of methods overestimated ETo-PM. These results agree with those observed by Mendonça (2001; 2003) and Fernandes (2006), assessments in the same area of study.

In the group of methods that are underestimated ETo FAO-24 Radiation, Jensen-Haise and Linacre. Observing Table 2, can be seen that the maximum error (EMAX) obtained similar behavior to EMA, decreasing as you increase the periods analyzed. The greater EMAX observed for a period of 1 day was the method of Linacre, equal 3.08 mm d^{-1}. Also for this method was the largest decrease in EMAX, when will the period 1 to10 days, with 1.79 mm d^{-1} for this period.

As for efficiency, it is clear that the methods of Hargreaves-Samaniand FAO-24 Radiation remained throughout the period evaluated with EF greater than 0.82. These methods also showed the best adjustment of the index of agreement ofWilmontt, and FAO-24 Radiation coming to D index of 0.99, for periods of 5 and 10 days.

The Simplified Penman method received a satisfactory performance for the estimation of ETo in the study region, for the daily period, with r^2 of 0.89, showing a small dispersion compared ETo_{PM}, a situation similar to that found by Villa Nova et al. (2006). The rate of agreement Wilmontt observed for this method was greater than 0.90 (D = 0.93).

The Simplified Penman and Makkink methods decreased efficiency over the period analyzed. EF were their best for the period of 1 day (0.72 to 0.60 mm d^{-1}, respectively) was lower for 10 days (0.66 and 0.47 mm d^{-1}, respectively.) as for the concordance index Wilmontt(D), Makkink values decreased from 0.89 (for the 1 day period) going to 0.86 for the period of 10 days. Simplified Penman method showed D constant for periods of 1, 5 and 10 days (0.93) Linacre was efficient and index D increased with the increase of the periods analyzed, as can be seen in table 1. Jensen-Haise showed EF decreasing with the increase of the evaluation period, with the worst rates of EF methods (ranging from 0.34 for 1 day, to 0.17 in 10 days). The index D remained above 0.86, D being its highest rate for the period of 1 day (0.89).

3.2. Comparison of methods for estimating the Kt

Table 3 shows the results of statistical analysis of different methodologies for determining the coefficient of the Pan (Kt) used in the Class "A" Pan method and Figure 3, the graphs of correlation in respect to the line 1:1

Periods (days)	Methods (TCA)	r^2	D	MAE ($mm\ d^{-1}$)	EF ($mm\ d^{-1}$)	EMAX ($mm\ d^{-1}$)	b
1	Allen	0.81	0.95	0.53	0.77	2.14	0.94
5	Allen	0.95	0.98	0.31	0.90	1.36	0.95
10	Allen	0.95	0.98	0.27	0.92	1.17	0.95
1	Bernardo	0.80	0.95	0.53	0.79	1.78	1.03
5	Bernardo	0.92	0.97	0.33	0.89	1.31	1.05
10	Bernardo	0.94	0.98	0.30	0.91	1.07	1.05
1	Cuenca	0.81	0.95	0.50	0.81	1.83	0.97
5	Cuenca	0.91	0.98	0.26	0.93	1.08	0.98
10	Cuenca	0.95	0.99	0.22	0.95	0.90	0.99
1	Synder	0.81	0,92	0,69	0,63	2,68	0.88
5	Synder	0.93	0,94	0,53	0,74	1,78	0.89
10	Synder	0.95	0,95	0,51	0,75	1,62	0.89

Table 3. Analysis of statistical methods of different methodologies for determining the coefficient of the Pan (Kt) for estimating evapotranspiration for averaging periods of 1, 5 and 10 days.

Looking at Table 3 and Figure 3 can see an increase in the coefficient of correlation (r^2) as it increases the evaluation period and all the methods used to determine the coefficient of the Pan (Kt) present good adjustment r^2, were above 0.80 in all periods. These results agree with those found by Conceição (2002; 2005) who compared the monthly ETo estimated by the Class "A" Pan with the Penman-Monteith-FAO and Mendonça et al. (2006) who compared the daily ETo estimated.

It is observed that the best method for determining Kt on a daily, and subsequent conversion of EV with ETO was proposed by Cuenca with EF = 0.81, followed by Bernardo, EF = 0.79. However, for the same period of 1 day, the index of agreement of Wilmontt of the two methods presented the same amount (D = 0.95).

http://dx.doi.org/10.5772/52278

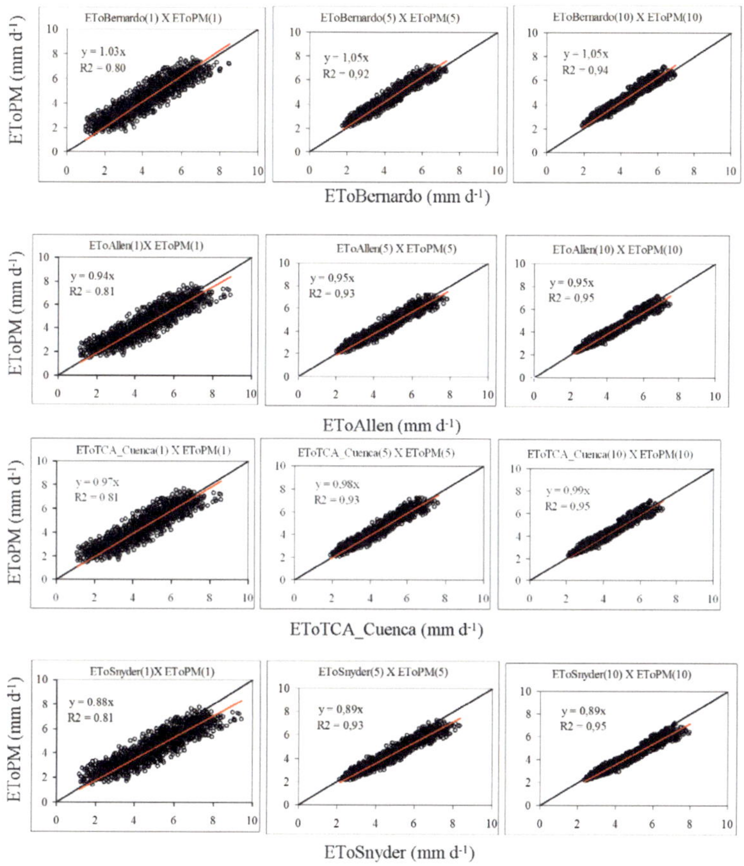

Figure 3. Graphs of correlation in respect to the line 1:1.

It was also found that all methods showed an increase in the value of efficiency (EF) as it increased the period and those proposed by Allen, Snyder and Cuenca, overestimated ETo-PM at all times. The methodology presented Allen, from 5 days a slope (b) constant at 0.95, as does the method of Snyder (b = 0.89).

The mean absolute error (MAE) of all methods decreased as the periods analyzed increased, and the method of Cuenca presented the lowest MAE (0.22 mm d^{-1}) for the period of 10 days. Methodologies for determining the assessed value of Kt, the worst results were obtained by the method of Snyder, the value of its efficiency (EF) 0.63, 0.74 and 0.75, respectively stop for a period of about 1, 5 and 10 days, while other methods reached levels higher than EF 0.90. This same method was maximum error (EMAX) decrease as you increased the periods analyzed, however, persisted for 10 days at 1.62 mm d^{-1}, while for the same period the methodology proposed by Allen reached 1.17 mm. d^{-1}, Bernardo et al., 1.07 mm d^{-1} and Cuenca, 0.90 mm d^{-1}.

4. Conclusions

Based on the results obtained in this work can be concluded that all indirect methods assessed showed improvements in their statistical indices as they increased the periods of analysis. It can be concluded that the methods of Makkink, Jansen-Haise, Linacre and Class "A" Pan using the methodology proposed by Snyder for obtaining Kt did not achieve satisfactory levels and should not be adopted for the estimation of reference evapotranspiration (ETo) in the study area. Otherwise, the methods of Hargreaves-Samani, FAO-24Solar Radiation, Simplified Penman Method and Class "A" Pan using the methodology proposed by Cuenca for obtaining Kt presented the best adjustment for the evaluation period and can be used satisfactorily for the estimation of ETo in the North Fluminense, Rio de Janeiro, Brazil.

Acknowledgements

The authors are grateful for the National Counsel for Scientific and Technological Development – CNPq; the Coordenação de Aperfeiçoamento de Pessoal de Nível superior – CAPES and the Fundação Carlos Chagas Filho de Amparo à Pesquisa do Estado do Rio de Janeiro – FAPERJ, for the financial support and logistics that made these studies possible.

Author details

José Carlos Mendonça*, Barbara dos Santos Esteves and Elias Fernandes de Sousa

*Address all correspondence to: mendonca@uenf.br; barbbarase@gmail.com; efs@uenf.br

Laboratory of Agricultural Engineering (LEAG / UENF), Rio de Janeiro, Brazil

References

[1] Allen, R. G.; Pereira, L. S.; Raes, D.; Smith, M. Crop evepotranspiration – guidelines for computing crop water requirements. Rome: FAO. FAO Irrigation and Drainage Paper 56, 300p. 1998.

[2] Bernardo, S., Sousa, E. F., & Carvalho, J. A. Estimativa da evapotranspiração de referencia (ETo), para as "Áreas de Baixada e de Tabuleiros" da Região Norte Fluminense. BoletimTécnico- Uenf, Campos dos Goytacazes, RJ. 1(1):14, 1996.

[3] Burman, R. D.; Nixon, P. R.; Wright, J. L.; Pruitt, W. O. Chapter 6, Water requirements. In: Design and operation of farm irrigation systems. ASCE Monography, No 3, Edited by JENSEN, M. E., St. Joseph, Michigan 49085, 829 p. 1983.

[4] Conceição, M. A. F. Reference Evapotranspiration Based on Class A Pan Evaporation. Scientia Agrícola, Piracicaba, SP, n.3, (2002). , 59, 417-420.

[5] Conceição, M. A. F., & Mandelli, F. Comparação Entre Métodos de Estimativa da Evapotranspiração de Referência em Bento Gonçalves, Rs. Revista Brasileira de Agrometeorologia, n.2, (2005). , 13, 303-307.

[6] Cuenca, R. H. Irrigation system design: an engineering approach. New Jersey: Prentice-Hall Englewood Cliffis. 1989. 133p.

[7] Doorembos, J., & Pruit, W. O. Guidelines for predicting crop water requirements. 2 ed. Rome: FAO (1977). p. (FAO- Irrigation and Drainage Paper 24)

[8] Ferreira, E., Toledo, J. H., Dantas, A. A. A., Pereira, R. M.: Cadastral Maps of Irrigated Areas by Center Pivots in Minas Gerais State, Using Cbers-2b/Ccd Satellite Imagery. Revista Engenharia Agrícola. In Press. Issn 0100-6916. 2011

[9] Fernandes, L. C. Avaliação de Diversas Equações Empíricas de Evapotranspiração. Estudo de Caso: Campos dos Goytacazes e Ilha do Fundão-RJ. (TeseMestrado)- Rio de Janeiro- RJ- Universidade Federal do Rio de Janeiro. (2006).

[10] Kumar, M., Raghuwanshi, N. S., Singh, R., Wallender, W. W., & Pruitt, W. O. (2002). Estimating evapotranspiration using artificial neural network. *Journal of Irrigation and Drainage Engineering*, 128, 224-233.

[11] Mendonça, J. C. Comparação Entre Métodos de Estimativa da Evapotranspiração Potencial de Referencia (ETo) para Região Norte Fluminense, RJ. Tese (MestradoemProdução Vegetal) Campos dos Goytacazes- RJ, Universida de Estadual do Norte Fluminense Uenf, 70, 2001.

[12] Mendonça, J. C. Comparação entre Métodos de Estimativa da Evapotranspiração Potencial de Referencia (ETo) para Região Norte Fluminense, RJ. Revista Brasileira de Engenharia Agrícola e Ambiental, Campina Grande, n.2, (2003). , 7, 275-279.

[13] Mendonça, J. C., Sousa, E. F., Andre, R. G. B., & Bernardo, S. Coeficientes do Tanque-Classe A para Estimativa de Evapotranspiração de Referência em Campos dos Goytacazes, RJ. Revista Brasileira de Agrometeorologia, n.1, (2006). , 14, 123-128.

[14] Penman, H. L. . Natural evaporation from open water, bare soil and grass."Proc. Roy. Society.London, A193, (1948). , 120-146.

[15] Pereira, A. R., Villa, Nova. N. A., & Sediyama, G. C. Evapotranspiração; Piracicaba & Fealq, 1997.

[16] Ramos, A. M.; Santos, L. A. R.; Fontes, L. T. G. (Organizadores). Normais Climatológicas do Brasil 1961-1990. Brasília, DF: Inmet, 465 p. 2009.

[17] Snyder, R. L. Equation for evaporation pan to evapotranspiration conversion. Journal of Irrigation and Drainage Engineeringof ASCE, New York, n.6, (1992). , 118, 977-980.

[18] Thornthwaite, C. W. An approach toward a rational classification of climate. Geographic Review, (1948). , 38.

[19] Villa, Nova. N. A., Miranda, J. H., Pereira, A. B., & Silva, K. O. Estimation of the Potential Evapotranspiration by a Simplified Penman Method. Revista Engenharia Agrícola, Jaboticabal, n.3, (2006). , 26, 713-721.

[20] Willmott, C. J. On the validation of model.*Physical Geography.* , n.2, , (1981). ., 2, 184-194.

[21] Zanetti, S. S.; Sousa, E. F.; Oliveira, V. P. S.; Almeida, F. T.; Bernardo, S. Estimating evapotranspiration using artificial neural network and minimum climatological data. Journal of Irrigation and Drainage Engineering, v.33, n.2, p.83-89, 2007.

[22] Zanetti, S.S.; Sousa, E. F. ; Carvalho, D. F.; Bernardo, S. Estimação da evapotranspiração de referência no Estado do Rio de Janeiro usando redes neurais artificiais. Revista Brasileira de EngenhariaAgrícola e Ambiental, v.12, n.2, p.174–180, 2008.